Game Theory

The New Palgrave Economics Collection
Editors: Steven N. Durlauf, University of Wisconsin-Madison, USA & Lawrence E. Blume, Cornell University, USA

Also in the series: Behavioural and Experimetal Economics
Economic Growth
Macroeconometrics and Time Series Analysis
Microeconometrics
Monetary Economics

Series Standing Order ISBN 978-0-230-24014-8 hardcover
Series Standing Order ISBN 978-0-230-24013-1 paperback

To receive future titles in this series as they are published quote one of the ISBNs listed above to set up a standing order: contact your bookseller; write to Customer Services Department, Macmillan Distribution Ltd, Houndmills, Basingstoke, Hampshire, RG21 6XS; or email orders@palgrave.com.

Game Theory

Edited by
Steven N. Durlauf
University of Wisconsin-Madison, USA

Lawrence E. Blume
Cornell University, USA

palgrave
macmillan

All articles first published in *The New Palgrave Dictionary of Economics*, 2nd Edition
Edited by Steven N. Durlauf and Lawrence E. Blume
in eight volumes, 2008

with the exception of Incentive Compatibility which first published in *The New Palgrave: A Dictionary of Economics*
Edited by John Eatwell, Murray Milgate and Peter Newman
in four volumes, 1987

First published 2010 by
PALGRAVE MACMILLAN

Palgrave Macmillan in the UK is an imprint of Macmillan Publishers Limited, registered in England, company number 785998, of Houndmills, Basingstoke, Hampshire RG21 6XS.

Palgrave Macmillan in the US is a division of St Martin's Press LLC, 175 Fifth Avenue, New York, NY 10010.

Palgrave Macmillan is the global academic imprint of the above companies and has companies and representatives throughout the world.

ISBN-13: 978-0-230-23889-3 hardback
ISBN-13: 978-0-230-23890-9 paperback

This book is printed on paper suitable for recycling and made from fully managed and sustained forest sources. Logging, pulping and manufacturing processes are expected to conform to the environmental regulations of the country of origin.

A catalogue record for this book is available from the British Library.

A catalog record for this book is available from the Library of Congress.

10 9 8 7 6 5 4 3 2 1
19 18 17 16 15 14 13 12 11 10

Printed in the United States of America

Contents

List of Contributors

SANDEEP BALIGA
Northwestern University, USA

SAMUEL BOWLES
University of Massachusetts, USA

ADAM BRANDENBURGER
New York University, USA

COLIN F. CAMERER
California Institute of Technology, USA

ROSS CRESSMAN
Wilfred Laurier University, Canada

MARTIN W.CRIPPS
University College London, UK

MARTIN DUFWENBERG
University of Arizona, USA

HERBERT GINTIS
University of Siena, Italy & University of Massachusetts, USA

SRIHARI GOVINDAN
University of Iowa, USA

FARUK GUL
Princeton University, USA

JOSEPH Y. HALPERN
Cornell University, USA

SERGIU HART
The Hebrew University Jerusalem, Israel

AVIAD HEIFETZ
The Open University, Israel

EHUD KALAI
Northwestern University, USA

MICHAEL KEARNS
University of Pennsylvania, USA

VIJAY KRISHNA
Penn State University, USA

JOHN O. LEDYARD
California Institute of Technology, USA

OLOF LEIMAR
Stockholm University, Sweden

KANDORI MICHIHIRO
University of Tokyo, Japan

JOHN MORGAN
University of California Berkeley, USA

STEPHEN MORRIS
Princeton University, USA

ROGER B. MYERSON
University of Chicago, USA

JOHN NACHBAR
Washington University, St Louis, USA

MARTIN J. OSBORNE
University of Toronto, Canada

PHILIP J. RENY
University of Chicago, USA

WILLIAM H. SANDHOLM
University of Wisconsin-Madison, USA

MARK A. SATTERTHWAITE
Northwestern University, USA

ROBERTO SERRANO
Brown University, USA

MARCIANO SINISCALCHI
Northwestern University, USA

TOMAS SJÖSTRÖM
Rutgers University, USA

XAVIER VIVES
University of Navarra, Spain

MARK WALKER
University of Arizona, USA

ROBERT B. WILSON
Stanford University, USA

JOHN WOODERS
University of Arizona, USA

H. PEYTON YOUNG
Johns Hopkins University, USA

General Preface

All economists of a certain age remember the "little green books". Many own a few. These are the offspring of *The New Palgrave: A Dictionary of Economics*; collections of reprints from *The New Palgrave* that were meant to deliver at least a sense of the *Dictionary* into the hands of those for whom access to the entire four volume, four million word set was inconvenient or difficult. *The New Palgrave Dictionary of Economics, Second Edition* largely resolves the accessibility problem through its online presence. But while the online search facility provides convenient access to specific topics in the now eight volume, six million word *Dictionary of Economics*, no interface has yet been devised that makes browsing from a large online source a pleasurable activity for a rainy afternoon. To our delight, *The New Palgrave*'s publisher shares our view of the joys of dictionary-surfing, and we are thus pleased to present a new series, the "little blue books", to make some part of the *Dictionary* accessible in the hand or lap for teachers, students, and those who want to browse. While the volumes in this series contain only articles that appeared in the 2008 print edition, readers can, of course, refer to the online *Dictionary* and its expanding list of entries.

The selections in these volumes were chosen with several desiderata in mind: to touch on important problems, to emphasize material that may be of more general interest to economics beginners and yet still touch on the analytical core of modern economics, and to balance important theoretical concerns with key empirical debates. The 1987 Eatwell, Milgate and Newman *The New Palgrave: A Dictionary of Economics* was chiefly concerned with economic theory, both the history of its evolution and its contemporary state. The second edition has taken a different approach. While much progress has been made across the board in the 21 years between the first and second editions, it is particularly the flowering of empirical economics which distinguishes the present interval from the 61 year interval between Henry Higgs' *Palgrave's Dictionary of Political Economy* and *The New Palgrave*. It is fair to say that, in the long run, doctrine evolves more slowly than the database of facts, and so some of the selections in these volumes will age more quickly than others. This problem will be solved in the online *Dictionary* through an ongoing process of revisions and updates. While no such solution is available for these volumes, we have tried to choose topics which will give these books utility for some time to come.

Steven N. Durlauf
Lawrence E. Blume

Introduction

Game theory was well-represented in *The New Palgrave: A Dictionary of Economics*, with important entries by Robert Aumann, John Harsanyi, Martin Shubik and others. However, timing was such that *The New Palgrave* just missed the renaissance of non-cooperative game theory that took place in the early 1980s. 1982 saw the publication of Kreps and Wilson on sequential equilibrium and the 'Gang-of-Four' papers, Rubinstein's alternating-offer bargaining model, and John Maynard Smith's *Evolution and the Theory of Games*. While the ideas behind these key works had been percolating at the important workshops and conferences for some time, their publication set off an intellectual land rush into an area that had heretofore been explored only by a small and specialized community. By 1989 game theory had a new journal, *Games and Economic Behaviour*, and the publication of David Kreps' *A Course in Microeconomic Theory* and then Mas-Colell, Whinston and Green's *Microeconomic Theory* squeezed general equilibrium theory aside to make game theory an integral part of the graduate microeconomics core.

The game theoretic explosion of the 1980s and 1990s is documented in this volume. The four entries titled 'Epistemic game theory: . . .' and the four entries 'Learning and evolution in games: . . .' survey the epist and evolutionary foundations of non-cooperative game theory. Important applications covered in this volume include bargaining and mechanism design. Cooperative game theory is represented by 'Games in coalitional form' and 'The Shapley value'. Game theory never has been simply a subfield of economic analysis. 'Game theory and biology' surveys the application of evolutionary game theory to biological phenomena. Among the most exciting game theoretic developments of the last several years is the growing interaction between computer scientists and game theory. 'Computer science and game theory' and 'Graphical games' sample this growing body of work.

The 1950s were a revolutionary decade for the theory of decision making by single individuals. Similarly, the last twenty five years have seen a revolution in the understanding of rational decision making by interacting agents. In his afterword to the 60th anniversary edition of *The Theory of Games and Economic Behaviour*, Ariel Rubinstein writes about the theory of interacting decision making that, 'it is my impression that the well of game theory is relatively dry.' The *Dictionary* entries reprinted in this volume are proof of how full that well has been over the past two decades. Game theory has opened economics to the possibilities of modelling directly the interaction of individual economic actors in social situations. In doing so it has dramatically enriched the neoclassical view of markets. The well of decision theory ran dry in the 1960s only to refill under the influence of behavioural studies of choice. If the well of game theory is now empty, this too is in all likelihood only a hiatus, as our understanding of agent interaction is refreshed with insights from biology, sociology and computer science.

Steven N. Durlauf
Lawrence E. Blume

bargaining

In its simplest definition, 'bargaining' is a socio-economic phenomenon involving two parties, who can cooperate towards the creation of a commonly desirable surplus, over whose distribution the parties are in conflict.

The nature of the cooperation in the agreement and the relative positions of the two parties in the status quo before agreement takes place will influence the way in which the created surplus is divided. Many social, political and economic interactions of relevance fit this definition: a buyer and a seller trying to transact a good for money, a firm and a union sitting at the negotiation table to sign a labour contract, a couple deciding how to split the intra-household chores, two unfriendly countries trying to reach a lasting peace agreement, or out-of-court negotiations between two litigating parties.

In all these cases three basic ingredients are present: (*a*) the status quo, or the disagreement point, that is, the arrangement that is expected to prevail if an agreement is not reached; (*b*) the presence of mutual gains from cooperation; and (*c*) the multiplicity of possible cooperative arrangements, which split the resulting surplus in different ways.

If the situation involves more than two parties, matters are different, as set out in von Neumann and Morgenstern (1944). Indeed, in addition to the possibilities already identified of either disagreement or agreement among all parties, it is conceivable that an agreement be reached among only some of the parties. In multilateral settings, we are therefore led to distinguish pure bargaining problems, in which partial agreements of this kind are not possible because subcoalitions have no more power than individuals alone, from coalitional bargaining problems (or simply coalitional problems), in which partial agreements become a real issue in formulating threats and predicting outcomes. An example of a pure bargaining problem would be a round of talks among countries in order to reach an international trade treaty in which each country has veto power, whereas an example of a coalitional bargaining problem would be voting in legislatures. In this article we concentrate on pure bargaining problems, leaving the description of coalitional problems to other articles in the dictionary. We are likewise not concerned with the vast informal literature on bargaining, which conducts case studies and tries to teach bargaining skills for the 'real world' (for this purpose, the reader is referred to Raiffa, 1982).

Approaches to bargaining before game theory

Before the adoption of game theoretic techniques, economists deemed bargaining problems (also called bilateral monopolies at the time) indeterminate. This was certainly the position adopted by important economic theorists, including Edgeworth (1881) and Hicks (1932). More specifically, it was believed that the solution to a bargaining problem must satisfy both individual rationality and collective rationality properties: the former means that neither party should end up worse than in the status quo and the latter refers to Pareto efficiency. Typically, the set of individually rational and Pareto-efficient agreements is very large in a bargaining problem, and these theorists were inclined to believe that theoretical arguments could go no further than this in obtaining a prediction. To be able to obtain such a prediction, one would have to rely on extra-economic variables, such as the bargaining power and abilities of either party, their state of mind in negotiations, their religious beliefs, the weather and so on.

A precursor to the game theoretic study of bargaining, at least in its attempt to provide a more determinate prediction, is the analysis of Zeuthen (1930). This Danish economist formulated a principle whereby the solution to a bargaining problem was dictated by the two parties' risk attitudes (given the probability of breakdown of negotiations following the adoption of a tough position at the bargaining table). The reader is referred to Harsanyi (1987) for a version of Zeuthen's principle and its connection with Nash's bargaining theory. The remainder of this article deals with game theoretic approaches to bargaining.

The axiomatic theory of bargaining

Nash (1950) and Nash (1953) are seminal papers that constitute the birth of the formal theory of bargaining. Two assumptions are central in Nash's theory. First, bargainers are assumed to be fully rational individuals, and the theory is intended to yield predictions based exclusively on data relevant to them (in particular, the agents are equally skilful in negotiations, and the other extraneous factors mentioned above do not play a role).

Second, a bargaining problem is represented as a pair (S, d) in the utility space, where S is a compact and convex subset of $\mathbf{IR}^2$ – the feasible set of utility pairs – and $d \in \mathbf{IR}^2$ is the disagreement utility point. Compactness follows from standard assumptions such as closed productions sets and bounded factor endowments, and convexity is obtained if one uses expected utility and lotteries over outcomes are allowed. Also, the set S must include points that dominate the disagreement point, that is, there is a positive surplus to be enjoyed if agreement is reached and the question is how this surplus should be divided. As in most of game theory, by 'utility' we mean von Neumann–Morgenstern expected utility; there may be underlying uncertainty, perhaps related to the probability of breakdown of negotiations. We shall normalize the disagreement utilities to 0 (this is without loss of generality if one uses expected utility because any positive affine transformation of utility functions represents the same preferences over lotteries). The resulting bargaining problem is called a normalized problem.

With this second assumption, Nash is implying that all information relevant to the solution of the problem must be subsumed in the pair (S, d). In other words, two bargaining situations that may include distinct details ought to be solved in the same way if both reduce to the same pair (S, d) in utility terms. In spite of this, it is sometimes convenient to distinguish between feasible utility pairs (points in S) and feasible outcomes in physical terms (such as the portions of a pie to be created after agreement).

Following the two papers by Nash (1950; 1953), bargaining theory is divided into two branches, the so-called axiomatic and strategic theories. The axiomatic theory, born with Nash (1950), which most authors identify with a normative approach to bargaining, proposes a number of properties that a solution to any bargaining problem should have, and proceeds to identify the solution that agrees with those principles. Meanwhile, the strategic theory, initiated in Nash (1953), is its positive counterpart: the usual approach here is the exact specification of the details of negotiation (timing of moves, information available, commitment devices, outside options and threats) and the identification of the behaviour that would occur in those negotiation protocols. Thus, while the axiomatic theory stresses how bargaining *should* be resolved between rational parties according to some desirable principles, the strategic theory describes how bargaining *could* evolve in a non-cooperative extensive form in the presence of common knowledge of rationality. Interestingly, the two theories connect and complement one another.

The Nash bargaining solution

The first contribution to axiomatic bargaining theory was made by John Nash in his path-breaking paper published in 1950. Nash wrote it as a term paper in an international trade course that he was taking as an undergraduate at Carnegie, at the age of 17. At the request of his Carnegie economics professor, Nash mailed his term paper to John von Neumann, who had just published his monumental book with Oskar Morgenstern. John von Neumann may not have paid enough attention to a paper sent by an undergraduate at a different university, and nothing happened with the paper until Nash arrived in Princeton to begin studying for his Ph.D. in mathematics.

According to Nash (1950), a solution to bargaining problems is simply a function that assigns to each normalized utility possibility set S one of its feasible points (recall that the normalization of the disagreement utilities has already been performed). The interpretation is that the solution dictates a specific agreement to each possible bargaining situation. Examples of solutions are: (*a*) the disagreement solution, which assigns to each normalized bargaining problem the point (0,0), a rather pessimistic solution; and (*b*) the dictatorial solution with bargainer 1 as the dictator, which assigns the point in the Pareto frontier of the utility possibility set in which agent 2 receives 0 utility. Surely, neither of these solutions looks very appealing: while the former is not Pareto efficient because it does not exploit the gains from cooperation associated with an agreement, the latter violates the most basic fairness principle by being so asymmetric.

Nash (1950) proceeds by proposing four desirable properties that a solution to bargaining problems should have.

1. *Scale invariance or independence of equivalent utility representations.* Since the bargaining problem is formulated in von Neumann–Morgenstern utilities, if utility functions are re-scaled but they represent the same preferences, the solution should be re-scaled in the same fashion. That is, no fundamental change in the recommended agreement will happen following a re-normalization of utility functions; the solution will simply re-scale utilities accordingly.
2. *Symmetry.* If a bargaining problem is symmetric with respect to the 45 degree line, the solution must pick a point on it: in a bargaining situation in which each of the threats made by one bargainer can be countered by the other with exactly the same threat, the two should be equally treated by the solution. This axiom is sometimes called 'equal treatment of equals' and it ensures that the solution yields 'fair' outcomes.
3. *Pareto efficiency.* The solution should pick a point of the Pareto frontier. As elsewhere in welfare economics, efficiency is the basic ingredient of a normative approach to bargaining; negotiations should yield an efficient outcome in which all gains from cooperation are exploited.
4. *Independence of irrelevant alternatives (IIA).* Suppose a solution picks a point from a given normalized bargaining problem. Consider now a new normalized problem, a subset of the original, but containing the point selected earlier by the solution. Then, the solution must still assign the same point. That is, the solution should be independent of 'irrelevant' alternatives: as in a constrained optimization programme, the deleted alternatives are deemed irrelevant because they were not chosen when they were present, so their absence should not alter the recommended agreement.

With the aid of these four axioms, Nash (1950) proves the following result:

Theorem 1. There is a unique solution to bargaining problems that satisfies properties (1–4): it is the one that assigns to each normalized bargaining problem the point that maximizes the product of utilities of the two bargainers.

Today we refer to this solution as the 'Nash solution'. Although some of the axioms have been the centre of some controversy – especially his fourth, IIA, axiom – the Nash solution has remained as the fundamental piece of this theory, and its use in applications is pervasive.

Some features of the Nash solution ought to be emphasized. First, the theory can be extended to the multilateral case, in which there are $n \geq 3$ parties present in bargaining: in a multilateral problem, it continues to be true that the unique solution that satisfies (1–4) is the one prescribing that agreement in which the product of utilities is maximized. See Lensberg (1988) for an important alternative axiomatization.

Second, the theory is independent of the details of the negotiation-specific protocols, since it is formulated directly in the space of utilities. In particular, it can be applied to problems where the utilities are derived from only one good or issue, as well as those where utility comes from multiple goods or issues.

Third, perhaps surprisingly because risk is not explicitly part of Nash's story, it is worth noting that the Nash solution punishes risk aversion. All other things equal, it will award a lower portion of the surplus to a risk-averse agent. This captures an old intuition in previous literature that risk aversion is detrimental to a bargainer: afraid of the bargaining breakdown, the more risk-averse a person is, the more he will concede in the final agreement. For example, suppose agents are bargaining over how to split a surplus of size 1. Let the utility functions be as follows: $u_1(x_1) = x_1^\alpha$ for $0 < \alpha \leq 1$, and $u_2(x_2) = x_2$, where x_1 and x_2 are the non-negative shares of the surplus, which add up to 1. The reader can calculate that the Pareto frontier of the utility possibility set corresponds to the agreements satisfying the equation $u_1^{1/\alpha} + u_2 = 1$. Therefore, the Nash solution awards the utility vector $(u_1^*, u_2^*) = ((\frac{\alpha}{\alpha+1})^\alpha, \frac{1}{\alpha+1})$, corresponding to shares of the surplus $(x_1, x_2) = (\frac{\alpha}{\alpha+1}, \frac{1}{\alpha+1})$. Note how the smaller α is, the more risk-averse bargainer 1 is.

Fourth, Zeuthen's principle turns out to be related to the Nash solution (see Harsanyi, 1987): in identifying the bargainer who must concede next, the Nash product of utilities of the two proposals plays a role. See Rubinstein, Safra and Thomson (1992) for a related novel interpretation of the Nash solution.

Fifth, the family of asymmetric Nash solutions has also been used in the literature as a way to capture unequal bargaining powers. If the bargaining power of player i is $\beta_i \in [0, 1]$, $\Sigma_i \beta_i = 1$, the asymmetric Nash solution with weights (β_1, β_2) is defined as the function that assigns to each normalized bargaining problem the point where $u_1^{\beta_1} u_2^{\beta_2}$ is maximized.

The Kalai–Smorodinsky bargaining solution

Several researchers have criticized some of Nash's axioms, IIA especially. To see why, think of the following example, which begins with the consideration of a symmetric right-angled triangle S with legs of length 1. Clearly, efficiency and symmetry alone determine that the solution must be the point (1/2,1/2). Next, chop off the top part of the triangle to get a problem $T \subset S$, in which all points where $u_2 > 1/2$ have been deleted. By IIA, the Nash solution applied to the problem T is still the point (1/2, 1/2).

Kalai and Smorodinsky (1975) propose to retain the first three axioms of Nash's, but drop IIA. Instead, they propose an individual monotonicity axiom. To understand it, let $a_i(S)$ be the highest utility that agent i can achieve in the normalized problem S, and let us call it agent i's aspiration level. Let $a(S) = (a_1(S), a_2(S))$ be the utopia point, typically not feasible.

5. *Individual monotonicity.* If $T \subset S$ are two normalized problems, and $a_j(T) = a_j(S)$, the solution must award i a utility in S at least as high as in T.

We can now state the Kalai–Smorodinsky theorem:

Theorem 2. There is a unique solution to bargaining problems that satisfies properties (1, 2, 3, 5): it is the one that assigns to each normalized bargaining problem the intersection point of the Pareto frontier and the straight line segment connecting 0 and the utopia point.

Note how the Kalai–Smorodinsky solution awards the point (2/3,1/3) to the problem T of the beginning of this subsection. In general, while the Nash solution pays attention to local arguments (it picks out the point of the smooth Pareto frontier where the utility elasticity $(du_2/u_2)/(du_1/u_1)$ is 1), the Kalai–Smorodinsky solution is mostly driven by 'global' considerations, such as the highest utility each bargainer can obtain in the problem.

Other solutions

Although the two major axiomatic solutions are Nash's and Kalai–Smorodinsky's, authors have derived a plethora of other solutions also axiomatically (see, for example, Thomson, 1994, for an excellent survey). Among them, one should perhaps mention the egalitarian solution, which picks out the point of the Pareto frontier where utilities are equal. This is based on very different principles, much more tied to ethics of a certain kind and less to the principles governing bargaining between two rational individuals. In particular, note how it is not invariant to equivalent utility representations, because of the strong interpersonal comparisons of utilities that it performs.

The strategic theory of bargaining

Now we are interested in specifying the details of negotiations. Thus, while we may lose the generality of the axiomatic approach, our goal is to study reasonable procedures and identify rational behaviour in them. For this and the next section, some major references include Osborne and Rubinstein (1990) and Binmore, Osborne and Rubinstein (1992).

Nash's demand game

Nash (1953) introduces the first bargaining model expressed as a non-cooperative game. Nash's demand game, as it is often called, captures in crude form the force of commitment in bargaining. Both bargainers must demand simultaneously a utility level. If the pair of utilities is feasible, it is implemented; otherwise, there is disagreement and both receive 0. This game admits a continuum of Nash equilibrium outcomes, including every point of the Pareto frontier, as well as disagreement. The first message that emerges from Nash's demand game is the indeterminacy of equilibrium outcomes, commonplace in non-cooperative game theory. In the same paper, advancing ideas that would be developed a couple of decades later, Nash proposed a refinement of the Nash equilibrium concept based on the possibility of uncertainty around the true feasible set. The result was a selection of one Nash equilibrium outcome, which converges to the Nash solution agreement as uncertainty vanishes.

The model just described is referred to as Nash's demand game with fixed threats: following an incompatible pair of demands, the outcome is the fixed disagreement point. Nash (1953) also analysed a variable threats model. In it, the stage of simultaneous demands is preceded by another stage, in which bargainers choose threats. Given a pair of threats chosen in the first stage, the refinement argument is used to obtain the Nash solution of the induced problem in the ensuing subgame (where the threats determine an endogenous disagreement point). Solving the entire game is possible by backward induction, appealing to logic similar to that in von Neumann's minimax theorem; see Abreu and Pearce (2002) for a connection between the variable threats model and repeated games.

The alternating offers bargaining procedure

The following game elegantly describes a stylized protocol of negotiations over time. It was studied by Stahl (1972) under the assumption of an exogenous deadline (finite horizon game) and by Rubinstein (1982) in the absence of a deadline (infinite horizon game). Players 1 and 2 are bargaining over a surplus of size 1. The bargaining protocol is one of alternating offers. In period 0, player 1 begins by making a proposal, a division of the surplus, say $(x, 1-x)$, where $0 \leq x \leq 1$ represents the part of the surplus that she demands for herself. Player 2 can then either accept or reject this proposal. If he accepts, the proposal is implemented; if he rejects, a period must elapse for them to come back to the negotiation table, and at that time (period 1) the roles are reversed so that player 2 will make a new proposal $(y, 1-y)$, where $0 \leq y \leq 1$ is the fraction of surplus that he offers to player 1. Player 1 must then either accept the new proposal, in which case bargaining ends with $(y, 1-y)$ as the agreement, or reject it, in which case a period must elapse before player 1 makes a new proposal. In period 2, player 1 proposes $(z, 1-z)$, to which player 2 must respond, and so on. The T-period finite horizon game imposes the disagreement outcome, with zero payoffs, after T proposals have been rejected. On the other hand, in the infinite horizon version, there is always a new proposal in the next period after a proposal is rejected.

Both players discount the future at a constant rate. Let $\delta \in [0, 1)$ be the per period discount factor. To simplify, let us assume that utility is linear in shares of the surplus. Therefore, from a share x agreed in period t, a player derives a utility of $\delta^{t-1}x$. Note how utility is increasing in the share of the surplus (monotonicity) and decreasing in the delay with which the agreement takes place (impatience).

A strategy for a player is a complete contingent plan of action to play the game. That is, a strategy specifies a feasible action every time a player is called upon to act in the game. In a dynamic game, Nash equilibrium does little to restrict the set of predictions: for example, it can be shown that in the alternating offers games, any agreement $(x, 1-x)$ in any period t, $0 \leq t \leq T < \infty$, can be supported by a Nash equilibrium; disagreement is also a Nash equilibrium outcome.

The prediction that game theory gives in a dynamic game of complete information is typically based on finding its subgame perfect equilibria. A subgame perfect equilibrium (SPE) in a two-player game is a pair of strategies, one for each player, such that the behaviour specified by them is a best response to each other at every point in time (not only at the beginning of the game). By stipulating that players must choose a best response to each other at every instance that they are supposed to act, SPE rules out incredible threats: that is, at an SPE players have an incentive to carry out the threat implicit in their equilibrium strategy because it is one of the best responses to the behaviour they expect the other player to follow at that point.

In the alternating offers games described above, there is a unique SPE, in both the finite and the infinite horizon versions. The SPE in the finite horizon game is found by backward induction. For example, in the one-period game, the so-called ultimatum game, the unique SPE outcome is the agreement on the split (1,0): since the outcome of a rejection is disagreement, the responder will surely accept any share of $\epsilon > 0$, which implies that in equilibrium the proposer ends up taking the entire surplus. Using this intuition, one can show that the outcome of the two-period game is the immediate agreement on the split $(1 - \delta, \delta)$: anticipating that if negotiations get to the final period, player 2 (the proposer in that final period) will take the entire surplus, player 1 persuades him not to get there simply by offering him the present discounted value of the entire surplus, that is, δ , while she takes the rest. This logic continues and can be extended to any finite horizon. The sequence of SPE outcomes so obtained as the deadline $T \to \infty$ is shown to converge to the unique SPE of the infinite horizon game. This game, more challenging to solve since one cannot go to its last period to begin inducting backwards, was studied in Rubinstein (1982). We proceed to state its main theorem and discuss the properties of the equilibrium (see Shaked and Sutton, 1984, for a simple proof).

Theorem 3. Consider the infinite horizon game of alternating offers, in which both players discount the future at a per period rate of $\delta \in [0, 1)$. There exists a unique SPE of this game: it prescribes immediate agreement on the division $(\frac{1}{1+\delta}, \frac{\delta}{1+\delta})$.

The first salient prediction of the equilibrium is that there will not be any delay in reaching an agreement. Complete information – each player knows the other player's preferences – and the simple structure of the game are key factors to explain this.

The equilibrium awards an advantage to the proposer, as expressed by the discount factor: note how the proposer's share exceeds the responder's by a factor of $1/\delta$. Given impatience, having to respond to a proposal puts an agent in a delicate position, since rejecting the offer entails time wasted until the next round of negotiations. This is the source of the proposer's advantage. Of course, this advantage is larger, the larger the impatience of the responder: note how if $\delta = 0$ (extreme impatience), the equilibrium awards all the surplus to the proposer because her offer is virtually an ultimatum; on the other hand, as $\delta \to 1$, the first-mover advantage disappears and the equilibrium tends to an equal split of the surplus.

To understand how the equilibrium works and in particular how the threats employed in it are credible, consider the SPE strategies. Both players use the same strategy, and it is the following: as a proposer, each player always asks for $1/(1 + \delta)$ and offers $\delta/(1 + \delta)$ to the other party; as a responder, a player accepts an offer as long as the share offered to the responder is at least $\delta/(1 + \delta)$. Note how rejecting a share lower than $\delta/(1 + \delta)$ is credible, in that its consequence, according to the equilibrium strategies, is to agree in the next period on a split that awards the rejecting player a share of $1/(1 + \delta)$, whose present discounted value at the time the rejection occurs is exactly $\delta/(1 + \delta)$.

To appreciate the difference from Nash equilibrium, let us argue, for example, that the split (0,1) cannot happen in an SPE. This agreement happens in a Nash equilibrium, supported by strategies that ask player 1 to offer the whole pie to player 2, and player 2 to reject any other offer. However, the threat embodied in player 2's strategy is not credible: when confronted with an offer $(\in, 1 - \in)$ for $\delta < 1 - \in < 1$, player 2 will have to accept it, contradicting his strategy. Can the reader argue why the Nash equilibrium split (1,0) is not an SPE outcome either (because to do so one would need to employ non-credible threats)? Rubinstein (1982) shows that the same non-credible threats are associated with any division of the pie other than the one identified in the theorem.

The Rubinstein–Stahl alternating offers game provides an elegant model of how negotiations may take place over time, and its applications are numerous, including bargaining problems pertaining to international trade, industrial organization, or political economy. However, unlike Nash's axiomatic theory, its predictions are sensitive to details. This is no doubt one of its strengths because one can calibrate how those details may influence the theory's prediction, but it is also its weakness in terms of lack of robustness in predictive power.

Incomplete information

In a static framework, Chatterjee and Samuelson (1983) study a double auction. A buyer and a seller are trying to transact a good. Each proposes a price, and trade takes place at the average of the two prices if and only if the buyer's price exceeds the seller's. Each trader knows his own valuation for the good. However, there is incomplete information on each side concerning the other side's valuation. It can be shown that in any equilibrium of this game there are inefficiencies: given certain *ex post* valuations of buyer and seller, there should be trade, yet it is precluded because of incomplete information, which leads traders to play 'too tough'.

Let us now turn to bargaining over time. As pointed out above, one prediction of the Rubinstein–Stahl model is immediate agreement. This may clash with casual observation; one may simply note the existence of strikes, lockouts and long periods of disagreement in many actual negotiations. As a consequence, researchers have suggested the construction of models in which inefficiencies, in the form of delay in agreement, occur in equilibrium. The main feature of bargaining models with this property is incomplete information. (For delay in agreement that does not rely on incomplete information, see Fernandez and Glazer, 1991, Avery and Zemsky, 1994, and Busch and Wen, 1995.)

If parties do not know each other's preferences (impatience rate, per period fixed cost of hiring a lawyer, profitability of the agreement, and so on), the actions taken by the parties in the bargaining game may be intended to elicit some of the information that they do not have, or perhaps to reveal or misrepresent some of the information privately held.

One technical remark is in order. The typical approach is to reduce the uncertainty to a game of imperfect information through the specification of types in the sense of

Harsanyi (1967–8). In such games, SPE no longer constitutes an appropriate refinement of Nash equilibrium. The relevant equilibrium notions are perfect Bayesian equilibrium and sequential equilibrium, and in them the off-equilibrium path beliefs play an important role in sustaining outcomes. Moreover, these concepts are often incapable of yielding a determinate prediction in many games, and authors have in these cases resorted to further refinements. One problem of the refinements literature, however, is that it lacks strong foundations. Often the successful use of a given refinement in a game is accompanied by a bizarre prediction when the same concept is used in other games. Therefore, one should interpret these findings as showing the possibilities that equilibrium can offer in these contexts, but the theory here is far from giving a determinate answer.

Rubinstein (1985) studies an alternating offers procedure in which there is one-sided incomplete information (that is, while player 1 has uncertainty regarding player 2's preferences, player 2 is fully informed). Suppose there are two types of player 2: one of them is 'weaker' than player 1, while the other is 'stronger' (in terms of impatience or per period costs). This game admits many equilibria, and they differ as a function of parameter configurations. There are pooling equilibria, in which an offer from player 1 is accepted immediately by both types of player 2. More relevant to the current discussion, there are also separating equilibria, in which player 1's offer is accepted by the weak type of player 2, while the strong type signals his true preferences by rejecting the offer and imposing delay in equilibrium. These equilibria are also used to construct other equilibria with more periods of delay in agreement. Some authors (Gul and Sonnenschein, 1988) argue that long delays in equilibrium are the product of strong non-stationary behaviour (that is, a player behaves very differently in and out of equilibrium, as a function of changes in his beliefs). They show that imposing stationary behaviour limits the delay in agreement quite significantly. One advantage of stationary equilibria is their simplicity, but one problem with them is that they impose stationary beliefs (players hold beliefs that are independent of the history of play).

The analysis is simpler and multiplicity of equilibrium is less of a problem in games in which the uninformed party makes all the offers. Consider, for example, a version of the model in Sobel and Takahashi (1983). The two players are a firm and a union. The firm is fully informed, while the union does not know the true profitability of the firm. The union makes all offers in these wage negotiations, and there is discounting across periods. In equilibrium, different types of the firm accept offers at different points in time: firms whose profitability is not very high can afford to reject the first high wage offers made by the union to signal their private information, while very profitable firms cannot because delay in agreement hurts them too much.

Most papers have studied the case of private values asymmetric information (if a player knows her type, she knows her preferences), although the correlated values case has also been analysed (where knowing one's type is not sufficient to know one's utility function); see Evans (1989) and Vincent (1989). The case of two-sided asymmetric information, in which neither party is fully informed, has been treated, for example, in Watson (1998). In all these results, one is able to find equilibria with significant delay in agreement, implying consequent inefficiencies. Uncertainty may also be about the rationality of the opponent: for example, one may be bargaining with a 'behavioural type' who has an unknown threshold below which he will reject all proposals (see Abreu and Gul, 2000).

A more general approach is adopted by studies of mechanism design. The focus is not simply on explaining delay as an equilibrium phenomenon in a given extensive form. Rather, the question is whether inefficiencies are a consequence of equilibrium behaviour in any bilateral bargaining game with incomplete information. The classic contribution to this problem is the paper by Myerson and Satterthwaite (1983). In a bilateral trading problem in which there is two-sided private values asymmetric information and the types of each trader are drawn independently from overlapping intervals, there does not exist any budget-balanced mechanism satisfying incentive compatibility, interim individual rationality and *ex post* efficiency. All these are desirable properties for a trading mechanism. Budget balance implies that payoffs cannot be increased with outside funds. Incentive compatibility requires that each type has no incentive to misrepresent his information. Interim individual rationality means that no type can be worse off trading than not trading. Finally, *ex post* efficiency imposes that trade takes place if and only if positive gains from trade exist. This impossibility result is a landmark of the limitations of bargaining under incomplete information, and has generated an important literature that explores ways to overcome it (see for example Gresik and Satterthwaite, 1989, and Satterthwaite and Williams, 1989).

Indivisibilities in the units

One important way in which Rubinstein's result is not robust happens when there is only a finite set of possible offers to be made (see van Damme, Selten and Winter, 1990, and Muthoo, 1991). Indivisibilities make it impossible for an exact adjustment of offers to leave the responder indifferent; as a result, multiple and inefficient equilibria appear. The issue concerns how fine the grid of possible instantaneous offers is with respect to the time grid in which bargaining takes place. If the former is finer than the latter, Rubinstein's uniqueness goes through; otherwise it does not. There will be circumstances for which one or the other specification of negotiation rules will be more appropriate.

Multi-issue bargaining

The following preliminary observation is worth making: if offers are made in utility space or all issues must be bundled in every offer, Rubinstein's result obtains. Thus, the literature on multi-issue bargaining has looked at procedures that depart from these assumptions.

The first generation of papers with multiple issues assumed that the agenda – that is, the order in which the different issues are brought to the table – was exogenously given. Since each issue is bargained over one at a time, Rubinstein's uniqueness and efficiency result obtains, simply proceeding by backward induction on the issues.

Fershtman (1990; 2000) and Busch and Horstmann (1997) study such games, from which one learns the comparative statics of equilibrium when agendas are exogenously fixed. The next group of papers studies more realistic games where the agenda is chosen endogenously by the players. The main lesson from this line of work is that restricting the issues that a proposer can bring to the table is a source of inefficiencies. Inderst (2000) and In and Serrano (2003) study a procedure where agenda is totally unrestricted, that is, the proposer can make offers on any subset of remaining issues and, by exploiting trade-offs in the marginal rates of substitution between issues, Rubinstein's efficiency result is also found. In contrast, Lang and Rosenthal (2001) and In and Serrano (2004) construct multiple and inefficient equilibria (including those with arbitrarily long delay in agreement) when agenda restrictions are imposed. Finally, Weinberger (2000) considers multi-issue bargaining when the responder can accept selectively subsets of proposals and also finds inefficiencies if issues are indivisible.

Multilateral bargaining

Even within the case of pure bargaining problems, one needs to make a distinction between different ways to model negotiations. The first extension of the Rubinstein game to this case is due to Shaked, as reported in Osborne and Rubinstein (1990, p. 63); see also Herrero (1985). Today we refer to the Shaked/Herrero game as the 'unanimity game'. In it, one of the players, say player 1, begins by making a public proposal to the others. A proposal is a division of the unit of surplus available when agreement is reached. Players $2, \ldots, n$ then must accept or reject this proposal. If all agree, it is implemented immediately. If at least one of them rejects it, time elapses and in the next period another player, say player 2, will make a new proposal, and so on. Note how these rules reduce to Rubinstein's when there are only two players. However, the prediction emerging from this game is dramatically different. For values of the discount factor that are sufficiently high (if $\delta \geq 1/(n-1)$), every feasible agreement can be supported by an SPE and, in addition, equilibria with an arbitrary number of periods of delay in agreement show up. The intuition for this extreme result is that the unanimity required by the rules in order to implement an agreement facilitates a plethora of equilibrium behaviours. For example, let us see how in the case of $n = 3$ it is possible to sustain an agreement where all the surplus goes to player 3. If player 2 rejects it, the same split will be repeated in the continuation, so it is pointless to reject it. If player 1 changes her proposal to try to obtain a gain, it will be rejected by that responder who in the proposal receives less than 1/2 (there must be at least one). This rejector can be bribed with receiving the entire surplus in the continuation, whose present discounted value is at least 1/2 (recall $\delta \geq 1/2$), thereby rendering his rejection credible. Of course, the choice of player 3 as the one receiving the entire surplus is entirely arbitrary and, therefore, one can see how extreme multiplicity of equilibrium is a phenomenon inherent to the unanimity game. This multiplicity relies on non-stationary strategies, as it can be shown that there is a unique stationary SPE.

An alternative extension of the Rubinstein rules to multilateral settings is given by exit games; see Jun (1987), Chae and Yang (1994), Krishna and Serrano (1996). As an illustration, let us describe the negotiation rules of the Krishna–Serrano game. Player 1 makes a public proposal, a division of the surplus, and the others must respond to it. Those who accept it leave the game with the shares awarded by the proposer, while the rejectors continue to bargain with the proposer over the part of the surplus that has not been committed to any player. A new proposal comes from one of the rejectors, and so on. These rules also reduce to Rubinstein's if $n = 2$, but now the possibility of exiting the game by accepting a proposal has important implications for the predictive power of the theory. Indeed, Rubinstein's uniqueness is restored and the equilibrium found inherits the properties of Rubinstein's, including its immediate agreement and the proposer's advantage (the equilibrium shares are $1/[1+(n-1)\delta]$ for the proposer and $\delta/[1+(n-1)\delta]$ for each responder). Note how, given that the others accept, each responder is *de facto* immersed in a two-player Rubinstein game, so in equilibrium he receives a share that makes him exactly indifferent between accepting and rejecting: this explains the ratio $1/\delta$ between the proposer's and each responder's equilibrium shares. The sensitivity of the result to the exact specification of details is emphasized in other papers. Vannetelbosch (1999) shows that uniqueness obtains in the exit game even with a notion of rationalizability, weaker than SPE; and Huang (2002) establishes that uniqueness is still the result in a model that combines unanimity and exit, since offers can be made both conditional and unconditional to each responder. Baliga and Serrano (1995; 2001) introduce imperfect information in the unanimity and exit games (offers are not public, but made in personalized envelopes), and multiplicity is found in both, based on multiple off-equilibrium path beliefs. Merlo and Wilson (1995) propose a stochastic specification and also find uniqueness of the equilibrium outcome. In a model often used in political applications, Baron and Ferejohn (1989) study a procedure with random proposers in which the proposals are adopted if approved by simple majority (between the unanimity and exit procedures described).

Bargaining and markets

Bargaining theory provides a natural approach to understand how prices may emerge in markets as a consequence of the direct interaction of agents. One can characterize the outcomes of models in which the interactions of small groups of agents are formulated as bargaining games, and compare them with market outcomes such as competitive equilibrium allocations. If a connection between the two is found, one is giving an answer to the long-standing question of the origin of competitive equilibrium prices without having to resort to the story of the Walrasian auctioneer. If not, one can learn the importance of the frictions in the model that may be preventing such a connection. Both kinds of results are valuable for economic theory.

Small markets

Models have been explored in which two agents are bargaining, but at least one of them may have an outside option (see Binmore, Shaked and Sutton, 1988). Thus, the bargaining pair is part of a larger economic context, which is not explicitly modelled. In the simplest specification, uniqueness and efficiency of the equilibrium is found. In the equilibrium, the outside option is used if it pays better than the Rubinstein equilibrium; otherwise it is ignored. Jehiel and Moldovanu (1995) show that delays may be part of the equilibrium when the agreement between a seller and several buyers is subject to externalities among the buyers: a buyer may have an incentive to reject an offer in the hope of making a different buyer accept the next offer and free-ride from that agreement. In general, these markets involving a small number of agents do not yield competitive allocations because market power is retained by some traders (see Rubinstein and Wolinsky, 1990).

Large markets under complete information

The standard model assumes a continuum of agents who are matched at random, typically in pairs, to perform trade of commodities. If a pair of agents agrees on a trade, they break the match. In simpler models, all traders leave the market after they trade once. In the more general models agents may choose either to leave and consume, or to stay in the market to be matched anew. Some authors have studied steady-state versions, in which the measure of traders leaving the market every period is offset exactly by the same measure of agents entering the market. In contrast, non-steady state models do not keep the measure of active traders constant (one prominent class of non-steady state models is that of one-time entry, in which after the initial period there is no new entry; certain transacting agents exit every period, so the market size dwindles over time). The analysis has been performed with discounting (where δ is the common discount factor that is thought of as being near 1) or without it: in both cases the idea is to describe frictionless or almost frictionless conditions (for example, Muthoo, 1993, considers several frictions and the outcomes that result when some, but not all, of them are removed).

The first models were introduced by Diamond and Maskin (1979), Diamond (1981), and Mortensen (1982), and they used the Nash solution to solve each bilateral bargaining encounter. Later each pairwise meeting has been modelled by adopting a procedure from the strategic theory.

The most general results in this area are provided by Gale (1986a; 1986b; 1986c; 1987). First, in a partial equilibrium set-up, a market for an indivisible good is analysed in Gale (1987), under both steady state and non-steady-state assumptions. The result is that all equilibrium outcomes yield trade at the competitive price when discounting is small: in all equilibria trade tends to take place at only one price, and that price must be the competitive price because it is the one that maximizes each trader's expected surplus. This generalizes a result of Binmore and Herrero (1988) and clarifies an earlier claim made by Rubinstein and Wolinsky (1985). Rubinstein and Wolinsky analysed the market in steady state and claimed that the market outcome was different from the competitive one. Their claim is justified if one measures the sets of traders in terms of the stocks present in the market, but Gale (1987) argues convincingly that, given the steady state imposed on the solution concept, it is the flow of agents into the market every period, not the total stock, that should comprise the relevant demand and supply curves. When this is taken into account, all prices are competitive because the measure of transacting sellers is the same as that of the transacting buyers.

In a more general model, Gale (1986a; 1986b; 1986c) studies an exchange economy with an arbitrary number of divisible goods. Now there is no discounting and agents can trade in as many periods as they wish before they leave the market place. Only after an agent rejects a proposal can he leave the market. Under a number of technical assumptions, Gale shows once again that all the equilibrium outcomes of his game are Walrasian:

Theorem 4. At every market equilibrium, each agent leaves the market with the bundle x_k with probability 1, where the list of such bundles is a Walrasian allocation of the economy.

Different versions of this result are proved in Gale (1986a; 1986c) and in Osborne and Rubinstein (1990). Also, Kunimoto and Serrano (2004) obtain the same result under substantially weaker assumptions on the economy, thereby emphasizing the robustness of the connection between the market equilibria of this decentralized exchange game and the Walrasian allocations of the economy. There are two key steps in this argument: first, one establishes that, since pairs are trading, pairwise efficiency obtains, which under some conditions leads to Pareto efficiency; and second, the equilibrium strategies imply budget balance so that each agent cannot end up with a bundle that is worth more than his initial endowment (given prices supporting the equilibrium allocation, already known to be efficient).

Dagan, Serrano and Volij (2000) also show a Walrasian result, but in their game the trading groups are coalitions of any finite size: in their proof, the force of the core equivalence theorem is exploited. One final comment is pertinent at this point. Some authors (for example, Gale, 2000) question the use of coalitions of any finite size in the trading procedure because the 'large' size of some of those groups seems to clash with the 'decentralized' spirit of these mechanisms. On the other hand, one can also argue that for the procedure to allow trade only in pairs, some market authority must be keeping track of this, making sure that coalitions of at least three agents are 'illegal'. Both trading technologies capture appealing aspects of decentralization, depending on the circumstances, and the finding is that either one yields a robust connection with the teachings of general equilibrium theory in frictionless environments. This is one more instance of the celebrated equivalence principle: in models involving a large number of agents, game theoretic predictions tend to converge, under some conditions, to the set of competitive allocations.

Large markets under incomplete information

If the asymmetric information is of the private values type, the same equivalence result is obtained between equilibria of matching and bargaining models and Walrasian allocations. This message is found, for example, in Rustichini, Satterthwaite and

Williams (1994), Gale (1987) and Serrano (2002). In the latter model, for instance, some non-Walrasian outcomes are still found in equilibrium, but they can be explained by features of the trading procedure that one could consider as frictions, such as a finite set of prices and finite sets of traders' types.

The result is quite different when asymmetric information goes beyond private values. For example, Wolinsky (1990) studies a market with pairwise meetings in which there is uncertainty regarding the true state of the world (which determines the true quality of the good being traded). Some traders know the state, while others do not, and there are uninformed traders among buyers and sellers (two-sided asymmetric information). The analysis is performed in steady state. To learn the true state, uninformed traders sample agents of the opposite side of the market. However, each additional meeting is costly due to discounting. The relevant question is whether information will be transmitted from the informed to the uninformed when discounting is removed. Wolinsky's answer is in the negative: as the discount factor $\delta \to 1$, a non-negligible fraction of uninformed traders transacts at a price that is not *ex post* individually rational. It follows that the equilibrium outcomes do not approximate those given by a fully revealing rational expectations equilibrium (REE). The reason for this result is that, while as $\delta \to 1$ sampling becomes cheaper and therefore each uninformed trader samples more agents, this is true on both sides, so that uninformed traders end up trying to learn from agents that are just as uninformed as they are. Serrano and Yosha (1993) overturn this result when asymmetric information is one-sided: in this case, although the noise force behind Wolinsky's result is not operative because of the absence of uninformed traders on one side, there is a negative force that works against learning, which is that misrepresenting information becomes cheaper for informed traders as $\delta \to 1$. The analysis in Serrano and Yosha's paper shows that, under steady state restrictions, the learning force is more powerful than the misrepresentation one, and convergence to REE is attained. Finally, Blouin and Serrano (2001) perform the analysis without the strong steady-state assumption, and show that with both information structures (one-sided and two-sided asymmetries) the result is negative: Wolinsky's noise force in the two-sided case continues to be crucial, while misrepresentation becomes very powerful in the one-sided model because of the lack of fresh uninformed traders. In these models, agents have no access to aggregate market signals; information is heavily restricted because agents observe only their own private history. It would be interesting to analyse other procedures where information may flow more easily.

ROBERTO SERRANO

See also **Nash program; Shapley value.**

Bibliography

Abreu, D. and Gul, F. 2000. Bargaining and reputation. *Econometrica* 68, 85–118.

Abreu, D. and Pearce, D. 2002. Bargaining, reputation and equilibrium selection in repeated games. Mimeo. Princeton University.

Avery, C. and Zemsky, P. 1994. Money burning and multiple equilibria in bargaining. *Games and Economic Behavior* 7, 154–68.

Baliga, S. and Serrano, R. 1995. Multilateral bargaining with imperfect information. *Journal of Economic Theory* 67, 578–89.

Baliga, S. and Serrano, R. 2001. Multilateral negotiations with private side-deals: a multiplicity example. *Economics Bulletin* 3(1), 1–7.

Baron, D. and Ferejohn, J. 1989. Bargaining in legislatures. *American Political Science Review* 83, 1181–206.

Binmore, K. and Herrero, M. 1988. Matching and bargaining in dynamic models. *Review of Economic Studies* 55, 17–31.

Binmore, K., Osborne, M. and Rubinstein, A. 1992. Non-cooperative models of bargaining. In *Handbook of Game Theory with Economic Applications*, vol. 1, ed. J. Aumann and S. Hart. New York: Elsevier.

Binmore, K., Shaked, A. and Sutton, J. 1988. An outside option experiment. *Quarterly Journal of Economics* 104, 753–70.

Blouin, M. and Serrano, R. 2001. A decentralized market with common values uncertainty: non-steady states. *Review of Economic Studies* 68, 323–46.

Busch, L.-A. and Horstmann, I. 1997. Bargaining frictions, bargaining procedures and implied costs in multiple-issue bargaining. *Economica* 64, 669–80.

Busch, L.-A. and Wen, Q. 1995. Perfect equilibria in a negotiation model. *Econometrica* 63, 545–65.

Chae, S. and Yang, J.-A. 1994. An *n*-person pure bargaining game. *Journal of Economic Theory* 62, 86–102.

Chatterjee, K. and Samuelson, W. 1983. Bargaining under incomplete information. *Operations Research* 31, 835–51.

Dagan, N., Serrano, R. and Volij, O. 2000. Bargaining, coalitions and competition. *Economic Theory* 15, 279–96.

Diamond, P. 1981. Mobility costs, frictional unemployment, and efficiency. *Journal of Political Economy* 89, 798–812.

Diamond, P. and Maskin, E. 1979. An equilibrium analysis of search and breach of contract I: steady states. *Bell Journal of Economics* 10, 282–316.

Edgeworth, F. 1881. Mathematical psychics. In *F. Y. Edgeworth's Mathematical Psychics and Further Papers on Political Economy*, ed. P. Newman. Oxford: Oxford University Press, 2003.

Evans, R. 1989. Sequential bargaining with correlated values. *Review of Economic Studies* 56, 499–510.

Fernandez, R. and Glazer, J. 1991. Striking for a bargain between two completely informed agents. *American Economic Review* 81, 240–52.

Fershtman, C. 1990. The importance of the agenda in bargaining. *Games and Economic Behavior* 2, 224–38.

Fershtman, C. 2000. A note on multi-issue two-sided bargaining: bilateral procedures. *Games and Economic Behavior* 30, 216–27.

Gale, D. 1986a. Bargaining and competition. Part I: characterization. *Econometrica* 54, 785–806.

Gale, D. 1986b. Bargaining and competition. Part II: existence. *Econometrica* 54, 807–18.

Gale, D. 1986c. A simple characterization of bargaining equilibrium in a large market without the assumption of dispersed characteristics. Working Paper 86–05. Philadelphia: CARESS, University of Pennsylvania.

Gale, D. 1987. Limit theorems for markets with sequential bargaining. *Journal of Economic Theory* 43, 20–54.

Gale, D. 2000. *Strategic Foundations of General Equilibrium – Dynamic Matching and Bargaining Games*. Cambridge: Cambridge University Press.

Gresik, T. and Satterthwaite, M. 1989. The rate at which a simple market converges to efficiency as the number of traders increases: an asymptotic result for optimal trading mechanisms. *Journal of Economic Theory* 48, 304–32.

Gul, F. and Sonnenschein, H. 1988. On delay in bargaining with one-sided uncertainty. *Econometrica* 56, 601–11.

Harsanyi, J. 1967–8. Games with incomplete information played by Bayesian players. Parts I, II and III. *Management Science* 14, 159–82, 320–34, 486–502.

Harsanyi, J. 1987. Bargaining. In *The New Palgrave Dictionary of Economics*, vol. 1, ed. J. Eatwell, M. Milgate and P. Newman. London: Macmillan.

Herrero, M. 1985. N-player bargaining and involuntary underemployment. In A strategic bargaining approach to market institutions. Ph.D. thesis. London: London School of Economics.

Hicks, J. 1932. *The Theory of Wages*. London: Macmillan; New York: St. Martin Press, 1963.

Huang, C.-Y. 2002. Multilateral bargaining: conditional and unconditional offers. *Economic Theory* 20, 401–12.

In, Y. and Serrano, R. 2003. Agenda restrictions in multi-issue bargaining II: unrestricted agendas. *Economics Letters* 79, 325–31.

In, Y. and Serrano, R. 2004. Agenda restrictions in multi-issue bargaining. *Journal of Economic Behavior and Organization* 53, 385–99.

Inderst, R. 2000. Multi-issue bargaining with endogenous agenda. *Games and Economic Behavior* 30, 64–82.

Jehiel, P. and Moldovanu, B. 1995. Cyclical delay in bargaining with externalities. *Review of Economic Studies* 62, 619–37.

Jun, B. 1987. A structural consideration on 3-person bargaining. In Essays on topics in economic theory. Ph. D. thesis. Philadelphia: Department of Economics, University of Pennsylvania.

Kalai, E. and Smorodinsky, M. 1975. Other solutions to Nash's bargaining problem. *Econometrica* 43, 513–18.

Krishna, V. and Serrano, R. 1996. Multilateral bargaining. *Review of Economic Studies* 63, 61–80.

Kunimoto, T. and Serrano, R. 2004. Bargaining and competition revisited. *Journal of Economic Theory* 115, 78–88.

Lang, K. and Rosenthal, R. 2001. Bargaining piecemeal or all at once? *Economic Journal* 111, 526–40.

Lensberg, T. 1988. Stability and the Nash solution. *Journal of Economic Theory* 45, 330–41.

Merlo, A. and Wilson, C. 1995. A stochastic model of sequential bargaining with complete information. *Econometrica* 63, 371–99.

Mortensen, D. 1982. Property rights and efficiency in mating, racing, and related games. *American Economic Review* 72, 968–79.

Muthoo, A. 1991. A note on bargaining over a finite set of feasible agreements. *Economic Theory* 1, 290–92.

Muthoo, A. 1993. Sequential bargaining and competition. *Economic Theory* 3, 353–63.

Myerson, R. and Satterthwaite, M. 1983. Efficient mechanisms for bilateral trading. *Journal of Economic Theory* 29, 265–81.

Nash, J. 1950. The bargaining problem. *Econometrica* 18, 155–62.

Nash, J. 1953. Two person cooperative games. *Econometrica* 21, 128–40.

Osborne, M. and Rubinstein, A. 1990. *Bargaining and Markets*. San Diego, CA: Academic Press. Online. Available at http://www.economics.utoronto.ca/osborne/bm, accessed 4 August 2005.

Raiffa, H. 1982. *The Art and Science of Negotiation*. Cambridge, MA: Harvard University Press.

Rubinstein, A. 1982. Perfect equilibrium in a bargaining model. *Econometrica* 50, 97–109.

Rubinstein, A. 1985. A bargaining model with incomplete information on time preferences. *Econometrica* 53, 1151–72.
Rubinstein, A., Safra, Z. and Thomson, W. 1992. On the interpretation of the Nash bargaining solution and its extension to non-expected utility preferences. *Econometrica* 60, 1171–86.
Rubinstein, A. and Wolinsky, A. 1985. Equilibrium in a market with sequential bargaining. *Econometrica* 53, 1133–50.
Rubinstein, A. and Wolinsky, A. 1990. Decentralized trading, strategic behavior and the Walrasian outcome. *Review of Economic Studies* 57, 63–78.
Rustichini, A., Satterthwaite, M. and Williams, S. 1994. Convergence to efficiency in a simple market with incomplete information. *Econometrica* 62, 1041–63.
Satterthwaite, M. and Williams, S. 1989. The rate of convergence to efficiency in the buyer's bid double auction as the market becomes large. *Review of Economic Studies* 56, 477–98.
Serrano, R. 2002. Decentralized information and the Walrasian outcome: a pairwise meetings market with private values. *Journal of Mathematical Economics* 38, 65–89.
Serrano, R. and Yosha, O. 1993. Information revelation in a market with pairwise meetings: the one-sided information case. *Economic Theory* 3, 481–99.
Shaked, A. and Sutton, J. 1984. Involuntary unemployment as a perfect equilibrium in a bargaining model. *Econometrica* 52, 1351–64.
Sobel, J. and Takahashi, I. 1983. A multi-stage model of bargaining. *Review of Economic Studies* 50, 411–26.
Stahl, I. 1972. *Bargaining Theory.* Stockholm: Stockholm School of Economics.
Thomson, W. 1994. Cooperative models of bargaining. In *Handbook of Game Theory with Economic Applications*, vol. 2, ed. R. Aumann and S. Hart. New York: Elsevier.
van Damme, E., Selten, R. and Winter, E. 1990. Alternating bid bargaining with a smallest money unit. *Games and Economic Behavior* 2, 188–201.
Vannetelbosch, V. 1999. Rationalizability and equilibrium in n-person sequential bargaining. *Economic Theory* 14, 353–71.
Vincent, D. 1989. Bargaining with common values. *Journal of Economic Theory* 48, 47–62.
von Neumann, J. and Morgenstern, O. 1944. *Theory of Games and Economic Behavior.* Princeton, NJ: Princeton University Press.
Watson, J. 1998. Alternating-offer bargaining with two-sided incomplete information. *Review of Economic Studies* 65, 573–94.
Weinberger, C. 2000. Selective acceptance and inefficiency in a two-issue complete information bargaining game. *Games and Economic Behavior* 31, 262–93.
Wolinsky, A. 1990. Information revelation in a market with pairwise meetings. *Econometrica* 58, 1–23.
Zeuthen, F. 1930. *Problems of Monopoly and Economic Warfare.* London: Kegan Paul.

behavioural economics and game theory

In traditional economic analysis, as well as in much of behavioural economics, the individual's motivations are summarized by a utility function (or a preference relation) over possible payoff-relevant outcomes while his cognitive limitations are described as incomplete information. Thus, the standard economic theory of the individual is couched in the language of constrained maximization and statistical inference.

The approach gains its power from the concise specification of payoff-relevant outcomes and payoffs as well as a host of auxiliary assumptions. For example, it is typically assumed that the individual's preferences are well behaved: that is, they can be represented by a function that satisfies conditions appropriate for the particular context such as continuity, monotonicity, quasi-concavity, and so on. When studying behaviour under uncertainty, it is often assumed that the individual's preference obeys the expected utility hypothesis. More importantly, it is assumed that the individual's subjective assessments of the underlying uncertainty are reasonably close to the observed distributions of the corresponding variables. Even after all these bold assumptions, the standard model would say little if the only relevant observation regarding the utility function is one particular choice outcome. Thus, economists will often assume that the same utility function is relevant for the individual's choices over some stretch of time during which a number of related choices are made. One hopes that these observations will generate enough variation to identify the decision-maker's (DM's) utility function. If not, the analyst may choose to utilize choice observations from different contexts to identify the individual's preferences or make parametric assumptions. The analyst may even pool information derived from observed choices of different individuals to arrive at a representative utility function.

1. Experimental challenges to the main axioms of choice theory

The simplest type of criticism of the standard theory accepts the usual economic abstractions and the standard framework but questions specific assumptions within this framework.

1.1 The independence axiom

Allais (1953) offers one of the earliest critiques of standard decision-theoretic assumptions. In his experiment, he provides two pairs of binary choices and shows that many subjects violate the expected utility hypothesis, in particular, the independence axiom. Allais's approach differs from the earlier criticisms: Allais questions an explicit axiom of choice theory rather than a perceived implicit assumption such as 'rationality'. Furthermore, he does so by providing a simple and clear experimental test of the particular assumption.

Subsequent research documents related violations of the independence axiom and classifies them. Researchers have responded to Allais's critique by developing a class of models that either abandons the independence axiom or replaces it with weaker alternatives. The agents in these models still maximize their preference and still reduce uncertainty to probabilistic assessments (that is, they are probabilistically sophisticated), but have preferences over lotteries that fail the independence axiom.

Non-expected utility preferences pose a difficulty for game theory: because many non-expected utility theories do not lead to quasi-concave utility functions, standard fixed point theorems cannot be used to establish the existence of Nash equilibrium. Crawford (1990) shows that if one interprets mixed strategies not as random behaviour but as the opponents' uncertainty regarding this behaviour, then the required convex-valuedness of the best response correspondence can be restored and existence of Nash equilibrium can be ensured.

In dynamic games, abandoning the independence axiom poses even more difficult problems. Without the independence axiom, conditional preferences at a given node of an extensive form game (or a decision-tree) depend on the unrealized payoffs earlier in the game. The literature has dealt with this problem in two ways: first, by assuming that the DM maximizes his conditional preference at each node (for a statement and defence of this approach, see Machina, 1989). This approach leads to dynamically consistent behaviour, since the DM ends up choosing the optimal strategy for the reduced (normal form) game. However, it is difficult to compute optimal strategies once conditional preference depends on the entire history of unrealized outcomes. The second approach rejects dynamic consistency and assumes that at each node the DM maximizes his unconditional preference given his prediction of future behaviour. Thus, in the second approach, each node is treated as a distinct player and a subgame perfect equilibrium of the extensive form game is computed. Game-theoretic models that abandon the independence axiom have favoured the second approach. Such models have been used to study auctions.

1.2 Redefining payoffs: altruism and fairness

The next set of behavioural criticisms question common assumptions regarding deterministic outcomes. Consider the *ultimatum game*: Player 1 chooses some amount $x \leq 100$ to offer to Player 2. If Player 2 accepts the offer, 2 receives x and 1 receives $100-x$; If 2 rejects, both players receive 0. Suppose the rewards are measured in dollars and Player 1 has to make his offer in multiples of a dollar. It is easy to verify that if the players care only about their own financial outcome, there is no subgame perfect Nash equilibrium of this game in which Player 1 chooses $x>1$. Moreover, in every equilibrium, any offer $x>0$ must be accepted with probability 1. Contrary to these predictions, experimental evidence indicates that small offers are often rejected. Hence, subjects in the Player 2 role resent either the unfairness of the (99,1) outcome, or Player 1's lack of generosity. Moreover, many experimental subjects anticipate this response and make more generous offers to ensure acceptance. Even in the version of this game in which Player 2 does not have the opportunity to reject (that is,

Player 1 is a *dictator*), Player 1 often acts altruistically and gives a significant share to Player 2.

More generally, there is empirical evidence that suggests that economic agents care not only about their physical outcomes but also about the outcomes of their opponents and how the two compare. Within game theory, this particular behavioural critique has been influential and has led to a significant theoretical literature on social preferences (see, for example, Fehr and Schmidt, 1999).

1.3 Redefining the objects of choice: ambiguity, timing of resolution of uncertainty, and preference for commitment

The next set of behavioural criticisms points out how the standard definition of outcome or consequence is inadequate. The literature on ambiguity questions probabilistic sophistication; that is, the idea that all uncertainty can be reduced to probability distributions. Ellsberg (1961) provides the original statement of this criticism. Consider the following choice problem: there are two urns; the first contains 50 red balls and 50 blue balls; the second contains 100 balls, each of which is either red or blue. The DM must select an urn and announce a colour. Then a ball will be drawn from the urn he selects. If the colour of the ball is the same as the colour the DM announces, he wins 100 dollars. Otherwise the DM gets zero. Experimental results indicate that many DMs are indifferent between (urn 1, red) and (urn 1, blue) but they strictly prefer either of these choices to (urn 2, red) and (urn 2, blue). If the DM were probabilistically sophisticated and assigned probability p to choosing a red ball from urn 1 and q to choosing a red ball from urn 2, the preferences above would indicate that $p = 1 - p$, $p > q$, and $p > 1 - q$, a contradiction. Hence, many DMs are not probabilistically sophisticated.

Ellsberg's experiment has lead to choice-theoretic models where agents are not probabilistically sophisticated and have an aversion to ambiguity; that is, the type of uncertainty associated with urn 2. Recent contributions have investigated auctions with ambiguity-averse bidders and mechanism design with ambiguity aversion.

Other developments in behavioural choice theory that fall into this category have had limited impact on game-theoretic research. For example, Kreps and Porteus (1978) introduce the notion of a temporal lottery to analyse economic agents' preference over the timing of resolution of uncertainty. The Kreps–Porteus model has been extremely influential in dynamic choice theory and asset pricing but has had less impact in strategic analysis.

Kreps (1979) takes as his primitive individuals' preferences over sets of objects. Hence, an object similar to the indirect utility function of demand theory defines the individual. Kreps uses this framework to analyse preference for flexibility. So far, there has been limited analysis of preference for flexibility in strategic problems.

Gul and Pesendorfer (2001) use preferences over sets to analyse agents who have a preference for commitment (an alternative approach to preference for commitment is discussed in Section (3.2)). The GP model has been used to analyse some mechanism design problems.

2. Limitations of the decision-maker

The work discussed in Section 1 explores alternative formulations of economic consequences to identify preference-relevant considerations that are ignored in standard economic analysis. The work discussed in this section provides a more fundamental challenge to standard economics. This research seeks alternatives to common assumptions regarding economic agents' understanding of their environments and their cognative/computational abilities.

2.1 Biases and heuristics

Many economic models are stated in subjectivist language. Hence probabilities, whether they represent the likelihood of future events or the individual's own ignorance of past events, are the DMs' personal beliefs rather than objective frequencies. Similarly, the DM's utility function is a description of his behaviour in a variety of contingencies rather than an assessment of the intrinsic value of the possible outcomes. Nevertheless, when economists use these models to analyse particular problems, the subjective probabilities (and sometimes other parameters) are often calibrated or estimated by measuring objective frequencies (or other objective variables).

Psychology and economics research has questioned the validity of this approach. Tversky and Kahneman (1974) identify systematic biases in how individuals make choices under uncertainty. This research has led to an extensive literature on heuristics and biases. Consider the following:

(a) Which number is larger $P(A|B)$ or $P(A \cap C|B)$? Clearly, $P(A|B)$ is the larger quantity; conditional on B or unconditionally, $A \cap C$ can never be more likely than A. Yet, when belonging to set C is considered 'typical' for a member of B, many subjects state that $A \cap C$ conditional on B is more likely than A conditional on B.

(b) Randomly selected subjects are tested for a particular condition. In the population, 95 per cent are healthy. The test is 90 per cent accurate; that is, a healthy subject tests negative and a subject having the condition tests positive with probability 0.9. If a randomly chosen person tests positive, what is the probability that he is ill? In such problems, subjects tend to ignore the low prior probability of having the condition and come up with larger estimates than the correct answer (less than one-third in this example).

Eyster and Rabin's (2005) analysis of auctions offers an example of a strategic model of biased decision-making. This work focuses on DMs' tendency to overemphasize their own (private) information at the expense of the information that is revealed through the strategic interaction.

2.2 Evolution and learning

As in decision theory, it is possible to state nearly all the assumptions of game theory in subjectivist language (see, for example, Aumann and Brandenburger, 1995). Hence, one can define Nash equilibrium as a property of players' beliefs. Of course, Nash equilibrium beliefs (together with utility maximization) will impose restrictions on observable behaviour, but these restrictions will fall short of demanding that the observed frequency of actions profiles constitute a Nash equilibrium. The theory of evolutionary games searches for dynamic mechanisms that lead to equilibrium behaviour, where equilibrium is identified with observable decisions (as opposed to beliefs) of individuals. The objective is to describe how equilibrium may emerge and which equilibria are more likely to emerge through repeated interaction in a setting where the typical epistemic assumptions of equilibrium analysis fail initially. Thus, such models are used both to justify Nash (or weaker) equilibrium notions and to justify refinements of these notions.

2.3 Cognitive limitations and game theory

Some game theoretic solution concepts require iterative procedures. For example, computing rationalizable outcomes in normal form games or finding backward induction solutions in extensive form games involves an iterative procedure that yields a smaller game after each step. The process ends when the final game, which consists exclusively of actions that constitute the desired solution, is reached. In principle, the number of steps needed to reach the solution can be arbitrarily large. Ho, Camerer and Weigelt (1998) observe that experimental subjects appear to carry out at most the first two steps of these procedures.

This line of work focuses both on organizing observed violations of standard game theoretic solutions concepts and interpreting the empirical regularities as the foundation of a behavioural notion of equilibrium.

3. Alternative models of the individual

The work discussed in this section poses the most fundamental challenge to the standard economic model of the individual. This work questions the usefulness of constrained maximization as a framework of economic analysis, or at least argues for a fundamentally different set of constraints.

3.1 Prospect theory and framing effects

Consider the following pair of choices (Tversky and Kahneman, 1981): an unusual disease is expected to kill 600 people. Two alternative programmes to combat the disease have been proposed.

Programme A will save 200; with Programme B, there is a one-third probability that 600 people will be saved, and a two-thirds probability that no one will be saved.

Next, consider the following restatement of what would appear to be the same options:

If Programme C is adopted 400 people will die; with Programme D, there is a one-third probability that nobody will die, and a two-thirds probability that 600 people will die.

Among subjects given a choice between A and B, most choose the safe option A, while the majority of the subjects facing the second pair of choices choose the risky option D.

Kahneman and Tversky's (1979) prospect theory combines issues discussed in Sections (1.1) and (2.2), with a more general critique of standard economic models, or at least of how such models are used in practice. Thus, while a standard model might favour a level of abstraction that ignores the framing issue above, Kahneman and Tversky (1979) argue that identifying the particular frame that the individual is likely to confront should be central to decision theory. In particular, these authors focus on the differential treatment of gains and losses. Prospect theory defines preferences not over lotteries of terminal wealth but over gains and losses, measured as differences from a status quo. In applications, the status quo is identified in a variety of ways.

For example, Köszegi and Rabin (2005–6) provide a theory of the status quo and utilize the resulting model to study a monopoly problem. In their theory, the DM's optimal choice becomes the status quo. Thus, the simplest form of the Köszegi–Rabin model defines optimal choices from a set A as $C(A) = \{x \in A | U(x, x) \geq U(y, x) \forall y \in A\}$. Hence, $x \in A$ is deemed to be a possible choice from A if the DM who views x as his reference point does not strictly prefer some other alternative y.

The three lines of work discussed below all represent a fundamental departure from the standard modelling of economic decisions: they describe behaviour as the outcome of a game even in a single person problem.

3.2 Preference reversals

Strotz (1955–6) introduces the idea of dynamic inconsistency: the possibility that a DM may prefer to consume x in period 2 to consuming y in period 1, if he makes the choice in period 0, but may have the opposite preference if he makes the choice in period 1. Strotz suggests that the appropriate way to model dynamically inconsistent behaviour is to assume that the period 0 individual treats his period 1 preference (and the implied behaviour) as a constraint on what he can achieve. Thus, suppose the period 0 DM has a choice between committing to z for period 2 consumption, or rejecting z and giving his period 1 self the choice between x in period 2 and y in period 1. Suppose also that the period 0 self prefers x to z and z to y while the period 1 self prefers y to x. Then, the Strotz model would imply that the DM ends up consuming z in period 2: the period 0 self realizes that if he does not commit to z, his period 1 self will choose y over x, which, for the period 0 self, is the least desirable outcome. Therefore, the period 0 self will commit to z. Hence, dynamic inconsistency leads to a preference for commitment.

Peleg and Yaari (1973) propose to reconcile the conflict among the different selves of a dynamically inconsistent DM with a strategic equilibrium concept. Their reformulation of Strotz's notion of consistent planning has facilitated the application of Strotz's ideas to more general settings, including dynamic games.

3.3 Imperfect recall

An explicit statement of the perfect recall assumption and analysis of its consequences (Kuhn, 1953) is one of the earliest contributions of extensive form game theory. In contrast, the analysis of forgetfulness, that is, extensive form games where the individual forgets his own past actions or information, is relatively recent (Piccione and Rubinstein, 1997).

Piccione and Rubinstein observe that defining optimal behaviour for players with imperfect recall is problematic and propose a few alternative definitions (1997). Subsequent work has focused on what they call the multi-selves approach. In the multi-selves approach to imperfect recall, as in dynamic inconsistency, each information set is treated as a separate player. Optimal behaviour is a profile of behavioural strategies and beliefs at information sets such that the beliefs are consistent with the strategy profile and each behavioural strategy maximizes the corresponding agent's payoff given his beliefs and the behaviour of the remaining agents. Hence, the multi-selves approach leads to a prediction of behaviour that is analogous to perfect Bayesian equilibrium.

3.4 Psychological games

Harsanyi (1967–8) introduces the notion of a type to facilitate analysis of the interaction of players' information in strategic problems. He argues that the notion of a type is flexible enough to accommodate all uncertainty and asymmetric information that is relevant in games. Geanakoplos, Pearce and Stacchetti (1989) observe that if payoffs are 'intrinsically' dependent on beliefs and beliefs are determined in equilibrium, then types cannot be defined independently of the particular equilibrium outcome. Their notion of a psychological game and type (for psychological games) allows for this interdependence between equilibrium expectations and payoffs.

Gul and Pesendorfer (2006) offer an alternative framework for dealing with interdependent preferences. In their analysis, players care not only about the physical consequences of their actions on their opponents, but also about their opponents' attitudes towards such consequences, and their opponents' attitudes towards others' attitudes towards such consequences, and so on. Gul and Pesendorfer provide a model of interdependent preference types similar to Harsanyi's interdependent belief types to analyse situations in which preference interdependence may arise not from the interaction of (subjective) information but from the interaction of the individuals' attitudes towards the well-being of others.

3.5 Neuroeconomics

The most comprehensive challenge to the standard economic modelling of the individual comes from research in neuroeconomics. Neuroeconomists argue that no matter how much the standard conventions are expanded to accommodate behavioural phenomena, it will not be enough: understanding economic behaviour requires studying the physiological, and in particular, neurological mechanisms behind choice. Recent experiments relate choice-theoretic variables to levels of brain activity, the type of choices to the parts of the brain that are engaged when making these choices, and hormone levels to behaviour (Camerer, 2006) provide a concise summary of recent research in neuroeconomics).

Neuroeconomists contend that 'neuroscience findings raise questions about the usefulness of some of the most common constructs that economists commonly use, such as risk aversion, time preference, and altruism' (Camerer, Loewenstein and Prelec, 2005). They argue that neuroscience evidence can be used directly to falsify or validate specific hypotheses about behaviour. Moreover, they claim that organizing choice theory and game theory around the abstractions of neuroscience will lead to better theories. Thus, neureconomics proposes to change both the language of game theory and what constitutes its evidence.

4. Conclusion

The interaction of behavioural economics and game theory has had two significant effects: first, it has broadened the subject matter and set of acceptable approaches to strategic analysis. New modelling techniques such as equilibrium notions that explicitly address biases have become acceptable and new questions such as the effect of ambiguity aversion in auctions have gained interest. More importantly, behavioural approaches have altered the set of empirical benchmarks – the stylized facts – that game theorists must address as they interpret their own conclusions.

FARUK GUL

Bibliography

Allais, M. 1953. Le comportement de l'homme rationnel devant le risque: critique des postulats et axiomes de l'ecole Americaine. *Econometrica* 21, 503–46.

Aumann, R.J. and Brandenburger, A. 1995. Epistemic conditions for Nash equilibrium. *Econometrica* 63, 1161–80.

Camerer, C.F., Loewenstein, G. and Prelec, D. 2005. Neuroeconomics: how neuroscience can inform economics. *Journal of Economic Literature* 43, 9–64.

Camerer, C.F. 2007. Neuroeconomics: using neuroscience to make economic predictions. *Economic Journal.*

Crawford, V.P. 1990. Equilibrium without independence. *Journal of Economic Theory* 50, 127–54.

Ellsberg, D. 1961. Risk, ambiguity, and the savage axioms. *Quarterly Journal of Economics* 75, 643–69.

Eyster, E. and Rabin, M. 2005. Cursed equilibrium. *Econometrica* 73, 1623–72.

Fehr, E. and Schimdt, K. 1999. A theory of fairness, competition and cooperation. *Quarterly Journal of Economics* 114, 817–68.

Geanakoplos, J., Pearce, D. and Stacchetti, E. 1989. Psychological games and sequential rationality. *Games and Economic Behavior* 1, 60–80.

Gul, F. and Pesendorfer, W. 2001. Temptation and self-control. *Econometrica* 2001, 1403–35.

Gul, F. and Pesendorfer, W. 2006. The canonical type space for interdependent preferences. Working paper. Princeton University.

Harsanyi, J. 1967–8. Games with incomplete information played by Bayesian players. *Management Science* 14, 159–82, 320–34, 486–502.

Ho, T., Camerer, C. and Weingelt, K. 1998. Iterated dominance and iterated best responses in p-beauty contests. *American Economic Review* 88, 947–69.

Kahneman, D. and Tversky, A. 1979. Prospect theory: an analysis of decision under risk. *Econometrica* 47, 263–92.
Köszegi, B. and Rabin, M. 2006. A model of reference-dependent preferences, *Quarterly Journal of Economics* 121, 1133–66.
Kreps, D.M. 1979. A preference for flexibility. *Econometrica* 47, 565–76.
Kreps, D.M. and Porteus, E.L. 1978. Temporal resolution of uncertainty and dynamic choice theory. *Econometrica* 46, 185–200.
Kuhn, H.W. 1953. Extensive games and the problem of information. In *Contributions to the Theory of Games*, vol. 2. Princeton, NJ: Princeton University Press.
Machina, M. 1989. Dynamic consistency and non-expected utility models of choice under uncertainty. *Journal of Economic Literature* 1622–68.
O'Donoghue, T. and Rabin, M. 1999. Doing it now or later. *American Economic Review* 103–24.
Peleg, B. and Yaari, M.E. 1973. On the existence of a consistent course of action when tastes are changing. *Review of Economic Studies* 40, 391–401.
Piccione, M. and Rubinstein, A. 1997. On the interpretation of decision problems with imperfect recall. *Games and Economic Behavior* 20, 3–24.
Strotz, R.H. 1955–6. Myopia and inconsistency in dynamic utility maximization. *Review of Economic Studies* 23, 165–80.
Tversky, A. and Kahnemann, D. 1974. Judgment under uncertainty: Heuristic and biases. *Science* 85, 1124–31.
Tversky, A. and Kahneman, D. 1981. The framing of decisions and the psychology of choice. *Science* 211, 453–8.

behavioural game theory

Analytical game theory assumes that players choose strategies which maximize the utility of game outcomes, based on their beliefs about what others players will do, given the economic structure of the game and history; in equilibrium, these beliefs are correct. Analytical game theory is enormously powerful, but it has two shortcomings as a complete model of behaviour by people (and other possible players, including non-human animals and organizations).

First, in complex naturally occurring games, equilibration of beliefs is unlikely to occur instantaneously. Models of choice under bounded rationality, predicting initial choices and equilibration with experience, are therefore useful.

Second, in empirical work, only received (or anticipated) payoffs are easily measured (for example, prices and valuations in auctions, or currency paid in an experiment). Since games are played over *utilities* for received payoffs, it is therefore necessary to have a theory of social preferences – that is, how measured payoffs determine players' utility evaluations – in order to make predictions.

The importance of understanding bounded rationality, equilibration and social preferences is provided by hundreds of experiments showing conditions under which predictions of analytical game theory are sometimes approximately satisfied, and sometimes badly rejected (Camerer, 2003). This article describes an emerging approach called 'behavioural game theory', which generalizes analytical game theory to explain experimentally observed violations. Behavioural game theory incorporates bounds on rationality, equilibrating forces, and theories of social preference, while retaining the mathematical formalism and generality across different games that has made analytical game theory so useful. While behavioural game theory is influenced by laboratory regularities, it is ultimately aimed at a broad range of applied questions such as worker reactions to employment terms, evolution of market institutions, design of auctions and contracts, animal behaviour, and differences in game-playing skill.

Social preferences

Let us start with a discussion of how preferences over outcomes of game can depart from pure material self-interest. In an ultimatum game a Proposer is endowed with a known sum, say ten dollars, and offers a share to another player, the Responder. If the Responder rejects the offer they both get nothing. The ultimatum game is a building block of more complex natural bargaining and a simple tool to measure numerically the price that Responders will pay to punish self-servingly unfair treatment.

Empirically, a large fraction of subjects rejects low offers of 20 per cent or so. Proposers fear these rejections reasonably accurately, and make offers around 40 per cent rather than very small offers predicted by perceived self-interest. (The earliest approximations of whether Proposers offer expected profit-maximizing offers, by Roth et al. 1991, suggested they did. However, those estimates were limited by the method of presenting Responders only with specific offers; since low offers are rare, it is hard to estimate the rejection rate of low offers accurately and hence hard to know conclusively whether offers are profit-maximizing. Different methods, and cross-population data used in Henrich et al., 2005, established that offers are too generous, even controlling for risk-aversion of the Proposers.) This basic pattern scales up to much higher stakes (the equivalent of months of wages) and does not change much when the experiment is repeated, so it is implausible to argue that subjects who reject offers (often highly intelligent college students) are confused.

It is crucial to note that rejecting two dollars out of ten dollars is a rejection of the *joint* hypothesis of utility-maximization and the auxiliary hypothesis that player *i*'s utility depends on only her own payoff x_i. An obvious place to repair the theory is to create a parsimonious theory of social preferences over (x_i, x_j) (and possibly of other features of the game) which predicts violations of self-interest across games with different structures. I will next mention some other empirical regularities, then turn to a discussion of such models of these regularities.

In ultimatum games, it appears that norms and judgements of fairness can depend on context and culture. For example, when Proposers earn the right to make the offer (rather than respond to an offer) by winning at a pre-play trivia game, they feel entitled to offer less – and Responders seem to accept less (Hoffman et al., 1994). Two comparative studies of small-scale societies show interesting variation across cultures. Subjects in a small Peruvian agricultural group, the Machiguenga, offer much less than those in other cultures (typically 15–25 per cent) and accept low offers. Across 15 societies, equality of average offers is positively related to the degree of cooperation in economic activity (for example, do men hunt collectively?) and to the degree of impersonal market trading (Henrich et al., 2005).

Ultimatum games tap negative reciprocity or vengeance. Other games suggest different psychological motives which correspond to different aspects of social preferences. In dictator games, a Proposer simply dictates an allocation of money and the Responder must accept it. In these games, Proposers offer less than in ultimatum games (about 15 per cent of the stakes on average), but offers vary widely with contextual labels and other variables (Camerer, 2003, ch. 2). In trust games, an Investor risks some of her endowment of money, which is increased by the experimenter (representing a return on social investment) and given to an anonymous Trustee. The Trustee pays back as much of the increased sum as she likes to the Investor (perhaps nothing) and keeps the rest. Trust games are models of opportunities to gain from investment with no legal protection against moral hazard by a business partner. Self-interested Trustees will never pay back money; self-interested Investors with equilibrium beliefs will anticipate this and invest nothing. In fact, Investors typically risk about half their money, and Trustees pay back slightly less than was risked (Camerer, 2003, ch. 2). Investments reflect expectations of repayment, along with altruism toward Investors (Ashraf, Bohnet and Piankov, 2006) and an aversion to 'betrayal' (Bohnet and Zeckhauser, 2004). Trustee payback is consistent with positive reciprocity, or a moral obligation to repay a player who risked money to benefit the group.

Importantly, competition has a strong effect in these games. If two or more Proposers make offers in an ultimatum game, and a single Responder accepts the highest offer, then the only equilibrium is for the Proposers to offer almost all the money to the Responder (the *opposite* of the prediction with one Proposer). In the laboratory this Proposer competition occurs rapidly, resulting in a very unfair allocation – almost no earnings for Proposers (for example, Camerer and Fehr, 2006). Similarly, when there is competition among Responders, at least one Responder accepts low offers and Proposers seem to anticipate this effect and offer much less. These regularities help explain an apparent paradox, why the competitive model based on self-interest works so well in explaining market prices in experiments with three or more traders on each side of the market. In these markets, traders with social preferences cannot make choices which reveal a trade-off of self-interest and concern for fairness. The parsimonious theory in which agents have social preferences can therefore explain both fairness-type effects in bilateral exchange and the absence of those effects in multilateral market exchange.

A good social preference theory should explain all these facts: rejections of substantial offers in ultimatum games, lower Proposer offers in dictator games than in ultimatum games, trust and repayment in trust games, and the effects of competition (which bring offers closer to the equilibrium self-interest prediction).

In 'inequality-aversion' theories of social preference, players prefer more money and also prefer that allocations be more equal (judged by differences in payoffs – Fehr and Schmidt, 1999 – or by deviations from payoff shares and equal shares – Bolton and Ockenfels, 2000). In a related 'Rawlsitarian' approach, players care about a combination of their own payoffs, the minimum payoff (à la Rawls) and the total payoff (utilitarian) (Charness and Rabin, 2002). These simple theories account relatively well for the regularities mentioned above across games, with suitable parameter values.

Missing from the inequality aversion and Rawlsitarian theories is a reaction to the intentions of players. Intentions seem to be important because players are much less likely to reject unequal offers that are created by a random device or third party than equivalently unequal offers proposed by a player who benefits from inequality (for example, Blount, 1995; Falk, Fehr and Fischbacher, 2007). In reciprocity theories which incorporate intentions, player A forms a judgement about whether another player B has sacrificed to benefit (or harm) her (for example, Rabin, 1993). A likes to reciprocate, repaying kindness with kindness, and meanness with vengeance. This idea can also explain the results mentioned above, and the effects of intentions shown in other studies.

A newer class of theories focused on 'social image' – that is, player A cares about whether another player B believes A adheres to a norm of fairness. For example, Dufwenberg and Gneezy (2000) show that Trustee repayments in a trust game are correlated with the Trustee's perception of what he or she thought the Investor expected to be repaid. These models hinge on delicate details of iterated beliefs (A's belief about B's belief about A's fairness), so they are more technically complicated but can also explain a wider range of results (see Bénabou and Tirole, 2006; Dillenberger and Sadowski, 2006). Models of this sort are also better equipped to explain deliberate avoidance of information. For example, in dictator games where the dictator can either keep nine dollars or can play a ten-dollar dictator game (knowing the Recipient will *not know* which path was chosen), players often choose the easy nine dollar payment (Dana, Cain and Dawes, 2006). Since they could just play the ten-dollar game and keep all ten dollars, the ten-dollars sacrifice is presumably the price paid to avoid knowing that another person knows you have been selfish (see also Dana, Weber and Kuang, 2007).

Social preference utility theories and social image concerns like these could be applied to explain charitable contribution, legal conflict and settlement, wage-setting and wage dispersion within firms, strikes, divorces, wars, tax policy, and bequests by parents to siblings. Explaining these phenomena with a single parsimonious theory would be very useful and important for policy and welfare economics.

Limited strategic thinking and quantal response equilibrium

In complex games, equilibrium analysis may predict poorly what players do in unique games, or in the first period of a repeated game. Disequilibrium behaviour is important to understand if equilibration takes a long time, and if initial behaviour is important in determining which of several multiple equilibria will emerge. Two types of theories are prominent: cognitive hierarchy theories of different limits on strategic thinking; and theories which retain the assumption of equilibrium beliefs but assume players make mistakes, choosing strategies with higher expected payoff deviations less often.

Cognitive hierarchy theories describe a 'hierarchy' of strategic thinking and constrain how the hierarchy works to make precise predictions. Iterated reasoning surely is limited in the human mind because of evolutionary inertia in promoting high-level thinking, because of constraints on working memory, and because of adaptive motives for overconfidence in judging relative skill (stopping after some steps of reasoning, believing others have reasoned less). Empirical evidence from many experiments with highly skilled subjects suggests that 0–2 steps of iterated reasoning are most likely in the first period of play. A simple illustration is the 'p-beauty contest' game (Nagel, 1995; Ho, Camerer and Weigelt, 1998). In this game, several players choose a number in the interval [0, 100]. The average of the numbers is computed, and multiplied by a value p (say 2/3). The player whose number is closest to p times the average wins a fixed prize.

In equilibrium players are never surprised what other players do. In the p-beauty contest game, this equilibrium condition implies that all players must be picking p times what others are choosing. This equilibrium condition only holds if everyone chooses 0 (the Nash equilibrium, consistent with iterated dominance). Figure 1 shows data from a game with $p=7$ and compares the Nash prediction (choosing 0) and the fit of a cognitive hierarchy model (Camerer, Ho and Chong, 2004). In this game, some players choose numbers scattered from 0 to 100, many others choose p times 50 (the average if others are expected to choose randomly) and others choose p^2 times 50. When the game is played repeatedly with the same players (who learn the average after each trial), numbers converge toward zero, a reminder that equilibrium concepts do reliably predict where an adaptive process leads, even if they do not predict the starting point of that process.

In cognitive hierarchy theories, players who do k steps of thinking anticipate that others do fewer steps. Fully specifying these theories requires specifying what 0-step players do, what higher-step players think, and the statistical distribution of players' thinking levels. One type of theory assumes players who do k steps of thinking believe others do k-steps (Nagel, 1995; Stahl and Wilson, 1995; Costa-Gomes, Crawford and Broseta, 2001). This specification is analytically tractable (especially in games with $n>$ two players) but implies that as players do more thinking their beliefs are further from reality. Another specification assumes increasingly rational expectations – k-level players truncate the actual distribution $f(k)$ of k-step thinkers and guess accurately the relative proportions of thinkers doing 0 to $k-1$ steps of thinking. Camerer, Ho and Chong (2004) and earlier studies show how these cognitive hierarchy theories can fit experimental data from a wide variety of games, with similar thinking-step parameters across games.

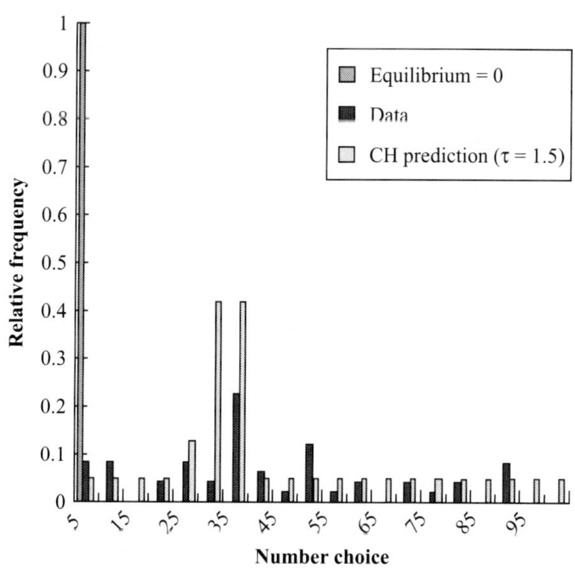

Figure 1 Number choices and theoretical predictions in beauty contest games. *Note*: Players choose numbers from 0 to 100 and the closest number to 0.7 times the average wins a fixed prize. *Source*: Camerer and Fehr (2006).

These cognitive hierarchy theories ignore the benefits and costs of thinking hard. Costs and benefits can be included by relaxing Nash equilibrium, so that players respond stochastically to expected payoffs and choose better responses more often then worse ones, but do not maximize. Denote player i's beliefs about the chance that other players j will choose strategy k by $P_i(s_j^k)$. The expected payoff of player i's strategy s_i^h is $E(s_i^h) = \sum_k P_i(s_j^k)\pi_i(s_i^h, s_j^k)$ (where $\pi_i(x, y)$ is i's payoff if i plays x and j plays y). If player i responds with a logit choice function, then $P_i(s_j^h) = \exp(\lambda E(s_i^h))/\sum_k \exp(\lambda E(s_i^k))$. In this kind of 'quantal response' equilibrium (QRE), each player's beliefs about choice probabilities of others are consistent with actual choice probabilities, but *players do not always choose the highest expected payoff strategy (and λ parameterizes the degree of responsiveness; larger λ implies better* response). QRE fits a wide variety of data better than Nash predictions (McKelvey and Palfrey, 1995; 1998; Goeree and Holt, 2001). It also circumvents some technical limits of Nash equilibrium because players always tremble but the degree of trembling in strategies is linked to expected payoff differences.

Learning

In complex games, it is unlikely that equilibrium beliefs arise from introspection or communication. Therefore, theorists have explored the mathematical properties of various rules under which equilibration might occur when rationality is bounded.

Much research is focused on population evolutionary rules, such as replicator dynamics, in which strategies which have a payoff advantage spread through the population (for example, Weibull, 1995). Schlag and Pollock (1999) show a link between imitation of successful players and replicator dynamics.

Several individual learning rules have been fit to many experimental data-sets. Most of these rules can be expressed as difference equations of underlying numerical propensities or attractions of stage-game strategies which are updated in response to experience. The simplest rule is choice reinforcement, which updates chosen strategies according to received payoffs (perhaps scaled by an aspiration level or reference point). These rules fit surprisingly well in some classes of games (for example, with mixed strategy equilibrium, so that all strategies are played and reinforced relatively often) and in environments with little information, where agents must learn payoffs from experience, but can fit quite poorly in other games. A more complex rule is weighted fictitious play (WFP), in which players form beliefs about what others will do in the future by taking a weighted average of past play, and then choose strategies with high expected payoffs given those beliefs (Cheung and Friedman, 1997). Camerer and Ho (1999) showed that WFP with geometrically declining weights is mathematically equivalent to generalized reinforcement in which unchosen strategies are reinforced as strongly as chosen ones. Building on this insight, they create a hybrid called experience weighted attraction (EWA). The original version of EWA has many parameters because it includes all the parameters used in the various special cases it hybridizes. The EWA form fits modestly better in some games (it adjusts carefully for overfitting by estimating parameters on part of the data and then forecasting out-of-sample), especially those with rapid learning across many strategies (such as pricing). In response to criticism about the number of free parameters, Ho, Camerer, and Chong (2007) created a version with zero *learning parameters* (*just a response sensitivity* λ *as in QRE*) by replacing parameters by 'self-tuning' functions of experience.

Some interesting learning rules do not fit neatly into the class of strategy-updating difference equations. Often it is plausible to think that players are reinforcing learning *rules* rather than strategies (for example, updating the reinforcement rule or the WFP rule; see Stahl, 2000). In many game it is also plausible that people update history-dependent strategies (like tit for tat; see Erev and Roth, 2001; McKelvey and Palfrey, 2001). Selten and Buchta (1999) discuss a concept of 'direction learning' in which players adjust based on experience in a 'direction' when strategies are numerically ordered.

All the rules described above are naive (called 'adaptive') in the sense that they do not incorporate the fact that other players are learning. Models which allow players to be 'sophisticated' and anticipate learning by other players (Stahl, 1999; Chong, Camerer and Ho, 2006) often fit better, especially with experienced subjects. Sophistication is particularly important if players are matched together repeatedly – as workers in firms, firms in strategic alliances, neighbours, spouses, and so forth. Then players have an incentive to take actions that 'strategically teach' an adaptive player what to do. Models of this sort have more moving parts but can explain some basic stylized facts (for example, differences in repeated-game play with fixed 'partner' and random 'stranger' matching of players) and fit a little better than equilibrium reputational models in trust and entry deterrence games (Chong, Camerer and Ho, 2006).

Conclusion

Behavioural game theory uses intuitions and experimental evidence to propose psychologically realistic models of strategic behaviour under rationality bounds and learning, and incorporates social motivations in valuation of outcomes. There are now many mathematical tools available in both of these domains that have been suggested by or fit closely to many different experimental games: cognitive hierarchy, quantal-response equilibrium, many types of learning models (for example, reinforcement, belief learning, EWA and self-tuning EWA), and many different theories of social preference based on inequality aversion, reciprocity, and social image. The primary challenge in the years ahead is to continue to compare and refine these models – in most areas, there is still lively debate about which simplifications are worth making, and why – and then apply them to the sorts of problems in contracting, auctions, and signalling that equilibrium analysis has been so powerfully applied to.

A relatively new challenge is to understand communication. Hardly any games in the world are played without some kind of pre-play messages (even in animal behaviour). However, communication is so rich that understanding how communication works by pure deduction is unlikely to succeed without help from careful empirical observation. A good illustration is Brandts and Cooper (2007), who show the nuanced ways in which communication and incentives, together, can influence coordination in a simple organizational team game.

COLIN F. CAMERER

Bibliography

Ashraf, N., Bohnet, I. and Piankov, N. 2006. Decomposing trust and trustworthiness. *Experimental Economics* 9, 193–208.

Bénabou, R. and Tirole, J. 2006. Incentives and prosocial behavior. *American Economic Review* 96, 1652–78.

Blount, S. 1995. When social outcomes aren't fair – the effect of causal attributions on preferences. *Organizational Behavior and Human Decision Processes* 63, 131–44.

Bohnet, I. and Zeckhauser, R. 2004. Trust, risk and betrayal. *Journal of Economic Behavior & Organization* 55, 467–84.

Bolton, G.E. and Ockenfels, A. 2000. ERC: a theory of equity, reciprocity, and competition. *American Economic Review* 90, 166–93.

Brandts, J. and Cooper, A. 2007. It's what you say, not what you pay: an experimental study of manager-employee relationships in overcoming coordination failure. *Journal of the European Economic Association* (forthcoming).

Camerer, C.F. 2003. *Behavioral Game Theory: Experiments on Strategic Interaction*. Princeton: Princeton University Press.

Camerer, C.F. and Fehr, E. 2006. When does 'economic man' dominate social behavior? *Science* 311, 47–52.

Camerer, C.F. and Ho, T.H. 1999. Experience-weighted attraction learning in normal form games. *Econometrica* 67, 827–74.

Camerer, C.F., Ho, T.-H. and Chong, J.-K. 2004. A cognitive hierarchy model of games. *Quarterly Journal of Economics* 119, 861–98.

Charness, G. and Rabin, M. 2002. Understanding social preferences with simple tests. *Quarterly Journal of Economics* 117, 817–69.

Cheung, Y.-W. and Friedman, D. 1997. Individual learning in normal form games: some laboratory results. *Games and Economic Behavior* 19, 46–76.

Chong, J.-K., Camerer, C.F. and Ho, T.H. 2006. A learning-based model of repeated games with incomplete information. *Games and Economic Behavior* 55, 340–71.

Costa-Gomes, M., Crawford, V.P. and Broseta, B. 2001. Cognition and behavior in normal-form games: an experimental study. *Econometrica* 69, 1193–235.

Dana, J., Cain, D.M. and Dawes, R.M. 2006. What you don't know won't hurt me: costly (but quiet) exit in dictator games. *Organizational Behavior and Human Decision Processes* 100, 193–201.

Dana, J.D., Weber, R.A. and Kuang, J.X. 2007. Exploiting moral wiggle room: behavior inconsistent with a preference for fair outcomes. *Economic Theory*. (forthcoming).

Dillenberger, D. and Sadowski, P. 2006. *Ashamed to Be Selfish*. Princeton: Princeton University Press.

Dufwenberg, M. and Gneezy, U. 2000. Measuring beliefs in an experimental lost wallet game. *Games and Economic Behavior* 30, 163–82.

Erev, I. and Roth, A.E. 2001. On simple reinforcement learning models and reciprocation in the prisoner dilemma game. In *The Adaptive Toolbox*, ed. G. Gigerenzer and R. Selten. Cambridge: MIT Press.

Falk, A., Fehr, E. and Fischbacher, U. 2007. Testing theories of intentions – fairness matters. *Games and Economic Behavior* (forthcoming).
Fehr, E. and Schmidt, K.M. 1999. A theory of fairness, competition, and cooperation. *Quarterly Journal of Economics* 114, 817–68.
Goeree, J.K. and Holt, C.A. 2001. Ten little treasures of game theory and ten intuitive contradictions. *American Economic Review* 91, 1402–22.
Henrich, J. et al. 2005. 'Economic man' in cross-cultural perspective: behavioral experiments in 15 small-scale societies. *Behavioral and Brain Sciences* 28, 795–815.
Ho, T.H., Camerer, C.F and Chong, J.-K. 2007. Self-tuning experience weighted attraction learning in games. *Journal of Economic Theory* 133, 177–98.
Ho, T.H., Camerer, C.F. and Weigelt, K. 1998. Iterated dominance and iterated best response in experimental 'p-beauty contests'. *American Economic Review* 88, 947–69.
Hoffman, E., McCabe, K., Shachat, K. and Smith, V. 1994. Preferences, property-rights, and anonymity in bargaining games. *Games and Economic Behavior* 7, 346–80.
McKelvey, R.D. and Palfrey, T.R. 1995. Quantal response equilibria for normal form games. *Games and Economic Behavior* 10, 6–38.
McKelvey, R.D. and Palfrey, T.R. 1998. Quantal response equilibria for extensive form games. *Experimental Economics* 1, 9–41.
McKelvey, R.D. and Palfrey, T.R. 2001. *Playing in the Dark: Information, Learning, and Coordination in Repeated Games.* Princeton: Princeton University Press.
Nagel, R. 1995. Unraveling in guessing games: an experimental study. *American Economic Review* 85, 1313–26.
Rabin, M. 1993. Incorporating fairness into game-theory and economics. *American Economic Review* 83, 1281–302.
Roth, A.E., Prasnikar, V., Okuno-Fujiwara, M. and Zamir, S. 1991. Bargaining and market behavior in Jerusalem, Ljubljana, Pittsburgh, and Tokyo: an experimental study. *American Economic Review* 81, 1068–95.
Schlag, K.H. and Pollock, G.B. 1999. Social roles as an effective learning mechanism. *Rationality and Society* 11, 371–97.
Selten, R. and Buchta, J. 1999. Experimental sealed bid first price auctions with directly observed bid functions. In *Games and Human Behavior: Essays in Honor of Amnon Rapoport*, ed. D.V. Budescu, I. Erev and R. Zwick. Mahwah, NJ: Lawrence Erlbaum.
Stahl, D. 1999. Sophisticated learning and learning sophistication. Working paper, University of Texas.
Stahl, D.O. 2000. Rule learning in symmetric normal-form games: theory and evidence. *Games and Economic Behavior* 32, 105–38.
Stahl, D.O. and Wilson, P. 1995. On players' models of other players: theory and experimental evidence. *Games and Economic Behavior* 10, 218–54.
Weibull, J. 1995. *Evolutionary Game Theory.* Cambridge, MA: MIT Press.

cheap talk

In the context of games of incomplete information, the term 'cheap talk' refers to direct and costless communication among players. Cheap-talk models should be contrasted with more standard signalling models. In the latter, informed agents communicate private information indirectly via their choices – concerning, say, levels of education attained – and these choices are costly. Indeed, signalling is credible precisely because choices are differentially costly – for instance, high-productivity workers may distinguish themselves from low-productivity workers by acquiring levels of education that would be too costly for the latter.

The central question addressed in cheap-talk models is the following. How much information, if any, can be credibly transmitted when communication is direct and costless? Interest in this question stems from the fact that with cheap talk there is always a 'babbling' equilibrium in which the participants deem all communication to be meaningless – after all, it has no direct payoff consequences – and as a result no one has any incentive to communicate anything meaningful. It is then natural to ask whether there are also equilibria in which communication is meaningful and informative.

We begin by examining the question posed above in the simplest possible setting: there is a single informed party – an expert – who offers information to a single uninformed decision maker. This simple model forms the basis of much work on cheap talk and was introduced in a now classic paper by Crawford and Sobel (1982). In what follows, we first outline the main finding of this paper, namely, that while there are informative equilibria, these entail a significant loss of information. We then examine various remedies that have been proposed to solve (or at least alleviate) the 'information problem'.

The information problem

We begin by considering the leading case in the model of Crawford and Sobel (henceforth CS). A decision maker must choose some decision y. Her payoff depends on y and on an unknown state of the world θ, which is distributed uniformly on the unit interval. The decision maker can base her decision on the costless message m sent by an expert who knows the precise value of θ. The decision maker's payoff is $U(y,\theta) = -(y-\theta)^2$, and the expert's payoff is $V(y,\theta,b) = -(y-(\theta+b))^2$, where $b \geq 0$ is a 'bias' parameter that measures how closely aligned the preferences of the two are. Because of the tractability of the 'uniform-quadratic' specification, this paper, and indeed much of the cheap talk literature, restricts attention to this case.

The sequence of play is as follows:

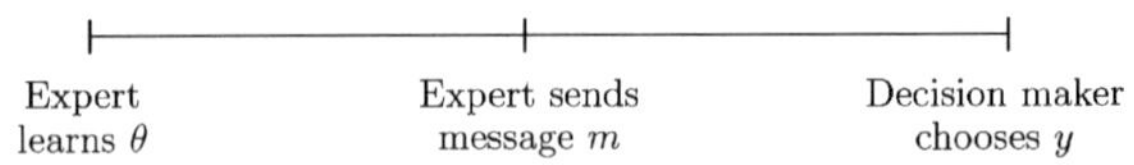

What can be said about (Bayesian-perfect) equilibria of this game? As noted above, there is always an equilibrium in which no information is conveyed, even in the case where preferences are perfectly aligned (that is, $b = 0$). In such a 'babbling' equilibrium, the decision maker believes (correctly it turns out) that there is no information content in the expert's message and hence chooses her decision only on the basis of her prior information. Given this, the expert has no incentive to convey any information – he may as well send random, uninformative messages – and hence the expert indeed 'babbles'. This reasoning is independent of any of the details of the model other than the fact that the expert's message is 'cheap talk'.

Are there equilibria in which all information is conveyed? When there is any misalignment of preferences, the answer turns out to be no. Specifically,

Proposition 1 If the expert is even slightly biased, all equilibria entail some information loss.

The proposition follows from the fact that, if the expert's message always revealed the true state and the decision maker believed him, then the expert would have the incentive to exaggerate the state – in some states θ, he would report $\theta + b$.

Are there equilibria in which some but not all information is shared? Suppose that, following message m, the decision maker holds posterior beliefs given by distribution function G. The action y is chosen to maximize her payoffs given G. Because payoffs are quadratic, this amounts to choosing a y satisfying:

$$y(m) = E[\theta|m] \tag{1}$$

Suppose that the expert faces a choice between sending a message m that induces action y or an alternative message, m', that induces an action $y' > y$. Suppose further that in state θ' the expert prefers y' to y and vice versa in state $\theta < \theta'$. Since the preferences satisfy the *single-crossing* condition, $V_{y\theta} > 0$, the expert would prefer y' to y in all states higher than θ'. This implies that there is a unique state a, satisfying $\theta < a < \theta'$, in which the expert is indifferent between the two actions. Equivalently, the distance between y and the expert's 'bliss' (ideal) action in state a is equal to the distance between action y' and the expert's bliss action in state a. Hence,

$$a + b - y = y' - (a + b) \tag{2}$$

Thus, message m is sent for all states $\theta < a$ and message m' for all states $\theta > a$.

To comprise an equilibrium where exactly two actions are induced, one would need to find values for a, y, and y' that simultaneously satisfy eqs. (1) and (2). Since m is sent in all states $\theta < a$, from eq. (1), $y = \frac{a}{2}$, Similarly, $y' = \frac{1+a}{2}$. Inserting these expression into eq. (2) yields

$$a = \frac{1}{2} - 2b \tag{3}$$

Equation (3) has several interesting properties. First, notice that a is uniquely determined for a given bias. Second, notice that, when the bias gets large ($b \geq \frac{1}{4}$), there is no feasible value of a, so no information is conveyed in any equilibrium. Finally, notice that, when the expert is unbiased ($b = 0$), there exists an equilibrium where the state space is equally divided into 'high' ($\theta > \frac{1}{2}$) and 'low' ($\theta < \frac{1}{2}$) regions and the optimal actions respond accordingly. As the bias increases, the low region shrinks in size while the high region grows; thus, the higher the bias is, the less the information conveyed.

For all $b < \frac{1}{4}$, we constructed an equilibrium that partitions the state space into two intervals. As the bias decreases, equilibria exist that partition the state space into more than two intervals. Indeed, Crawford and Sobel (1982) showed that:

Proposition 2 All equilibria partition the state space into a finite number of intervals. The information conveyed in the most informative equilibrium is decreasing in the bias of the expert.

If the expert were able to commit to fully reveal what he knows, *both* parties would be better off than in any equilibrium of the game described above. With full revelation, the decision maker would choose $y = \theta$ and earn a payoff of zero, while the expert would earn a payoff of $-b^2$. It is easily verified that in any equilibrium the payoffs of both parties are lower than this. The overall message of the CS model is that, absent any commitment possibilities, cheap talk inevitably leads to information loss, which is increasing in the bias of the expert. The remainder of the article studies various 'remedies' for the information loss problem: more extensive communication, delegation, contracts, and multiple experts.

Remedies

Extensive communication

In the CS model, the form of the communication between the two parties was one-sided – the expert simply offered a report to the decision maker, who then acted on it. Of course, communication can be much richer than this, and it is natural to ask whether its form affects information transmission. One might think that it would not. First, one-sided communication where the expert speaks two or more times is no better than having him speak once, since any information the expert might convey in many messages can be encoded in a single message. Now, suppose the communication is two-sided – it is a conversation – so the decision maker also speaks. Since she has no information of her own to contribute, all she can do is to send random messages, and at first glance this seems to add little. As we will show, however, random messages improve information transmission by acting as *coordinating devices.*

To see this, suppose the expert has bias $b = \frac{1}{12}$. As we previously showed, when only he speaks, the best equilibrium is where the expert reveals whether the state is above or below $\frac{1}{3}$. Suppose instead that we allow for *face-to-face* conversation – a simultaneous exchange of messages – and that the sequence of play is:

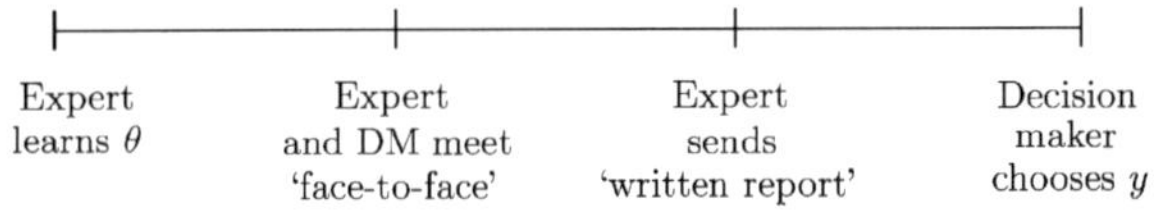

The following strategies constitute an equilibrium. The expert reveals some information at the face-to-face meeting, but there is also some randomness in what transpires. Depending on how the conversation goes, the meeting is deemed by both parties to be a 'success' or a 'failure'. After the meeting, and depending on its outcome, the expert may send an additional 'written report' to the decision maker.

During the meeting, the expert reveals whether θ is above or below $\frac{1}{6}$; he also sends some additional messages that affect the success or failure of the meeting. If he reveals that $\theta \leq \frac{1}{6}$, the meeting is adjourned, no more communication takes place, and the decision maker chooses a low action $y_L = \frac{1}{12}$ that is optimal given the information that $\theta \leq \frac{1}{6}$.

If, however, he reveals that $\theta > \frac{1}{6}$, then the written report depends on whether the meeting was a success or a failure. If the meeting is a failure, no more communication takes place, and the decision maker chooses the 'pooling' action $y_P = \frac{7}{12}$ that is optimal given that $\theta > \frac{1}{6}$. If the meeting is a success, however, the written report further divides the interval $[\frac{1}{6}, 1]$ into $[\frac{1}{6}, \frac{5}{12}]$ and $[\frac{5}{12}, 1]$. In the first sub-interval, the medium action $y_M = \frac{7}{24}$ is taken and in the second sub-interval the high action $y_H = \frac{17}{24}$ is taken. The actions taken in different states are depicted in Figure 1. The dotted line depicts the actions, $\theta + \frac{1}{12}$, that are 'ideal' for the expert.

Notice that in state $\frac{1}{6}$, the expert prefers y_L to y_P (y_L is closer to the dotted line than is y_P) and prefers y_M to y_L. Thus, if there were no uncertainty about the outcome of the meeting – for instance, if all meetings were 'successes' – then the expert would not be willing to reveal whether the state is above or below $\frac{1}{6}$; for states $\theta = \frac{1}{6} - \varepsilon$, the expert would say $\theta \in [\frac{1}{6}, \frac{5}{12}]$, thereby inducing y_M instead of y_L. If all meetings were failures, then for states $\theta = \frac{1}{6} + \varepsilon$, the expert would say $\theta < \frac{1}{6}$ thereby inducing y_L instead of y_P

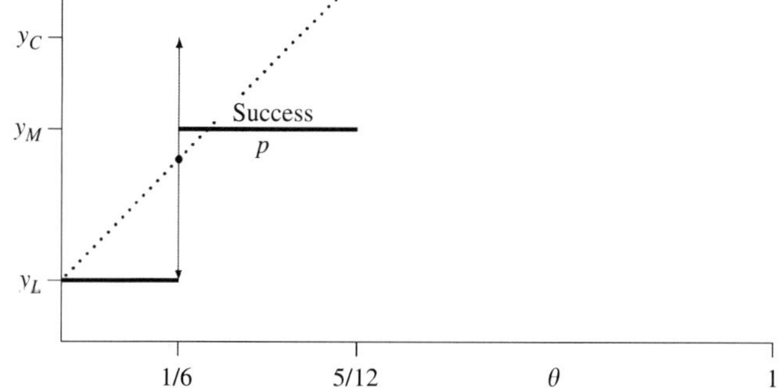

Figure 1 Equilibrium with face-to-face meeting

There exists a probability $p = \frac{16}{21}$ such that when $\theta = \frac{1}{6}$ the expert is indifferent between y_L and a $(p, 1-p)$ lottery between y_M and y_P (whose certainty equivalent is labelled y_C in the figure). Also, when $\theta < \frac{1}{6}$, the expert prefers y_L to a $(p, 1-p)$ lottery between y_M and y_P, and when $\theta > \frac{1}{6}$, the expert prefers a $(p, 1-p)$ lottery between y_M and y_P to y_L.

It remains to specify a conversation such that the meeting is successful with probability $p = \frac{16}{21}$. Suppose the expert sends a message (*Low*, A_i) or (*High*, A_i) and the decision maker sends a message A_j, where $i, j \in \{1, 2, ..., 21\}$. These messages are interpreted as follows. *Low* signals that $\theta \leq \frac{1}{6}$ and *High* signals that $\theta > \frac{1}{6}$. The A_i and A_j messages play the role of a coordinating device and determine whether the meeting is successful. The expert chooses A_i at random and each A_i is equally likely. Similarly, the decision maker chooses A_j at random. Given these choices, the meeting is a

Success if $0 \leq i - j < 16$ or $j - i > 5$

Failure otherwise

For example, if the messages of the expert and the decision maker are (*High*, A_{17}) and A_5, respectively, then it is inferred that $\theta > \frac{1}{6}$ and, since $i - j = 12 < 16$, the meeting is a success. Observe that with these strategies, given any A_i or A_j, the probability that the meeting is a success is exactly $\frac{16}{21}$.

The equilibrium constructed above conveys more information than any equilibria of the CS game. The remarkable fact about the equilibrium is that this improvement in information transmission is achieved by adding a stage in which the *uninformed* decision maker also participates. While the analysis above concerns itself with the case where $b = \frac{1}{12}$, informational improvement through a 'conversation' is a general phenomenon (Krishna and Morgan, 2004a):

Proposition 3 Multiple stages of communication together with active participation by the decision maker always improve information transmission.

What happens if the two parties converse more than once? Does every additional stage of communication lead to more information transmission? In a closely related setting, Aumann and Hart (2003) obtain a precise but abstract characterization of the set of equilibrium payoffs that emerge in sender–receiver games with a finite number of states and actions when the number of stages of communication is infinite. Because the CS model has a continuum of states and actions, their characterization does not directly apply. Nevertheless, it can be shown that, even with an unlimited conversation, full revelation is impossible. A full characterization of the set of equilibrium payoffs with multiple stages remains an open qst.

Delegation

A key tenet of organizational theory is the 'delegation principle', which says that the power to make decisions should reside in the hands of those with the relevant information (Milgrom and Roberts, 1992). Thus, one approach to solving the information problem is simply to delegate the decision to the expert. However, the expert's bias will distort the chosen action from the decision maker's perspective. Delegation this leads to a trade-off between an optimal decision by an uninformed party and a biased decision by an informed party.

Is delegation worthwhile? Consider again an expert with bias $b = \frac{1}{12}$. The decision maker's payoff from the most informative partition equilibrium is $-\frac{1}{36}$. Under delegation, the action chosen is $y = \theta + b$ and the payoff is $-b^2 = -\frac{1}{144}$. Thus delegation is preferred. Dessein (2002) shows that this is always true:

Proposition 4 If the expert's bias is not too large ($b \leq \frac{1}{4}$), delegation is better than all equilibria of the CS model.

In fact, by exerting only slightly more control, the decision maker can do even better. As first pointed out by Holmström (1984), the optimal delegation scheme involves limiting the scope of actions from which the expert can choose. Under the uniform-quadratic specification, the decision maker should optimally limit the expert's choice of actions to $y \in [0, 1 - b]$. When $b = \frac{1}{12}$, limiting actions in this way raises the decision maker's payoff from $-\frac{1}{144}$ to $-\frac{1}{162}$.

Optimal delegation still leads to information loss. When the expert's choice is 'capped', in high states the action is unresponsive to the state.

An application of the delegation principle arises in the US House of Representatives. Typically a specialized committee – analogous to an informed expert – sends a bill to the floor of the House – the decision maker. How it may then be amended depends on the legislative rule under effect. Under the so-called *closed rule* the floor is limited in its ability to amend the bill, while under the *open rule* the floor may freely amend the bill. Thus, operating under a closed rule is similar to delegation, while an open rule is similar to the CS model. The proposition above suggests, and Gilligan and Krehbiel (1987; 1989) have shown, that in some circumstances the floor may benefit by adopting a closed rule.

Contracts

Up until now we have assumed that the decision maker did not compensate the expert for his advice. Can compensation, via an incentive contract, solve the information problem? To examine this, we amend the model to allow for compensation and use mechanism design to find the optimal contract. Suppose that the payoffs are now given by

$$U(y, \theta, t) = -(y - \theta)^2 - t \quad V(y, \theta, b, t) = -(y - \theta - b)^2 + t$$

where $t \geq 0$ is the amount of compensation.

Using the revelation principle, we can restrict attention to a direct mechanism where both t and y depend on the state θ reported by the expert. Notice that such mechanisms directly link the expert's reports to payoffs – talk is no longer cheap.

Contracts are powerful instruments. A contract that leads to full information revelation and first-best actions is:

$$t(\hat{\theta}) = 2b(1 - \hat{\theta}) \quad y(\hat{\theta}) = \hat{\theta}$$

where $\hat{\theta}$ is the state reported by the expert. Under this contract, the expert can do no better than to tell the truth, that is, to set $\hat{\theta} = \theta$, and, as a consequence, the action undertaken in this scheme is the 'bliss' action for the decision maker. Full revelation is expensive, however. When $b = \frac{1}{12}$, the decision maker's payoff from this scheme is $-\frac{1}{12}$. Notice that this is worse than the payoff of $-\frac{1}{36}$ in the best CS equilibrium, which can be obtained with no contract at all. The costs of implementing the fully revealing contract outweigh the benefits.

In general, Krishna and Morgan (2004b) show:

Proposition 5 With contracts, full revelation is always feasible but never optimal.

The proposition above shows that full revelation is never optimal. No contract at all is also not optimal – delegation is preferable. What is the structure of the optimal contract? A typical optimal contract is depicted as the dark line in Figure 2. First, notice that, even though the decision maker could induce his bliss action for some states, it is never optimal to do so. Instead, for low states ($\theta < b$) the decision maker implements a 'compromise' action – an action that lies between θ and $\theta + b$. When $\theta > b$, the optimal contract simply consists of capped delegation.

Multiple senders

Thus far we have focused attention on how a decision maker should consult a single expert. In many instances, decision makers consult multiple experts – often with similar information but differing ideologies (biases). Political leaders often form cabinets of advisors with overlapping expertise. How should a cabinet be constituted? Is a balanced cabinet – one with advisors with opposing ideologies – helpful? How should the decision maker structure the 'debate' among her advisors?

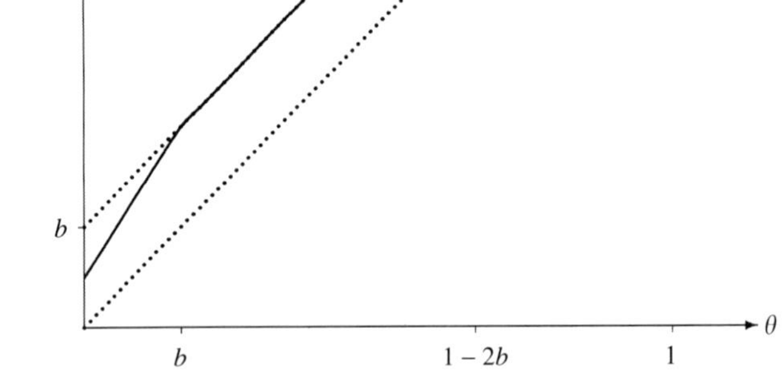

Figure 2 An optimal contract, $b \leq \frac{1}{3}$

To study these issues, we add a second expert having identical information to the CS model. To incorporate ideological differences, suppose the experts have differing biases. When both b_1 and b_2 are positive, the experts have *like bias* – both prefer higher actions than does the decision maker. In contrast, if $b_1 > 0$ and $b_2 < 0$, then the experts have *opposing bias* – expert 1 prefers a higher action and expert 2 a lower action than does the decision maker.

Simultaneous talk

When both experts report to the decision maker simultaneously, the information problem is apparently solved – full revelation is now an equilibrium. To see this, suppose the experts have like bias and consider the following strategy for the decision maker: choose the action that is the more 'conservative' of the two recommendations. Precisely, if $m_1 < m_2$, choose action m_1 and vice versa if $m_2 < m_1$. Under this strategy, each expert can do no better than to report θ honestly if the other does likewise. If expert 2 reports $m_2 = \theta$, then a report $m_1 > \theta$ has no effect on the action. However, reporting $m_1 < \theta$ changes the action to $y = m_1$, but this is worse for expert 1. Thus, expert 1 is content to simply tell the truth. Opposing bias requires a more complicated construction, but the effect is the same: full revelation is an equilibrium (see Krishna and Morgan, 2001b).

Notice that the above construction is fragile because truth-telling is a weakly dominated strategy. Each expert is at least as well off by reporting $m_i = \theta + b_i$ and strictly better off in some cases. Battaglini (2002) defines an equilibrium refinement for such games which, like the notion of perfect equilibrium in finite games, incorporates the usual idea that players may make mistakes. He then shows that such a refinement rules out all equilibria with full revelation regardless of the direction of the biases. While the set of equilibria satisfying the refinement is unknown, the fact that full revelation is ruled out means that simply adding a second expert does not solve the information problem satisfactorily.

Sequential talk

Finally, we turn to the case where the experts offer advice in sequence:

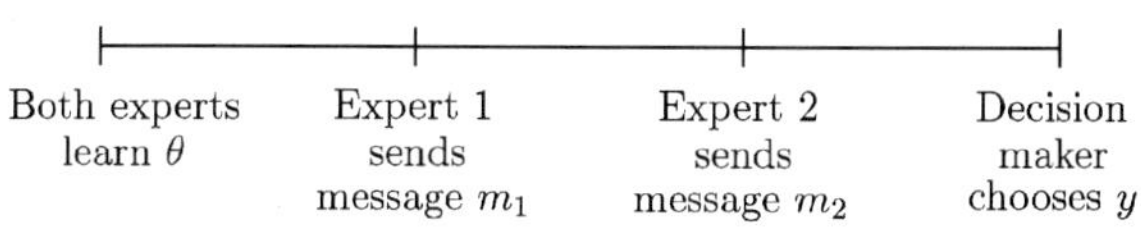

Suppose that the two experts have biases $b_1 = \frac{1}{18}$ and $b_2 = \frac{1}{12}$, respectively. It is easy to verify (with the use of (2)) that, if only expert 1 were consulted, then the most informative equilibrium entails his revealing that the state is below $\frac{1}{9}$, or between $\frac{1}{9}$ and $\frac{4}{9}$, or above $\frac{4}{9}$. If only expert 2 were consulted, then the most informative equilibrium is where he reveals whether the state is below or above $\frac{1}{3}$. If the decision maker were able to consult only one of the two experts, she would be better off consulting the more loyal expert 1.

But what happens if she consults both? It turns out that, if both experts actively contribute information, then the decision maker can do no better than the following equilibrium. Expert 1 speaks first and reveals whether or not the state is above or below $\frac{11}{27}$. If expert 1 reveals that the state is above $\frac{11}{27}$, expert 2 reveals nothing further. If, however, expert 1 reveals that the state is below $\frac{11}{27}$, then expert 2 reveals further whether or not it is above or below $\frac{1}{27}$. That this is an equilibrium may be verified again by using (2) and recognizing that, in state $\frac{1}{27}$, expert 2 must be indifferent between the optimal action in the interval $[0, \frac{1}{27}]$ and the optimal action in $[\frac{1}{27}, \frac{11}{27}]$. In state $\frac{11}{27}$, expert 1 must be indifferent between the optimal action in $[\frac{1}{27}, \frac{11}{27}]$ and the optimal action in $[\frac{11}{27}, 1]$.

Sadly, by actively consulting both experts, the decision maker is worse off than if she simply ignored expert 2 and consulted only her more loyal advisor, expert 1. This result is quite general, as shown by Krishna and Morgan (2001a):

Proposition 6 When experts have like biases, actively consulting the less loyal expert never helps the decision maker.

The situation is quite different when experts have opposing biases, that is, when the cabinet is balanced. To see this, suppose that the cabinet is comprised of two equally loyal experts biases $b_1 = \frac{1}{12}$ and $b_2 = -\frac{1}{12}$. Consulting expert 1 alone leads to a partition $[0, \frac{1}{3}], [\frac{1}{3}, 1]$ while consulting expert 2 alone leads to the partition $[0, \frac{2}{3}], [\frac{2}{3}, 1]$. If instead the decision maker asked both experts for advice, the following is an equilibrium: expert 1 reveals whether θ is above or below $\frac{2}{9}$. If he reveals that the state is below $\frac{2}{9}$, the discussion ends. If, however, expert 1 indicates that the state is above $\frac{2}{9}$, expert 2 is actively consulted and reveals further whether the state is above or below $\frac{7}{9}$. Based on this, the decision maker takes the appropriate action. One may readily verify that this is an improvement over consulting either expert alone. Once again the example readily generalizes:

Proposition 7 When experts have opposing biases, actively consulting both experts always helps the decision maker.

Indeed, the decision maker can be more clever than this. One can show that, with experts of opposing bias, there exist equilibria where a portion of the state space is *fully revealed*. By allowing for a 'rebuttal' stage in the debate, there exists an equilibrium where *all* information is fully revealed.

VIJAY KRISHNA AND JOHN MORGAN

Bibliography

Aumann, R. and Hart, S. 2003. Long cheap talk. *Econometrica* 71, 1619–60.

Austen-Smith, D. 1993. Interested experts and policy advice: multiple referrals under open rule. *Games and Economic Behavior* 5, 3–43.

Baron, D. 2000. Legislative organization with informational committees. *American Journal of Political Science* 44, 485–505.

Battaglini, M. 2002. Multiple referrals and multidimensional cheap talk. *Econometrica* 70, 1379–401.

Crawford, V. and Sobel, J. 1982. Strategic information transmission. *Econometrica* 50, 1431–51.

Dessein, W. 2002. Authority and communication in organizations. *Review of Economic Studies* 69, 811–38.
Gilligan, T. and Krehbiel, K. 1987. Collective decision-making and standing committees: an informational rationale for restrictive amendment procedures. *Journal of Law, Economics, and Organization* 3, 287–335.
Gilligan, T. and Krehbiel, K. 1989. Asymmetric information and legislative rules with a heterogeneous committee. *American Journal of Political Science* 33, 459–90.
Grossman, G. and Helpman, E. 2001. *Special Interest Politics.* Cambridge, MA: MIT Press.
Holmström, B. 1984. On the theory of delegation. In *Bayesian Models in Economic Theory*, ed. M. Boyer and R. Kihlstrom. Amsterdam: North-Holland.
Krishna, V. and Morgan, J. 2001a. A model of expertise. *Quarterly Journal of Economics* 116, 747–75.
Krishna, V. and Morgan, J. 2001b. Asymmetric information and legislative rules: some amendments. *American Political Science Review* 95, 435–52.
Krishna, V. and Morgan, J. 2004a. The art of conversation: eliciting information from experts through multi-stage communication. *Journal of Economic Theory* 117, 147–79.
Krishna, V. and Morgan, J. 2004b. Contracting for information under imperfect commitment. Paper CPC05-051. Competition Policy Center, University of California, Berkeley.
Milgrom, P. and Roberts, J. 1992. *Economics, Organization and Management.* Englewood Cliffs, NJ: Prentice Hall.
Wolinsky, A. 2002. Eliciting information from multiple experts. *Games and Economic Behavior* 41, 141–60.

computer science and game theory

1. Introduction

There has been a remarkable increase in work at the interface of computer science and game theory in the past decade. Game theory forms a significant component of some major computer science conferences (see, for example, Kearns and Reiter, 2005; Sandholm and Yokoo, 2003); leading computer scientists are often invited to speak at major game theory conferences, such as the World Congress on Game Theory 2000 and 2004. In this article I survey some of the main themes of work in the area, with a focus on the work in computer science. Given the length constraints, I make no attempt at being comprehensive, especially since other surveys are also available, including Halpern (2003), Linial (1994), Papadimitriou (2001), and a comprehensive survey book (Nisan et al., 2007).

The survey is organized as follows. I look at the various roles of computational complexity in game theory in Section 2, including its use in modelling bounded rationality, its role in mechanism design, and the problem of computing Nash equilibria. In Section 3, I consider a game-theoretic problem that originated in the computer science literature, but should be of interest to the game theory community: computing the *price of anarchy*, that is, the cost of using a decentralizing solution to a problem. In Section 4, I consider interactions between distributed computing and game theory. In Section 5, I consider the problem of implementing mediators, which has been studied extensively in both computer science and game theory. I conclude in Section 6 with a discussion of a few other topics of interest.

2. Complexity considerations

The influence of computer science in game theory has perhaps been most strongly felt through complexity theory. I consider some of the strands of this research here. There are a numerous basic texts on complexity theory that the reader can consult for more background on notions like NP-completeness and finite automata, including Hopcroft and Ullman (1979) and Papadimitriou (1994a).

2.1 Bounded rationality

One way of capturing bounded rationality is in terms of agents who have limited computational power. In economics, this line of research goes back to the work of Neyman (1985) and Rubinstein (1986), who focused on finitely repeated Prisoner's Dilemma. In n-round finitely repeated Prisoner's Dilemma, there are 2^{2^n-1} strategies (since a strategy is a function from histories to {cooperate, defect}, and there are clearly 2^n–1 histories of length $< n$). Finding a best response to a particular move can thus potentially be difficult. Clearly people do not find best responses by doing extensive computation. Rather, they typically rely on simple heuristics, such as 'tit for tat' (Axelrod, 1984). Such heuristics can often be captured by finite automata; both Neyman and Rubinstein thus focus on finite automata playing repeated Prisoner's Dilemma. Two computer scientists, Papadimitriou and Yannakakis (1994), showed that if both players in an n-round Prisoner's Dilemma are finite automata with at least 2^n–1 states, then the only equilibrium is the one where they defect in every round. This result says that a finite automaton with exponentially many states can compute best responses in Prisoner's Dilemma.

We can then model bounded rationality by restricting the number of states of the automaton. Neyman (1985) showed, roughly speaking, that if the two players in n-round Prisoner's Dilemma are modelled by finite automata with a number of states in the interval $[n^{1/k}, n^k]$ for some k, then collaboration can be approximated in equilibrium; more precisely, if the payoff for (cooperate, cooperate) is (3, 3) there is an equilibrium in the repeated game where the average payoff per round is greater than $3 - \frac{1}{k}$ for each player. Papadimitriou and Yannakakis (1994) sharpen this result by showing that if at least one of the players has fewer than $2^{c_\varepsilon n}$ states, where $c_\varepsilon = \frac{\varepsilon}{12(1+\varepsilon)}$, then for sufficiently large n, there is an equilibrium where each player's average payoff per round is greater than 3–ε. Thus, computational limitations can lead to cooperation in Prisoner's Dilemma.

There have been a number of other attempts to use complexity-theoretic ideas from computer science to model bounded rationality (see Rubinstein, 1998, for some examples). However, it seems that there is much more work to be done here.

2.2 Computing Nash equilibrium

Nash (1950) showed every finite game has a Nash equilibrium in mixed strategies. But how hard is it to actually find that equilibrium? On the positive side, there are well known algorithms for computing Nash equilibrium, going back to the classic Lemke–Howson (1964) algorithm, with a spate of recent improvements (see, for example, Govindan and Wilson, 2003; Blum, Shelton and Koller, 2003; Porter, Nudelman and Shoham, 2004). Moreover, for certain classes of games (for example, symmetric games, Papadimitriou and Roughgarden, 2005), there are known to be polynomial-time algorithms. On the negative side, many qsts about Nash equilibrium are known to be NP-hard. For example, Gilboa and Zemel (1989) showed that, for a game presented in normal form, deciding whether there exists a Nash equilibrium where each player gets a payoff of at least r is NP-complete. Interestingly, Gilboa and Zemel also show that computing whether there exists a *correlated* equilibrium (Aumann, 1987) where each player gets a payoff of at least r is computable in polynomial time. In general, questions regarding correlated equilibrium seem easier than the analogous qsts for Nash equilibrium; see Papadimitriou (2005) and Papadimitriou and Roughgarden (2005) for further examples. Chu and Halpern (2001) prove similar NP-completeness results if the game is represented in extensive form, even if all players have the same payoffs (a situation that arises frequently in computer science applications, where we can view the players as agents of some designer, and take the payoffs to be the designer's payoffs). Conitzer and Sandholm (2003) give a compendium of hardness results for various questions regarding Nash equilibria.

Nevertheless, there is a sense in which it seems that the problem of finding a Nash equilibrium is easier than typical NP-complete problems, because every game is guaranteed to have a Nash equilibrium. By way of contrast, for a typical NP-complete problem like propositional satisfiability, whether or not a propositional formula is satisfiable is not known. Using this observation, it can be shown that if finding a Nash equilibrium is NP-complete, then NP = coNP. Recent work has in a sense completely characterized the complexity of finding a Nash equilibrium in normal-form games: it is a *PPAD-complete* problem (Chen and Deng, 2006; Daskalis, Goldberg and Papadimitriou, 2006). PPAD stands for 'polynomial parity argument (directed case)'; see Papadimitriou (1994b) for a formal definition and examples of other PPAD problems. It is believed that PPAD-complete problems are not solvable in polynomial time, but are simpler than NP-complete problems, although this remains an open problem. See Papadimitriou (2007) for an overview of this work.

2.3 Algorithmic mechanism design

The problem of mechanism design is to design a game such that the agents playing the game, motivated only by self-interest, achieve the designer's goals. This problem has much in common with the standard computer science problem of designing protocols that satisfy certain specifications (for example, designing a distributed protocol that achieves Byzantine agreement; see Section 4). Work on mechanism design has traditionally ignored computational concerns. But Kfir-Dahav, Monderer and Tennenholtz (2000) show that, even in simple settings, optimizing social welfare is NP-hard, so that perhaps the most common approach to designing mechanisms, applying the Vickrey–Groves–Clarke (VCG) procedure (Clarke, 1971; Groves, 1973; Vickrey, 1961), is not going to work in large systems. We might hope that, even if we cannot compute an optimal mechanism, we might be able to compute a reasonable approximation to it. However, as Nisan and Ronen (2000; 2001) show, in general, replacing a VCG mechanism by an approximation does not preserve truthfulness. That is, even though truthfully revealing one's type is an optimal strategy in a VCG mechanism, it may no longer be optimal in an approximation. Following Nisan and Ronen's work, there has been a spate of papers either describing computationally tractable mechanisms or showing that no computationally tractable mechanism exists for a number of problems, ranging from task allocation (Archer and Tardos, 2001; Nisan and Ronen, 2001) to cost-sharing for multicast trees (Feigenbaum, Papadimitriou and Shenker, 2000) (where the problem is to share the cost of sending, for example, a movie over a network among the agents who actually want the movie) to finding low-cost paths between nodes in a network (Archer and Tardos, 2002).

The problem that has attracted perhaps the most attention is *combinatorial auctions*, where bidders can bid on bundles of items. This becomes of particular interest in situations where the value to a bidder of a bundle of goods cannot be determined by simply summing the value of each good in isolation. To take a simple example, the value of a pair of shoes is much higher than that of the individual shoes; perhaps more interestingly, an owner of radio stations may value having a licence in two adjacent cities more than the sum of the individual licences. Combinatorial auctions are of great interest in a variety of settings including spectrum auctions, airport time slots (that is, take-off and landing slots), and industrial procurement. There are many complexity-theoretic issues related to combinatorial auctions. For a detailed discussion and references see Cramton, Shoham and Steinberg (2006); I briefly discuss a few of the issues involved here.

Suppose that there are n items being auctioned. Simply for a bidder to communicate her bids to the auctioneer can take, in general, exponential time, since there are 2^n bundles. In many cases, we can identify a bid on a bundle with the bidder's valuation of the bundle. Thus, we can try to carefully design a bidding language in which a bidder can communicate her valuations succinctly. Simple information-theoretic arguments can be used to show that, for every bidding language, there will be valuations that will require length at least 2^n to express in that language. Thus, the best we can hope for is to design a language that can represent the 'interesting' bids succinctly. See Nisan (2006) for an overview of various bidding languages and their expressive power.

Given bids from each of the bidders in a combinatorial auction, the auctioneer would like to then determine the winners. More precisely, the auctioneer would like to allocate the m items in an auction so as to maximize his revenue. This problem, called the *winner determination problem*, is NP-complete in general, even in relatively simple classes of combinatorial auctions with only two bidders making rather restricted bids. Moreover, it is not even polynomial-time approximable, in the sense that there is no constant d and polynomial-time algorithm such that the algorithm produces an allocation that gives revenue that is at least $1/d$ of optimal. On the other hand, there are algorithms that provably find a good solution, seem to work well in practice, and, if they seem to be taking too long, can be terminated early, usually with a good feasible solution in hand. See Lehmann, Müller and Sandholm, (2006) for an overview of the results in this area.

In most mechanism design problems, computational complexity is seen as the enemy. There is one class of problems in which it may be a friend: voting. One problem with voting mechanisms is that of *manipulation* by voters. That is, voters may be tempted to vote strategically rather than ranking the candidates according to their true preferences, in the hope that the final outcome will be more favourable. This situation arises frequently in practice; in the 2000 US presidential election, American voters who preferred Nader to Gore to Bush were encouraged to vote for Gore, rather than 'wasting' a vote on Nader. The classic Gibbard–Satterthwaite theorem (Gibbard, 1973; Satterthwaite, 1975) shows that, if there are at least three alternatives, then in any nondictatorial voting scheme (that is, one where it is *not* the case that one particular voter dictates the final outcome, irrespective of how the others vote), there are preferences under which an agent is better off voting strategically. The hope is that, by constructing the voting mechanism appropriately, it may be computationally intractable to find a manipulation that will be beneficial. While finding manipulations for the plurality protocol (the candidate with the most votes wins) is easy, there are well-known voting protocols for which manipulation is hard in the presence of three or more candidates. See Conitzer, Sandholm and Lang (2007) for a summary of results and further pointers to the literature.

2.4 Communication complexity

Most mechanisms in the economics literature are designed so that agents truthfully reveal their preferences. However, in some settings, revealing one's full preferences can require a prohibitive amount of communication. For example, in a combinatorial auction of m items, revealing one's full preferences may require revealing what one would be willing to pay for each of the 2^m-1 possible bundles of items. Even if $m=30$, this requires revealing more than one billion numbers. This leads to an obvious question: how much communication is required by various mechanisms? Formal work on this question in the economics community goes back to Hurwicz (1977) and Mount and Reiter (1974); their definitions focused on the dimension of the message space. Independently (and later), there was active work in computer science on *communication complexity*, the number of bits of communication needed for a set of n agents to compute the value of a function $f : x_{i=1}^{n}\Theta_i \rightarrow X$, where each agent i knows $\theta_i \in \Theta_i$. (Think of θ_i as representing agent i's type.) Recently there has been an explosion of work, leading to a better understanding of the communication complexity for many important economic allocation problems; see Segal (2006) for an overview. Two important themes in this work are understanding the role of price-based market mechanisms in solving allocation problems with minimal communication, and designing mechanisms that provide agents with incentives to communicate truthfully while having low communication requirements.

3. The price of anarchy

In a computer system, there are situations where we may have a choice between a centralized and a decentralized solution to a problem. By 'centralized' here, I mean that each agent in the system is told exactly what to do and must do so; in the decentralized solution, each agent tries to optimize his own selfish interests. Of course, centralization comes at a cost. For one thing, there is a problem of enforcement. For another, centralized solutions tend to be more vulnerable to failure. On the other hand, a centralized solution may be more socially beneficial. How much more beneficial can it be?

Koutsoupias and Papadimitriou (1999) formalized this question by considering the ratio of the social welfare of the centralized solution to the social welfare of the Nash equilibrium with the worst social welfare (assuming that the social welfare function is always positive). They called this ratio the *price of anarchy*, and proved a number of results regarding the price of anarchy for a scheduling problem on parallel machines. Since the original paper, the price of anarchy has been studied in many settings, including traffic routing (Roughgarden and Tardos, 2002), facility location games (for example, where is the best place to put a factory) (Vetta, 2002), and spectrum sharing (how should channels in a WiFi network be assigned) (Halldórsson et al., 2004).

To give a sense of the results, consider the traffic-routing context of Roughgarden and Tardos (2002). Suppose that the travel time on a road increases in a known way with the congestion on the road. The goal is to minimize the average travel time for all drivers. Given a road network and a given traffic load, a centralized solution would tell each driver which road to take. For example, there could be a rule that cars with odd-numbered licence plates take road 1, while those with even-numbered plates take road 2, to minimize congestion on either road. Roughgarden and Tardos show that the price of anarchy is unbounded if the travel time can be a nonlinear function of the congestion. On the other hand, if it is linear, they show that the price of anarchy is at most 4/3.

The price of anarchy is but one way of computing the 'cost' of using a Nash equilibrium. Others have been considered in the computer science literature. For example, Tennenholtz (2002) compares the *safety level* of a game – the optimal amount that an agent can guarantee himself, independent of what the other agents do – to what the agent gets in a Nash equilibrium, and shows, for interesting classes of games, including load-balancing games and first-price auctions, that the ratio between the safety level and the Nash equilibrium is bounded. For example, in the case of first-price auctions, it is bounded by the constant e.

4. Game theory and distributed computing

Distributed computing and game theory are interested in much the same problems: dealing with systems where there are many agents, facing uncertainty and having possibly different goals. In practice, however, there has been a significant difference in emphasis between the two areas. In distributed computing, the focus has been on problems such as fault tolerance, asynchrony, scalability, and proving correctness of algorithms; in game theory, the focus has been on strategic concerns. I discuss here some issues of common interest. Most of the discussion in the remainder of this section is taken from Halpern (2003).

To understand the relevance of fault tolerance and asynchrony, consider the *Byzantine agreement* problem, a paradigmatic problem in the distributed systems literature. In this problem, there are assumed to be n soldiers, up to t of which may be faulty (the t stands for *traitor*); n and t are assumed to be common knowledge. Each soldier starts with an initial preference, to either attack or retreat. (More precisely, there are two types of nonfaulty agents – those that prefer to attack, and those that prefer to retreat.) We want a protocol that guarantees that (1) all *nonfaulty* soldiers reach the same decision, and (2) if all the soldiers are nonfaulty and their initial preferences are identical, then the final decision agrees with their initial preferences. (The condition simply prevents the obvious trivial solutions, where the soldiers attack no matter what, or retreat no matter what.)

The problem was introduced by Pease, Shostak and Lamport (1980), and has been studied in detail since then; Chor and Dwork (1989), Fischer (1983), and Linial (1994) provide overviews. Whether the Byzantine agreement problem is solvable depends in part on what types of failures are considered, on whether the system is *synchronous* or *asynchronous*, and on the ratio of n to t. Roughly speaking, a system is synchronous if there is a global clock and agents move in lockstep; a 'step' in the system corresponds to a tick of the clock. In an asynchronous system, there is no global clock. The agents in the system can run at arbitrary rates relative to each other. One step for agent 1 can correspond to an arbitrary number of steps for agent 2 and vice versa. Synchrony is an implicit assumption in essentially all games. Although it is certainly possible to model games where player 2 has no idea how many moves player 1 has taken when player 2 is called upon to move, it is not typical to focus on the effects of synchrony (and its lack) in games. On the other hand, in distributed systems, it is typically a major focus.

Suppose for now that we restrict to *crash failures*, where a faulty agent behaves according to the protocol, except that it might crash at some point, after which it sends no messages. In the round in which an agent fails, the agent may send only a subset of the messages that it is supposed to send according to its protocol. Further suppose that the system is synchronous. In this case, the following rather simple protocol achieves Byzantine agreement:

- In the first round, each agent tells every other agent its initial preference.
- For rounds 2 to $t+1$, each agent tells every other agent everything it has heard in the previous round. Thus, for example, in round 3, agent 1 may tell agent 2 that it heard from agent 3 that its initial preference was to attack, and that it (agent 3) heard from agent 2 that its initial preference was to attack, and it heard from agent 4 that its initial preferences was to retreat, and so on. This means that messages get exponentially long, but it is not difficult to represent this information in a compact way so that the total communication is polynomial in n, the number of agents.
- At the end of round $t+1$, if an agent has heard from any other agent (including itself) that its initial preference was to attack, it decides to attack; otherwise, it decides to retreat.

Why is this correct? Clearly, if all agents are correct and want to retreat (resp., attack), then the final decision will be to retreat (resp., attack), since that is the only preference that agents hear about (recall that for now we are considering only crash failures). It remains to show that if some agents prefer to attack and others to retreat, then all the nonfaulty agents reach the same final decision. So suppose that i and j are nonfaulty and i decides to attack. That means that i heard that some agent's initial preference was to attack. If it heard this first at some round $t' < t+1$, then i will forward this message to j, who will receive it and thus also attack. On the other hand, suppose that i heard it first at round $t+1$ in a message from i_{t+1}. Thus, this message must be of the form 'i_t said at round t that … that i_2 said at round 2 that i_1 said at round 1 that its initial preference was to attack.' Moreover, the agents $i_1, \ldots, i_{t+1}$ must all be distinct. Indeed, it is easy to see that i_k must crash in round k before sending its message to i (but after sending its message to i_{k+1}), for $k=1,\ldots, t$, for otherwise i must have gotten the message from i_k, contradicting the assumption that i first heard at round $t+1$ that some agent's initial preference was to attack. Since at most t agents can crash, it follows that i_{t+1}, the agent that sent the message to i, is not faulty, and thus sends the message to j. Thus, j also decides to attack. A symmetric argument shows that if j decides to attack, then so does i.

It should be clear that the correctness of this protocol depends on both the assumptions made: crash failures and synchrony. Suppose instead that *Byzantine* failures are allowed, so that faulty agents can deviate in arbitrary ways from the protocol; they may 'lie', send deceiving messages, and collude to fool the nonfaulty agents in the most malicious ways. In this case, the protocol will not work at all. In fact, it is known that agreement can be reached in the presence of Byzantine failures iff $t<n/3$, that is, iff fewer than a third of the agents can be faulty (Pease, Shostak and Lamport, 1980). The effect of asynchrony is even more devastating: in an asynchronous system, it is impossible to reach agreement using a deterministic protocol even if $t=1$ (so that there is at most one failure) and only crash failures are allowed (Fischer, Lynch and Paterson, 1985). The problem in the asynchronous setting is that if none of the agents have heard from, say, agent 1, they have no way of knowing whether agent 1 is faulty or just slow. Interestingly, there are randomized algorithms (that is, behavioural strategies) that achieve agreement with arbitrarily high probability in an asynchronous setting [Ben-Or, 1983; Rabin, 1983].

Byzantine agreement can be viewed as a game where, at each step, an agent can either send a message or decide to attack or retreat. It is essentially a game between two teams, the nonfaulty agents and the faulty agents, whose composition is unknown (at least by the correct agents). To model it as a game in the more traditional sense, we could imagine that the nonfaulty agents are playing against a new player, the 'adversary'. One of the adversary's moves is that of 'corrupting' an agent: changing its type from 'nonfaulty' to 'faulty.' Once an agent is corrupted, what the adversary can do depends on the failure type being considered. In the case of crash failures, the adversary can decide which of a corrupted agent's messages will be delivered in the round in which the agent is corrupted; however, it cannot modify the messages themselves. In the case of Byzantine failures, the adversary essentially gets to make the moves for agents that have been corrupted; in particular, it can send arbitrary messages.

Why has the distributed systems literature not considered strategic behaviour in this game? Crash failures are used to model hardware and software failures; Byzantine failures are used to model random behaviour on the part of a system (for example, messages getting garbled in transit), software errors, and malicious adversaries (for example, hackers). With crash failures, it does not make sense to view the adversary's behaviour as strategic, since the adversary is not really viewed as having strategic interests. While it would certainly make sense, at least in principle, to consider the probability of failure (that is, the probability that the adversary corrupts an agent), this approach has by and large been avoided in the literature because it has proved difficult to characterize the probability distribution of failures over time. Computer components can perhaps be characterized as failing according to an exponential distribution (see Babaoglu, 1987, for an analysis of Byzantine agreement in such a setting), but crash failures can be caused by things other than component failures (faulty software, for example); these can be extremely difficult to characterize probabilistically. The problems are even worse when it comes to modelling random Byzantine behaviour.

With malicious Byzantine behaviour, it may well be reasonable to impute strategic behaviour to agents (or to an adversary controlling them). However, it is often difficult to characterize the payoffs of a malicious agent. The goals of the agents may vary from that of simply trying to delay a decision to that of causing disagreement. It is not clear what the appropriate payoffs should be for attaining these goals. Thus, the distributed systems literature has chosen to focus instead on algorithms that are guaranteed to satisfy the specification without making assumptions about the adversary's payoffs (or nature's probabilities, in the case of crash failures).

Recently, there has been some work on adding strategic concerns to standard problems in distributed computing; see, for example, Alvisi et al. (2005) and Halpern and Teague (2004). Moving in the other direction, there has also been some work on adding concerns of fault tolerance and asynchrony to standard problems in game theory; see, for example, Eliaz (2002), Monderer and Tennenholtz (1999a; 1999b) and the definitions in the next section. This seems to be an area that is ripe for further developments. One such development is the subject of the next section.

5. Implementing mediators

The question of whether a problem in a multiagent system that can be solved with a trusted mediator can be solved by just the agents in the system, without the mediator, has attracted a great deal of attention in both computer science (particularly in the cryptography community) and game theory. In cryptography, the focus on the problem has been on *secure multiparty computation*. Here it is assumed that each agent i has some private information x_i. Fix functions $f_1, \ldots, f_n$. The goal is to have agent i learn $f_i(x_1, \ldots, x_n)$ without learning anything about x_j for $j \neq i$ beyond what is revealed by the value of $f_i(x_1, \ldots, x_n)$. With a trusted mediator, this is trivial: each agent i just gives the mediator its private value x_i; the mediator then sends each agent i the value $f_i(x_1, \ldots, x_n)$. Work on multiparty computation (Goldreich, Micali and Wigderson, 1987; Shamir, Rivest and Adelman, 1981; Yao, 1982) provides conditions under which this can be done. In game theory, the focus has been on whether an equilibrium in a game with a mediator can be implemented using what is called *cheap talk* – that is, just by players communicating among themselves (cf. Barany, 1992; Ben-Porath, 2003; Forges, 1990; Gerardi, 2004; Heller, 2005; Urbano and Vila, 2004). As suggested in the previous section, the focus in the computer science literature has been in doing multiparty computation in the presence of possibly malicious adversaries, who do everything they can to subvert the computation, while in the game theory literature the focus has been on strategic agents. In recent work, Abraham et al. (2006) and Abraham, Dolev and Halpern (2007) considered deviations by both rational players, who have preferences and try to maximize them, and players who can viewed as malicious, although it is perhaps better to think of them as rational players whose utilities are not known by the other players or mechanism designer. I briefly sketch their results here; the following discussion is taken from Abraham, Dolev and Halpern (2007).

The idea of tolerating deviations by coalitions of players goes back to Aumann (1959); more recent refinements have been considered by Moreno and Wooders (1996). Aumann's definition is essentially the following.

Definition 1 $\vec{\sigma}$ is a k-resilient′ equilibrium *if, for all sets C of players with* $|C| \leq k$*, it is not the case that there exists a strategy* $\vec{\tau}$ *such that* $u_i(\vec{\tau}_C, \vec{\sigma}_{-C}) > u_i(\vec{\sigma})$ *for all* $i \in C$.

As usual, the strategy $(\vec{\tau}_C, \vec{\sigma}_{-C})$ is the one where each player $i \in C$ plays τ_i and each player $i \notin C$ plays σ_i. As the prime notation suggests, this is not quite the definition we want to work with. The trouble with this definition is that it suggests that coalition members cannot communicate with each other during the game. Perhaps surprisingly, allowing communication can *prevent* certain equilibria (see Abraham, Dolev and Halpern, 2007, for an example). Since we should expect coalition members to communicate, the following definition seems to capture a more reasonable notion of resilient equilibrium. Let the cheap-talk extension of a game Γ be, roughly speaking, the game where players are allowed to communicate among themselves in addition to performing the actions of Γ and the payoffs are just as in Γ.

Definition 2 $\vec{\sigma}$ is a *k-resilient equilibrium* in a game Γ if $\vec{\sigma}$ is a k-resilient′ equilibrium in the cheap-talk extension of Γ (where we identify the strategy σ_i in the game Γ with the strategy in the cheap-talk game where player i never sends any messages beyond those sent according to σ_i).

A standard assumption in game theory is that utilities are (commonly) known; when we are given a game we are also given each player's utility. When players make decisions, they can take other players' utilities into account. However, in large systems it seems almost invariably the case that there will be some fraction of users who do not respond to incentives the way we expect. For example, in a peer-to-peer network like Kazaa or Gnutella, it would seem that no rational agent should share files. Whether or not you can get a file depends only on whether other people share files. Moreover, there are disincentives for sharing (the possibility of lawsuits, use of bandwidth, and so on). Nevertheless, people do share files. However, studies of the Gnutella network have shown almost 70 per cent of users share no files and nearly 50 per cent of responses are from the top one per cent of sharing hosts (Adar and Huberman, 2000).

One reason that people might not respond as we expect is that they have utilities that are different from those we expect. Alternatively, the players may be irrational, or (if moves are made using a computer) they may be playing using a faulty computer and thus not able to make the move they would like, or they may not understand how to get the computer to make the move they would like. Whatever the reason, it seems important to design strategies that tolerate such unanticipated behaviours, so that the payoffs of the users with 'standard' utilities do not get affected by the nonstandard players using different strategies. This can be viewed as a way of adding fault tolerance to equilibrium notions.

Definition 3 A joint strategy $\vec{\sigma}$ is *t-immune* if, for all $T \subseteq N$ with $|T| \leq t$, all joint strategies $\vec{\tau}$, and all $i \notin T$, we have $u_i(\vec{\sigma}_{-T}, \vec{\tau}_T) \geq u_i(\vec{\sigma})$.

The notion of t-immunity and k-resilience address different concerns. For t immunity, we consider the payoffs of the players not in T, and require that they are not worse due to deviation; for resilience, we consider the payoffs of players in C, and require that they are not better due to deviation. It is natural to combine both notions. Given a game Γ, let $\Gamma_{\vec{\tau}}^{T}$ be the game that is identical to Γ except that the players in T are fixed to playing strategy $\vec{\tau}$.

Definition 4 $\vec{\sigma}$ is a (k, t) *-robust* equilibrium if $\vec{\sigma}$ is t-immune and, for all $T \subseteq N$ such that $|T| \leq t$ and all joint strategies $\vec{\tau}$, $\vec{\sigma}_{-T}$ is a k-resilient strategy of $\Gamma_{T}^{\vec{\tau}}$.

To state the results of Abraham et al. (2006) and Abraham, Dolev and Halpern (2007) on implementing mediators, three games need to be considered: an *underlying game* Γ, an extension Γ_d of Γ with a mediator, and a cheap-talk extension Γ_{CT} of Γ. Assume that Γ is a *normal-form Bayesian game*: each player has a type from some type space with a known distribution over types, and the utilities of the agents depend on the types and actions taken. Roughly speaking, a cheap talk game *implements* a game with a mediator if it induces the same distribution over actions in the underlying game, for each type vector of the players. With this background, I can summarize the results of Abraham et al. (2006) and Abraham, Dolev and Halpern (2007).

- If $n > 3k + 3t$, a (k, t) -robust strategy $\vec{\sigma}$ with a mediator can be implemented using cheap talk (that is, there is a (k, t)-robust strategy $\vec{\sigma}'$ in a cheap talk game such that $\vec{\sigma}$ and $\vec{\sigma}'$ induce the same distribution over actions in the underlying game). Moreover, the implementation requires no knowledge of other agents' utilities, and the cheap talk protocol has bounded running time that does not depend on the utilities.
- If $n \leq 3k + 3t$, then, in general, mediators cannot be implemented using cheap talk without knowledge of other agents' utilities. Moreover, even if other agents' utilities are known, mediators cannot, in general, be implemented without having a $(k+t)$-punishment strategy (that is, a strategy that, if used by all but at most $(k+t)$ players, guarantees that every player gets a worse outcome than they do with the equilibrium strategy) nor with bounded running time.
- If $n > 2k+3t$, then mediators can be implemented using cheap talk if there is a punishment strategy (and utilities are known) in finite expected running time that does not depend on the utilities.
- If $n \leq 2k + 3t$ then mediators cannot, in general, be implemented, even if there is a punishment strategy and utilities are known.

- If $n > 2k + 2t$ and there are broadcast channels then, for all ε, mediators can be ε-implemented (intuitively, there is an implementation where players get utility within ε of what they could get by deviating) using cheap talk, with bounded expected running time that does not depend on the utilities.
- If $n \leq 2k + 2t$, then mediators cannot, in general, be ε-implemented, even with broadcast channels. Moreover, even assuming cryptography and polynomially bounded players, the expected running time of an implementation depends on the utility functions of the players and ε.
- If $n > k + 3t$, then, assuming cryptography and polynomially bounded players, mediators can be ε-implemented using cheap talk, but if $n \leq 2k + 2t$, then the running time depends on the utilities in the game and ε.
- If $n \leq k + 3t$, then even assuming cryptography, polynomially bounded players, and a $(k + t)$-punishment strategy, mediators cannot, in general, be ε-implemented using cheap talk.
- If $n > k + t$, then, assuming cryptography, polynomially bounded players, and a public-key infrastructure (PKI), we can ε-implement a mediator.

The proof of these results makes heavy use of techniques from computer science. All the possibility results showing that mediators can be implemented use techniques from secure multiparty computation. The results showing that if $n \leq 3k + 3t$, then we cannot implement a mediator without knowing utilities, and that, even if utilities are known, a punishment strategy is required, use the fact that Byzantine agreement cannot be reached if $t < n/3$; the impossibility result for $n \leq 2k + 3t$ also uses a variant of Byzantine agreement.

A related line of work considers implementing mediators assuming stronger primitives (which cannot be implemented in computer networks); see Izmalkov, Micali and Lepinski (2005) and Lepinski et al. (2004) for details.

6. Other topics

There are many more areas of interaction between computer science than I have indicated in this brief survey. I briefly mention a few others here.

6.1 Interactive epistemology

Since the publication of Aumann's (1976) seminal paper, there has been a great deal of activity in trying to understand the role of knowledge in games, and providing epistemic analyses of solution concepts; see Battigalli and Bonanno (1999) for a survey. In computer science, there has been a parallel literature applying epistemic logic to reason about distributed computation. One focus of this work has been on characterizing the level of knowledge needed to solve certain problems. For example, to achieve Byzantine agreement common knowledge among the nonfaulty agents of an initial value is necessary and sufficient. More generally, in a precise sense, common knowledge is necessary and sufficient for coordination. Another focus has been on defining logics that capture the reasoning of resource-bounded agents. A number of approaches have been considered. Perhaps the most common considers logics for reasoning about *awareness*, where an agent may not be aware of certain concepts, and can know something only if he is aware of it. This topic has been explored in both computer science and game theory; see Dekel, Lipman and Rustichini (1998), Fagin and Halpern (1988), Halpern (2001), Halpern and Rêgo (2007), Heifetz, Meier and Schipper (2006), and Modica and Rustichini (1994; 1999) for some of the work in this active area. Another approach, so far considered only by computer scientists, involves *algorithmic knowledge*, which takes seriously the assumption that agents must explicitly compute what they know. See Fagin et al. (1995) for an overview of the work in epistemic logic in computer science.

6.2 Network growth

If we view networks as being built by selfish players (who decide whether or not to build links), what will the resulting network look like? How does the growth of the network affect its functionality? For example, how easily will influence spread through the network? How easy is it to route traffic? See Fabrikant et al. (2003) and Kempe, Kleinberg and Tardos (2003) for some recent computer science work in this burgeoning area.

6.3 Efficient representation of games

Game theory has typically focused on 'small' games, often two- or three-player games, that are easy to describe, such as Prisoner's Dilemma, in order to understand subtleties regarding basic issues such as rationality. To the extent that game theory is used to tackle larger, more practical problems, it will become important to find efficient techniques for describing and analysing games. By way of analogy, $2^n - 1$ numbers are needed to describe a probability distribution on a space characterized by n binary random variables. For $n = 100$ (not an unreasonable number in practical situations), it is impossible to write down the probability distribution in the obvious way, let alone do computations with it. The same issues will surely arise in large games. Computer scientists use graphical approaches, such as *Bayesian networks* and *Markov networks* (Pearl, 1988), for representing and manipulating probability measures on large spaces. Similar techniques seem applicable to games; see, for example, Kearns, Littman and Singh (2001), Koller and Milch (2001), and La Mura (2000) for specific approaches, and Kearns (2007) for a recent overview. Note that representation is also an issue when we consider the complexity of problems such as computing Nash or correlated equilibria. The complexity of a problem is a function of the size of the input, and the size of the input (which in this case is a description of the game) depends on how the input is represented.

6.4 Learning in games

There has been a great deal of work in both computer science and game theory on learning to play well in different settings (see Fudenberg and Levine, 1998, for an overview of the work in game theory). One line of research in computer science has involved learning to play optimally in a reinforcement-learning setting, where an agent interacts with an unknown (but fixed) environment. The agent then faces a fundamental tradeoff between *exploration* and *exploitation*. The question is how long it takes to learn to play well (to get a reward within some fixed ε of optimal); see Brafman and Tennenholtz (2002) and Kearns and Singh (1998) for the current state of the art. A related question is efficiently finding a strategy minimizes *regret* – that is, finding a strategy that is guaranteed to do not much worse than the best strategy would have done in hindsight (that is, even knowing what the opponent would have done). See Blum and Mansour (2007) for a recent overview of work on this problem.

JOSEPH Y. HALPERN

See also **epistemic game theory: an overview; epistemic game theory: beliefs and types; mechanism design: new developments.**

The work for this article was supported in part by NSF under grants CTC-0208535 and ITR-0325453, by ONR under grant N00014-02-1-0455, by the DoD Multidisciplinary University Research Initiative (MURI) program administered by the ONR under grants N00014-01-1-0795 and N00014-04-1-0725, and by AFOSR under grant F49620-02-1-0101. Thanks to Larry Blume, Christos Papadimitriou, Ilya Segal, Éva Tardos, and Moshe Tennenholtz for useful comments.

Bibliography

Abraham, I., Dolev, D., Gonen, R. and Halpern, J. 2006. Distributed computing meets game theory: robust mechanisms for rational secret sharing and multiparty computation. In *Proceedings of the 25th ACM Symposium on Principles of Distributed Computing*, 53–62.

Abraham, I., Dolev, D. and Halpern, J. 2007. Lower bounds on implementing robust and resilient mediators. Available at http://arxiv.org/abs/0704.3646, accessed 19 June 2007.

Adar, E. and Huberman, B. 2000. Free riding on Gnutella. *First Monday* 5(10).

Alvisi, L., Ayer, A.S., Clement, A., Dahlin, M., Martin, J.P. and Porth, C. 2005. BAR fault tolerance for cooperative services. In *Proceedings of the 20th ACM Symposium on Operating Systems Principles (SOSP 2005)*, 45–58.

Archer, A. and Tardos, É. 2001. Truthful mechanisms for one-parameter agents. In *Proceedings of the 42nd IEEE Symposium on Foundations of Computer Science*, 482–91.

Archer, A. and Tardos, É. 2002. Frugal path mechanisms. In *Proceedings of the 13th ACM-SIAM Symposium on Discrete Algorithms*, 991–9.

Aumann, R. 1959. Acceptable points in general cooperative *n*-person games. In *Contributions to the Theory of Games IV*, ed. A. Tucker and R. Luce. Princeton, NJ: Princeton University Press, pp. 287–324.

Aumann, R.J. 1976. Agreeing to disagree. *Annals of Statistics* 4, 1236–9.

Aumann, R.J. 1987. Correlated equilibrium as an expression of Bayesian rationality. *Econometrica* 55, 1–18.

Axelrod, R. 1984. *The Evolution of Cooperation*. New York: Basic Books.

Babaoglu, O. 1987. On the reliability of consensus-based fault-tolerant distributed computing systems. *ACM Translation on Computer Systems* 5, 394–416.

Barany, I. 1992. Fair distribution protocols or how the players replace fortune. *Mathematics of Operations Research* 17, 327–40.

Battigalli, P. and Bonanno, G. 1999. Recent results on belief, knowledge and the epistemic foundations of game theory. *Research in Economics* 53, 149–225.

Ben-Or, M. 1983. Another advantage of free choice: completely asynchronous agreement protocols. In *Proceedings of the Second ACM Symposium on Principles of Distributed Computing*, 27–30.

Ben-Porath, E. 2003. Cheap talk in games with incomplete information. *Journal of Economic Theory* 108, 45–71.

Blum, A. and Mansour, Y. 2007. Learning, regret minimization, and equilibria. In *Algorithmic Game Theory*, ed. N. Nisan, T. Roughgarden, É. Tardos, and V. Vazirani. Cambridge: Cambridge University Press.

Blum, B., Shelton, C.R. and Koller, D. 2003. A continuation method for Nash equilibria in structured games. In *Proceedings of the 18th International Joint Conference on Artificial Intelligence (IJCAI '03)*, 757–64.

Brafman, R.I. and Tennenholtz, M. 2002. R-MAX: a general polynomial time algorithm for near-optimal reinforcement learning. *Journal of Machine Learning Research* 3, 213–31.

Chen, X. and Deng, X. 2006. Settling the complexity of 2-player Nash equilibrium. In *Proceedings of the 47th IEEE Symposium on Foundations of Computer Science*, 261–72.

Chor, B. and Dwork, C. 1989. Randomization in Byzantine agreement. In *Advances in Computing Research 5: Randomness and Computation*, ed. S. Micali. Greenwich, CT: JAI Press, pp. 443–97.

Chu, F. and Halpern, J.Y. 2001. A decision-theoretic approach to reliable message delivery. *Distributed Computing* 14, 359–89.

Clarke, E.H. 1971. Multipart pricing of public goods. *Public Choice* 11, 17–33.

Conitzer, V. and Sandholm, T. 2003. Complexity results about Nash equilibria. In *Proceedings of the 18th International Joint Conference on Artificial Intelligence (IJCAI '03)*, 765–71.

Conitzer, V., Sandholm, T. and Lang, J. 2007. When are elections with few candidates hard to manipulate? *Journal of the ACM* 54(3), Article 14.

Cramton, P., Shoham, Y. and Steinberg, R. 2006. *Combinatorial Auctions*. Cambridge, MA: MIT Press.

Daskalis, C., Goldberg, P.W. and Papadimitriou, C.H. 2006. The complexity of computing a Nash equilibrium. In *Proceedings of the 38th ACM Symposium on Theory of Computing*, 71–8.

Dekel, E., Lipman, B. and Rustichini, A. 1998. Standard state-space models preclude unawareness. *Econometrica* 66, 159–73.

Eliaz, K. 2002. Fault-tolerant implementation. *Review of Economic Studies* 69, 589–610.

Fabrikant, A., Luthra, A., Maneva, E., Papadimitriou, C.H. and Shenker, S. 2003. On a network creation game. In *Proceedings of the 22nd ACM Symposium on Principles of Distributed Computing*, 347–51.

Fagin, R. and Halpern, J.Y. 1988. Belief, awareness, and limited reasoning. *Artificial Intelligence* 34, 39–76.

Fagin, R., Halpern, J.Y., Moses, Y. and Vardi, M.Y. 1995. *Reasoning about Knowledge*. Cambridge, MA: MIT Press. A slightly revised paperback version was published in 2003.

Feigenbaum, J., Papadimitriou, C. and Shenker, S. 2000. Sharing the cost of muliticast transmissions (preliminary version). In *Proceedings of the 32nd ACM Symposium on Theory of Computing*, 218–27.

Fischer, M.J. 1983. The consensus problem in unreliable distributed systems. In *Foundations of Computation Theory*, vol. 185, ed. M. Karpinski. Lecture Notes in Computer Science. Berlin/New York: Springer, pp. 127–40.

Fischer, M.J., Lynch, N.A. and Paterson, M.S. 1985. Impossibility of distributed consensus with one faulty processor. *Journal of the ACM* 32, 374–82.

Forges, F. 1990. Universal mechanisms. *Econometrica* 58, 1341–64.

Fudenberg, D. and Levine, D. 1998. *The Theory of Learning in Games*. Cambridge, MA: MIT Press.

Gerardi, D. 2004. Unmediated communication in games with complete and incomplete information. *Journal of Economic Theory* 114, 104–31.

Gibbard, A. 1973. Manipulation of voting schemes. *Econometrica* 41, 587–602.
Gilboa, I. and Zemel, E. 1989. Nash and correlated equilibrium: some complexity considerations. *Games and Economic Behavior* 1, 80–93.
Goldreich, O., Micali, S. and Wigderson, A. 1987. How to play any mental game. In *Proceedings of the 19th ACM Symposium on Theory of Computing*, 218–29.
Govindan, S. and Wilson, R. 2003. A global Newton method to compute Nash equilibria. *Journal of Economic Theory* 110, 65–86.
Groves, T. 1973. Incentives in teams. *Econometrica* 41, 617–31.
Halldórsson, M.M., Halpern, J.Y., Li, L. and Mirrokni, V. 2004. On spectrum sharing games. In *Proceedings of the 23rd ACM Symposium on Principles of Distributed Computing*, 107–14.
Halpern, J.Y. 2001. Alternative semantics for unawareness. *Games and Economic Behavior* 37, 321–39.
Halpern, J.Y. 2003. A computer scientist looks at game theory. *Games and Economic Behavior* 45, 114–32.
Halpern, J.Y. and Rêgo, L.C. 2007. Reasoning about knowledge of unawareness. *Games and Economic Behavior.* Also available at http://arxiv.org/abs/cs.LO/0603020, accessed 24 June 2007.
Halpern, J.Y. and Teague, V. 2004. Rational secret sharing and multiparty computation: extended abstract. In *Proceedings of the 36th ACM Symposium on Theory of Computing*, 623–32.
Heifetz, A., Meier, M. and Schipper, B. 2006. Interactive unawareness. *Journal of Economic Theory* 130, 78–94.
Heller, Y. 2005. A minority-proof cheap-talk protocol. Unpublished manuscript.
Hopcroft, J.E. and Ullman, J.D. 1979. *Introduction to Automata Theory, Languages and Computation.* New York: Addison-Wesley.
Hurwicz, L. 1977. On the dimensional requirements of informationally decentralized Pareto satisfactory processes. In *Studies in Resource AllocationProcesses*, ed. K.J. Arrow and L. Hurwicz. New York: Cambridge University Press, pp. 413–24.
Izmalkov, S., Micali, S. and Lepinski, M. 2005. Rational secure computation and ideal mechanism design. In *Proceedings of the 46th IEEE Symposium Foundations of Computer Science*, 585–95.
Kearns, M. 2007. Graphical games. In *Algorithmic Game Theory*, ed. N. Nisan, T. Roughgarden, É. Tardos and V. Vazirani. Cambridge: Cambridge University Press.
Kearns, M., Littman, M.L. and Singh, S.P. 2001. Graphical models for game theory. In *Proceedings of the 17th Conference on Uncertainty in Artificial Intelligence (UAI 2001)*, 253–60.
Kearns, M.J. and Reiter, M.K. 2005. *Proceedings of the Sixth ACM Conference on Electronic Commerce (EC '05).* New York: ACM. Table of contents available at http://www.informatik.unitrier.de/~ley/db/conf/sigecom/sigecom2005.html, accessed 19 June 2007.
Kearns, M. and Singh, S.P. 1998. Near-optimal reinforcement learning in polynomial time. In *Proceedings of the 15th International Conference on Machine Learning*, pp. 260–8.
Kempe, D., Kleinberg, J. and Tardos, É. 2003. Maximizing the spread of influence through a social network. In *Proceedings of the Ninth ACM SIGKDD International Conference on Knowledge Discovery and Data Mining*, 137–46.
Kfir-Dahav, N.E., Monderer, D. and Tennenholtz, M. 2000. Mechanism design for resource-bounded agents. In *International Conference on Multiagent Systems*, 309–16.
Koller, D. and Milch, B. 2001. Structured models for multiagent interactions. In *Theoretical Aspects of Rationality and Knowledge: Proceedings of the Eighth Conference (TARK 2001)*, 233–48.
Koutsoupias, E. and Papadimitriou, C.H. 1999. Worst-case equilibria. In *Proceedings of the 16th Conference on Theoretical Aspects of Computer Science*, vol. 1563, Lecture Notes in Computer Science. Berlin: Springer, pp. 404–13.
Kushilevitz, E. and Nisan, N. 1997. *Communication Complexity.* Cambridge: Cambridge University Press.

La Mura, P. 2000. Game networks. In *Proceedings of the 16th Conference on Uncertainty in Artificial Intelligence (UAI 2000)*, 335–42.

Lehmann, D., Müller, R. and Sandholm, T. 2006. The winner determination problem. In *Combinatorial Auctions*, ed. P. Cramton, Y. Shoham, and R. Steinberg. Cambridge, MA: MIT Press.

Lemke, C.E. and Howson, J.J.T. 1964. Equilibrium points of bimatrix games. *Journal of the Society for Industrial and Applied Mathematics* 12, 413–23.

Lepinski, M., Micali, S., Peikert, C. and Shelat, A. 2004. Completely fair SFE and coalition-safe cheap talk. In *Proceedings of the 23rd ACM Symposium Principles of Distributed Computing*, 1–10.

Linial, N. 1994. Games computers play: Game-theoretic aspects of computing. In *Handbook of Game Theory*, vol. 2, ed. R.J. Aumann and S. Hart. Amsterdam: North-Holland.

Modica, S. and Rustichini, A. 1994. Awareness and partitional information structures. *Theory and Decision* 37, 107–24.

Modica, S. and Rustichini, A. 1999. Unawareness and partitional information structures. *Games and Economic Behavior* 27, 265–98.

Monderer, D. and Tennenholtz, M. 1999a. Distributed games. *Games and Economic Behavior* 28, 55–72.

Monderer, D. and Tennenholtz, M. 1999b. Distributed Games: From Mechanisms to Protocols. In *Proceedings of the 16th National Conference on Artificial Intelligence (AAAI '99)*, 32–7.

Moreno, D. and Wooders, J. 1996. Coalition-proof equilibrium. *Games and Economic Behavior* 17, 80–112.

Mount, K. and Reiter, S. 1974. The informational size of message spaces. *Journal of Economic Theory* 8, 161–92.

Nash, J. 1950. Equilibrium points in *n*-person games. *Proceedings of the National Academy of Sciences* 36, 48–9.

Neyman, A. 1985. Bounded complexity justifies cooperation in finitely repeated Prisoner's Dilemma. *Economic Letters* 19, 227–9.

Nisan, N. 2006. Bidding languages for combinatorial auctions. In *Combinatorial Auctions*. Cambridge, MA: MIT Press.

Nisan, N. and Ronen, A. 2000. Computationally feasible VCG mechanisms. In *Proceedings of the Second ACM Conference on Electronic Commerce (EC '00)*, 242–52.

Nisan, N. and Ronen, A. 2001. Algorithmic mechanism design. *Games and Economic Behavior* 35, 166–96.

Nisan, N., Roughgarden, T., Tardos, É. and Vazirani, V. 2007. *Algorithmic Game Theory*. Cambridge, UK: Cambridge University Press.

Papadimitriou, C.H. 1994a. *Computational Complexity*. Reading, MA: Adison-Wesley.

Papadimitriou, C.H. 1994b. On the complexity of the parity argument and other inefficient proofs of existence. *Journal of Computer and System Sciences* 48, 498–532.

Papadimitriou, C.H. 2001. Algorithms, games, and the internet. In *Proceedings of the 33rd ACM Symposium on Theory of Computing*. 749–53.

Papadimitriou, C.H. 2005. Computing correlated equilibria in multiplayer games. In *Proceedings of the 37th ACM Symposium on Theory of Computing*, 49–56.

Papadimitriou, C.H. 2007. The complexity of finding Nash equilibria. In *Algorithmic Game Theory*, ed. N. Nisan, T. Roughgarden, É. Tardos and V. Vazirani. Cambridge: Cambridge University Press.

Papadimitriou, C.H. and Roughgarden, T. 2005. Computing equilibria in multi-player games. In *Proceedings of the 16th ACM–SIAM Symposium on Discrete Algorithms*, 82–91.

Papadimitriou, C.H. and Yannakakis, M. 1994. On complexity as bounded rationality. In *Proceedings of the 26th ACM Symposium on Theory of Computing*, 726–33.
Pearl, J. 1988. *Probabilistic Reasoning in Intelligent Systems*. San Francisco: Morgan Kaufmann.
Pease, M., Shostak, R. and Lamport, L. 1980. Reaching agreement in the presence of faults. *Journal of the ACM* 27, 228–34.
Porter, R., Nudelman, E. and Shoham, Y. 2004. Simple search methods for finding a Nash equilibrium. In *Proceedings of the 21st National Conference on Artificial Intelligence (AAAI '04)*, 664–9.
Rabin, M.O. 1983. Randomized Byzantine generals. In *Proceedings of the 24th IEEE Symposium on Foundations of Computer Science*, 403–9.
Roughgarden, T. and Tardos, É. 2002. How bad is selfish routing? *Journal of the ACM* 49, 236–59.
Rubinstein, A. 1986. Finite automata play the repeated prisoner's dilemma. *Journal of Economic Theory* 39, 83–96.
Rubinstein, A. 1998. *Modeling Bounded Rationality*. Cambridge, MA: MIT Press.
Sandholm, T. and Yokoo, M. 2003. *The Second International Joint Conference on Autonomous Agents and Multiagent Systems (AAMAS 2003)*. Table of contents available at http://www.informatik.uni-trier.de/~ley/db/conf/atal/aamas2003.html, accessed 25 June 2007.
Satterthwaite, M. 1975. Strategy-proofness and Arrow's conditions: existence and correspondence theorems for voting procedures and social welfare functions. *Journal of Economic Theory* 10, 187–217.
Segal, I. 2006. Communication in economic mechanisms. In *Advances in Economics and Econometrics: Theory and Application, Ninth World Congress (Econometric Society Monographs*, ed. R. Blundell, W.K. Newey and T. Persson. Cambridge: Cambridge University Press, pp. 222–68.
Shamir, A., Rivest, R.L. and Adelman, L. 1981. Mental poker. In *The Mathematical Gardner*, ed. D.A. Klarner. Boston, MA: Prindle, Weber, and Schmidt, pp. 37–43.
Tennenholtz, M. 2002. Competitive safety analysis: robust decision-making in multi-agent systems. *Journal of A.I. Research* 17, 363–78.
Urbano, A. and Vila, J.E. 2002. Computational complexity and communication: coordination in two-player games. *Econometrica* 70, 1893–927.
Urbano, A. and Vila, J.E. 2004. Computationally restricted unmediated talk under incomplete information. *Economic Theory* 23, 283–320.
Vetta, A. 2002. Nash equilibria in competitive societies, with applications to facility location, traffic routing and auctions. In *Proceedings of the 43rd IEEE Symposium on Foundations of Computer Science*, 416–25.
Vickrey, W. 1961. Counterspeculation, auctions and competitive sealed tenders. *Journal of Finance* 16, 8–37.
Yao, A. 1982. Protocols for secure computation (extended abstract). In *Proceedings of the 23rd IEEE Symposium on Foundations of Computer Science*, 160–4.

cooperation

Cooperation is said to occur when two or more individuals engage in joint actions that result in mutual benefits. Examples include the mutually beneficial exchange of goods, the payment of taxes to finance public goods, team production, common pool resource management, collusion among firms, voting for income redistribution to others, participating in collective actions such as demonstrations, and adhering to socially beneficial norms.

A major goal of economic theory has been to explain how wide-scale cooperation among self-regarding individuals occurs in a decentralized setting. The first thrust of this endeavour involved Walras's general equilibrium model, culminating in the celebrated 'invisible hand' theorem of Arrow and Debreu (Arrow and Debreu, 1954; Debreu, 1959; Arrow and Hahn, 1971). But, the assumption that contracts could completely specify all relevant aspects of all exchanges and could be enforced at zero cost to the exchanging parties is not applicable to many important forms of cooperation. Indeed, such economic institutions as firms, financial institutions, and state agencies depend on incentive mechanisms involving strategic interaction in addition to explicit contracts (Blau, 1964; Gintis, 1976; Stiglitz, 1987; Tirole, 1988; Laffont, 2000).

The second major thrust in explaining cooperation eschewed complete contracting and developed sophisticated repeated game-theoretic models of strategic interaction. These models are based on the insights of Shubik (1959), Taylor (1976), Axelrod and Hamilton (1981) and others that repetition of social interactions plus retaliation against defectors by withdrawal of cooperation may enforce cooperation among self-regarding individuals. A statement of this line of thinking, applied towards understanding the broad historical and anthropological sweep of human experience is the work of Ken Binmore (1993; 1998; 2005). For Binmore, a society's moral rules are instructions for behaviour in conformity with one of the myriad of Nash equilibria of a repeated n-player social interaction. Because the interactions are repeated, and these rules form a Nash equilibrium, the self-regarding individuals who comprise the social order will conform to the moral rules.

We begin by reviewing models of repeated dyadic interaction in which cooperation may occur among players who initially cooperate and in the next round adopt the action of the other player in the previous round, called *tit for tat*. These models show that as long as the probability of game repetition is sufficiently great and individuals are sufficiently patient, a cooperative equilibrium can be sustained once it is implemented. This reasoning applies to a wide range of similar strategies. We then analyse *reputation maintenance* models of dyadic interaction, which are relevant when individuals interact with many different individuals, and hence the number of periods before a repeat encounter with any given individual may be too great to support the tit-for-tat strategy.

We then turn to models of cooperation in larger groups, arguably the most relevant case, given the scale on which cooperation frequently takes place. The folk theorem (Fudenberg and Maskin, 1986) shows that, in groups of any size, cooperation can be maintained on the assumption that the players are sufficiently future-oriented and termination of the interaction is sufficiently unlikely. We will see, however, that these models do not successfully extend the intuitions of the dyadic models to many-person interactions. The reason is that the level of cooperation that may be supported in this way deteriorates as group size increases and the probability of either behavioural or perceptual error rises, and because the theory lacks a plausible account of how individuals would discover and coordinate on the complicated strategies necessary for cooperation to be sustained in these models. This difficulty bids us investigate how other-regarding preferences, *strong reciprocity* in particular, may sustain a high level of cooperation, even with substantial errors and in large groups.

Repetition allows cooperation in groups of size two

Consider a pair of individuals who play the following *stage game* repeatedly: each can *cooperate* (that is, help the other) at a cost $c>0$ to himself, providing a benefit to the other of $b>c$. Alternatively, each player can *defect*, incurring no cost and providing no benefit. Clearly, both would gain by cooperating in the stage game, each receiving a net gain of $b-c>0$. However, the structure of the game is that of a *Prisoner's Dilemma*, in which a self-regarding player earns higher payoff by defecting, no matter what his partner does.

The behaviour whereby each individual provides aid as long as this aid has been reciprocated by the other in the previous encounter, is called *tit for tat*. Although termed 'reciprocal altruism' by biologists, this behaviour is self-regarding, because each individual's decisions depend only on the expected net benefit the individual enjoys from the long-term relationship.

On the assumption that after each round of play the interaction will be continued with probability δ, and that players have discount factor d (so $d=1/(1+r)$, where r is the rate of time preference), then provided

$$\delta db>c, \tag{1}$$

each individual paired with a tit-for-tat player does better by cooperating (that is, playing tit for tat) rather than by defecting. Thus tit for tat is a best response to itself. To see this, let **v** be the present value of cooperating when paired with a tit-for-tat player. Then

$$\mathbf{v}=b-c+\delta\, d\, \mathbf{v}, \tag{2}$$

which gives

$$\mathbf{v}=\frac{b-c}{1-\delta d}. \tag{3}$$

The present value of defecting for ever on a tit-for-tat playing partner is b (the single period gain of b being followed by zero gains in every subsequent period as a result of the tit-for-tat player's defection), so playing tit-for-tat is a best response to itself if and only if $(b-c)(1-\delta d) > b$, which reduces to (1). Under these conditions unconditional defect is also a best response to itself, so either cooperation or defection can be sustained.

But suppose that, instead of defection for ever, the alternative to tit for tat is for a player to defect for a certain number of rounds, before returning to cooperation on round $k > 0$. The payoff to this strategy against tit for tat is $b-(\delta d)^k c + (\delta d)^{k+1}$ **v**. This payoff must not be greater than **v** if tit for tat is to be a best response to itself. It is an easy exercise in algebra to show that the inequality

$$\mathbf{v} \geq b - (\delta d)^k c + (\delta d)^{k+1}\mathbf{v}$$

simplifies to (1), no matter what the value of k. A similar argument shows that when (1) holds, defecting for ever (that is, $k = \infty$) does not have a higher payoff than cooperating.

Cooperation through reputation maintenance

Tit for tat takes the form of frequent repetition of the Prisoner's Dilemma stage game inducing a pair of self-regarding individuals to cooperate. In a sizable group, an individual may interact frequently with a large number of partners but infrequently with any single one, say on the average of once every k periods. Players then discount future gains so that a payoff of **v** in k periods from now is worth d^k **v** now. Then, an argument parallel to that of the previous section shows that cooperating is a best response if and only if

$$\frac{b-c}{1-\delta d^k} > b$$

which reduces to

$$\delta d^k b > c. \qquad (4)$$

Note that this is the same equation as (1) except that the effective discount factor falls from d to d^k. For sufficiently large k, it will not pay to cooperate. Therefore, the conditions for tit-for-tat reciprocity will not obtain.

But cooperation may be sustained in this situation if each individual keeps a mental model of exactly which group members cooperated in the previous period and which did not. In this case, players may cooperate in order to cultivate a *reputation for cooperation*. When individuals tend to cooperate with others who have a reputation for cooperation, a process called *indirect reciprocity* can sustain cooperation. Let us say that an individual who cooperated in the previous period *in good standing*, and specify that the only way an individual can fall into *bad standing* is by defecting on a partner who is in good standing. Note that an individual can always defect when his partner is in bad standing without losing his good standing status. In this more general setting the tit-for-tat strategy is replaced by the following *standing strategy*: cooperate if and only if your current partner is in good standing, except that, if you accidentally defected the previous period, cooperate this period unconditionally, thereby restoring your status as a member in good standing. This *standing model* is due to Sugden (1986).

Panchanathan and Boyd (2004) have proposed an ingenious deployment of indirect reciprocity, assuming that there is an ongoing dyadic helping game in society based on the indirect reciprocity information and incentive structure, and there is also an n-player public goods game, played relatively infrequently by the same individuals. In the dyadic helping game, two individuals are paired and each member of the pair may confer a benefit b upon his partner at a cost c to himself, an individual remaining in good standing so long as he does not defect on a partner who is in good standing. This random pairing is repeated with probability δ and with discount factor d. In the public goods game, an individual produces a benefit b_g that is shared equally by all the other members, at a cost c_g to himself. The two games are linked by defectors in the public goods game being considered in bad standing at the start of the helping game that directly follows. Then, cooperation can be sustained in both the public goods game and in the dyadic helping game so long as

$$c_g \leq \frac{b(1-\varepsilon)-c}{1-\delta d}, \tag{5}$$

where ε is the rate at which cooperators unintentionally fail to produce the benefit. Parameters favouring this solution are that the cost c_g of cooperating in the public goods game be low, the factor δd is close to unity, and the net benefit $b(1-\varepsilon)-c$ of cooperating in the reputation-building reciprocity game be large.

The major weakness of the standing model is its demanding informational requirements. Each individual must know the current standing of each member of the group, the identity of each member's current partner, and whether each individual cooperated or defected against his current partner. Since dyadic interactions are generally private, and hence are unlikely to be observed by more than a small number of others, errors in determining the standing of individuals may be frequent. This contrasts sharply with the repeated game models of the previous section, which require only that an individual know how many of his current partners defected in the previous period. Especially serious is that warranted non-cooperation (because in one's own mental accounting one's partner is in bad standing) may be perceived to be unwarranted defection by some third parties but not by others. This will occur with high frequency if information partially private rather than public (not everyone has the same information). It has been proposed that gossip and other forms of communication can transform private into public information, but how this might occur among self-regarding individuals has not been (and probably cannot be) shown, because in any practical setting individuals may benefit by reporting dishonestly on what they have observed, and self-regarding individuals do not care about the harm to others induced by false information. Under such conditions, disagreements among individuals about who ought to be punished can reach extremely high levels, with the unravelling of cooperation as a result.

In response to this weakness of the standing model, Nowak and Sigmund (1998) developed an indirect reciprocity model which they term *image scoring*. Players in the image scoring need not know the standing of recipients of aid, so the informational requirements of indirect reciprocity are considerably reduced. Nowak and Sigmund show that the strategy of cooperating with others who have cooperated in the past, *independent of the reputation of the cooperator's partner*, is stable against invasion by defectors, and weakly stable against invasion by unconditional cooperators once defectors are eliminated from the population. Leimar and Hammerstein (2001), Panchanathan and Boyd (2003), and Brandt and Sigmund (2004; 2005), explore the applicability of image scoring.

Cooperation in large groups of self-regarding individuals

Repeated game theory has extended the above two-player results to a general n-player stage game, the so-called *public goods game*. In this game each player cooperates at cost $c>0$, contributing an amount $b>c$ that is shared equally among the other $n-1$ players. We define the *feasible payoff set* as the set of possible payoffs to the various players, assuming each cooperates with a certain probability, and each player does at least as well as the payoffs obtaining under mutual defection. The set of feasible payoffs for a two-player public goods game is given in Figure 1 by the four-sided figure ABCD. For the n-player game, the figure ABCD is replaced by a similar n-dimensional polytope.

Repeated game models have demonstrated the so-called folk theorem, which asserts that any distribution of payoffs to the n players that lies in the feasible payoff set can be supported by an equilibrium in the repeated public goods game, provided the discount factor times the probability of continuation, δd, is sufficiently close to unity. The equilibrium concept employed is a refinement of subgame perfect equilibrium. Significant contributions to this literature include Fudenberg and Maskin (1986), assuming perfect information, Fudenberg, Levine and Maskin (1994), assuming imperfect information, so that cooperation is sometimes inaccurately reported as

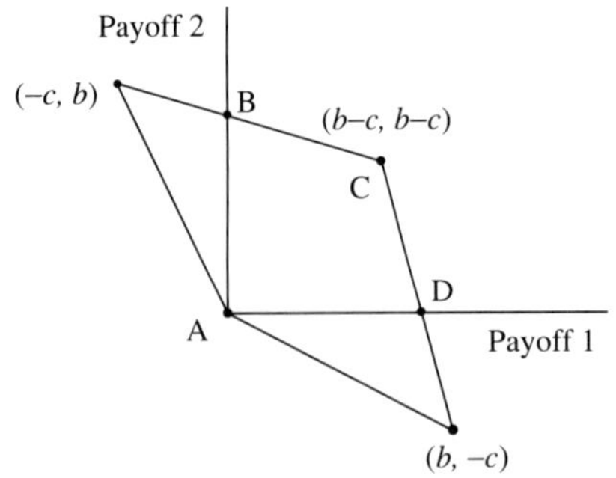

Figure 1 Two-player public goods game

defection, and Sekiguchi (1997), Piccione (2002), Ely and Välimäki (2002), Bhaskar and Obara (2002) and Mailath and Morris (2006), who assume that different players receive different, possibly inaccurate, information concerning the behaviour of the other players.

The folk theorem is an *existence theorem* affirming that any outcome that is a Pareto improvement over universal defection may be supported by a Nash equilibrium, including point C (full cooperation) in the figure and outcomes barely superior to A (universal defection). The theorem is silent on which of this vast number of equilibria is more likely to be observed or how they might be attained. When these issues are addressed two problems are immediately apparent: first, equilibria in the public goods game supported in this manner exhibit very little cooperation if large numbers of individuals are involved or errors in execution and perception are large, and second, the equilibria are not robust because they require some mechanism allowing coordination on highly complex strategies. While such a mechanism could be provided by centralized authority, decentralized mechanisms, as we will see, are not sustainable in a plausible dynamic.

The dynamics of cooperation

The first difficulty, the inability to support high levels of cooperation in large groups or with significant behavioural or perceptual noise, stems from the fact that the only way players may punish defectors is *to withdraw their own cooperation*. In the two-person case, defectors are thus targeted for punishment. But for large n, withdrawal of cooperation to punish a single defector punishes all group members equally, most of whom, in the neighbourhood of a cooperative equilibrium, will be cooperators. Moreover, in large groups, the rate at which erroneous signals are propagated will generally increase with group size, and the larger the group, the larger the fraction of time group members will spend punishing (miscreants and fellow cooperators alike). For instance, suppose the rate at which cooperators accidentally fail to produce b, and hence signal defection, is five per cent. Then, in a group of size two, a perceived defection will occur in about ten per cent of all periods, while in a group of size 20, at least one perceived defection will occur in about 64 per cent of all periods.

As a result of these difficulties, the folk theorem assertion that we can approximate the per-period expected payoff as close to the efficient level (point C in Figure 1) as desired as long as the discount factor δ is sufficiently close to unity is of little practical relevance. The reason is that as $\delta \to 1$, the current payoff approximates zero, and the expected payoff is deferred to future periods at very little cost, since future returns are discounted at a very low rate. Indeed, with the discount factor δ held constant, the efficiency of cooperation in the Fudenberg, Levine and Maskin model declines at an exponential rate with increasing group size (Bowles and Gintis, 2007, ch. 13). Moreover, in an agent-based simulation of the public goods with punishment model, on the assumption of a benefit/cost ratio of $b/c = 2$ (that is, contributing to the public good costs half of the benefit conferred on members of the group) and a discount factor times probability of repetition of $d\delta = 0.96$, even for an error rate as low as $\varepsilon = 0.04$, fewer than half of the members contribute to the public good in groups of size $n = 4$, and less that 20 per cent contribute in groups of size $n = 6$ (Bowles and Gintis, 2007, ch. 5).

The second limitation of the folk theorem analysis is that it has not been shown (and probably cannot be shown) that the equilibria supporting cooperation are dynamically robust, that is, asymptotically stable with a large basin of attraction in the relevant dynamic. Equilibria for which this is not the case will seldom be observed because they are unlikely to be attained and if attained unlikely to persist for long.

The Nash equilibrium concept applies when each individual expects all others to play their parts in the equilibrium. But, when there are multiple equilibria, as in the case of the folk theorem, where there are many possible patterns of response to given pattern of defection, each imposing distinct costs and requiring distinct, possibly stochastic, behaviours on the part of players, there is no single set of beliefs and expectations that group members can settle upon to coordinate their actions (Aumann and Brandenburger, 1995).

While game theory does not provide an analysis of how beliefs and expectations are aligned in a manner allowing cooperation to occur, sociologists (Durkheim, 1902; Parsons and Shils, 1951) and anthropologists (Benedict, 1934; Boyd and Richerson, 1985; Brown, 1991) have found that virtually every society has such processes, and that they are key to understanding strategic interaction. Borrowing a page from sociological theory, we posit that groups may have *focal rules* that are common knowledge among group members. Focal rules could suggest which of a countless number of strategies that could constitute a Nash equilibrium should all individuals adopt them, thereby providing the coordination necessary to support cooperation. These focal rules do not ensure equilibrium, because error, mutation, migration, and other dynamical forces ensure that on average not all individuals conform to the focal rules of the groups to which they belong. Moreover, a group's focal rules are themselves subject to dynamical forces, those producing better outcomes for their members displacing less effective focal rules.

In the case of the repeated public goods game, which is the appropriate model for many forms of large-scale cooperation, Gintis (2007) shows that focal rules capable of supporting the kinds of cooperative equilibria identified by the folk theorem are not evolutionarily stable, meaning that groups whose focal rules support highly cooperative equilibria do worse than groups with less stringent focal rules, and as a result the focal rules necessary for cooperation are eventually eliminated.

The mechanism behind this result can be easily explained. Suppose a large population consists of many smaller groups playing n-person public goods games, with considerable migration across groups, and with the focal rules of successful groups being copied by less successful groups. To maintain a high level of cooperation in a group, focal rules should foster punishing defectors by withdrawing cooperation. However, such punishment is both costly and provides an external benefit to other groups by reducing the frequency of defection-prone individuals who might migrate elsewhere. Hence, groups that 'free ride' by not punishing defectors harshly will support higher payoffs for its members than groups that punish assiduously. Such groups will then be copied by other groups, leading to a secular decline in the frequency of punishment suggested by focal rules in all groups. Thus, suppose that the groups in question were competitive firms whose profits depend on the degree of cooperation among firm members. If all adopted a zero-tolerance rule (all would defect if even a single defection was perceived), then a firm adopting a rule that tolerated a single defection would sustain higher profits and replace the zero-tolerance firms. But this firm would in turn be replaced by a firm adopting a rule that tolerates two defections.

These two problems – the inability to support efficient levels of cooperation in large groups with noisy information, and dynamic instability – have been shown for the case where information is public. Private information, in general the more relevant case, considerably exacerbates these problems.

Cooperation with other-regarding individuals

The models reviewed thus far have assumed that individuals are entirely self-regarding. But cooperation in sizable groups is possible if there exist other-regarding individuals in the form of *strong reciprocators*, who cooperate with one another and punish defectors, even if they sustain net costs. Strong reciprocators are altruistic in the standard sense that they confer benefits on other members of their group (in this case, because their altruistic punishment of defectors sustains cooperation) but would increase their own payoffs by adopting self-regarding behaviours. A model with *social preferences* of this type can explain large-scale decentralized cooperation with noisy information as long as the information structure is such that defectors expect a level of punishment greater than costs of cooperating.

Cooperation is not a puzzle if a sufficient number of individuals with social preferences are involved. The puzzle that arises is how such altruistic behaviour could have become common, given that bearing costs to support the benefits of others reduces payoffs, and both cultural and genetic updating of behaviours is likely to favour traits with higher payoffs. This evolutionary puzzle applies to strong reciprocity. Since punishment is costly to the individual, and an individual could escape punishment by cooperating, while avoiding the costs of punishment by not punishing, we are obliged to exhibit a mechanism whereby strong reciprocators could proliferate when rare and be sustained in equilibrium, despite their altruistic behaviour.

This is carried out in Sethi and Somanathan (2001), Gintis (2000), Boyd et al. (2003), Gintis (2003) and Bowles and Gintis (2004). The evolutionary viability of other types of altruistic cooperation is demonstrated in Bowles, Jung-Kyoo and Hopfensitz (2003), Boyd et al. (2003), Bergstrom (1995) and Salomonsson and Weibull (2006). The critical condition allowing the evolution of strong reciprocity and other forms of altruistic social preferences is that individuals with social preferences are more likely than random to interact with others with social preferences. Positive assortment arises in these models due to deliberate exclusion of those who have defected in the past (by ostracism, for example), random differences in the composition of groups (due to small group size and limited between-group mobility), limited dispersion of close kin who share common genetic and cultural inheritance, and processes of social learning such as conformism or group level socialization contributing to homogeneity within groups. As in the repeated game models, smaller groups favour cooperation, but in this case for a different reason: positive assortment tends to decline with group size. But the group sizes that sustain the altruistic preferences that support cooperative outcomes in these models are at least an order of magnitude larger than those indicated for the repeated game models studied above.

In sum, we think that other-regarding preferences provide a compelling account of many forms of human cooperation that are not well explained by repeated game models with self-regarding preferences. Moreover, a number of studies have shown that strong reciprocity and other social preferences are a common human behaviour (Fehr and Gächter, 2000; Henrich et al., 2005) and could have emerged and been sustained in a gene-culture co-evolutionary dynamic under conditions experienced by ancestral humans (Bowles, 2006). The above models also show that strong reciprocity and other social preferences that support cooperation can evolve and persist even when there are many self-regarding players, where group sizes are substantial, and when behavioural or perception errors are significant.

Conclusion: economics and the missing choreographer

The shortcomings of the economic theory of cooperation based on repeated games strikingly replicate those of economists' other main contribution to the study of decentralized cooperation, namely, general equilibrium theory. Both prove the existence of equilibria with socially desirable properties, while leaving the question of how such equilibria are achieved as an afterthought, thereby exhibiting a curious lack of attention to dynamics and out-of equilibrium behaviour. Both purport to model decentralized interactions but on close inspection require a level of coordination that is not explained, but rather posited as a *deus ex machina*. To ensure that only equilibrium trades are executed, general equilibrium theory resorts to a fictive 'auctioneer'. No counterpart to the auctioneer has been made explicit in the repeated-game approach to cooperation. Highly choreographed coordination on complex strategies capable of deterring defection are supposed to materialize quite without the need for a choreographer.

Humans are unique among living organisms in the degree and range of cooperation among large numbers of substantially unrelated individuals. The global division of labour and exchange, the modern democratic welfare state, and contemporary warfare alike evidence our distinctiveness. These forms of cooperation emerged historically and are today sustained as a result of the interplay of self-regarding and social preferences operating under the influence of group-level institutions of governance and socialization that favour cooperators, in part by protecting them from exploitation by defectors.

The norms and institutions that have accomplished this evolved over millennia through trial and error. Consider how real-world institutions addressed two of the shoals on which the economic models foundered. First, the private nature of information, as we have seen, makes it virtually impossible to coordinate the targeted punishment of miscreants. Converting private information about transgressions into public information that can provide the basis of punishment often involves civil or criminal trials, elaborate processes that rely on commonly agreed upon rules of evidence and ethical norms of appropriate behaviour. Even these complex institutions frequently fail to transform the private protestations of innocence and guilt into common knowledge. Second, as in the standing models with private information, cooperation often unravels when the withdrawal of cooperation by the civic-minded intending to punish a defector is interpreted by others as a violation of a cooperative norm, inviting further defections. In all successful modern societies, this problem was eventually addressed by the creation of a corps of specialists entrusted with carrying out the more severe of society's punishments, whose uniforms conveyed the civic purpose of the punishments they meted out, and whose professional norms, it was hoped, would ensure that the power to punish was not used for personal gain. Like court proceedings, this institution works imperfectly. It is hardly surprising then that economists have encountered difficulty in devising simple models of how large numbers of self-regarding individuals might sustain cooperation in a truly decentralized setting.

Modelling this complex process is a major challenge of contemporary science. Economic theory, favouring parsimony over realism, has instead sought to explain cooperation without reference to other-regarding preferences and with minimalist or fictive descriptions of social institutions.

SAMUEL BOWLES AND HERBERT GINTIS

See also **behavioural economics and game theory; behavioural game theory; repeated games.**

Bibliography

Arrow, K.J. and Debreu, G. 1954. Existence of an equilibrium for a competitive economy. *Econometrica* 22, 265–90.

Arrow, K.J. and Hahn, F. 1971. *General Competitive Analysis.* San Francisco: Holden-Day.

Aumann, R.J. and Brandenburger, A. 1995. Epistemic conditions for Nash equilibrium. *Econometrica* 65, 1161–80.

Axelrod, R. and Hamilton, W.D. 1981. The evolution of cooperation. *Science* 211, 1390–6.

Benedict, R. 1934. *Patterns of Culture.* Boston: Houghton Mifflin.

Bergstrom, T.C. 1995. On the evolution of altruistic ethical rules for siblings. *American Economic Review* 85, 58–81.

Bhaskar, V. and Obara, I. 2002. Belief-based equilibria the repeated Prisoner's Dilemma with private monitoring. *Journal of Economic Theory* 102, 40–69.

Binmore, K. 1993. *Game Theory and the Social Contract: Playing Fair.* Cambridge, MA: MIT Press.

Binmore, K. 1998. *Game Theory and the Social Contract: Just Playing.* Cambridge, MA: MIT Press.

Binmore, K.G. 2005. *Natural Justice*. Oxford: Oxford University Press.

Blau, P. 1964. *Exchange and Power in Social Life*. New York: Wiley.

Bowles, S. 2006. Group competition, reproductive leveling, and the evolution of human altruism. *Science* 314, 1669–72.

Bowles, S. and Gintis, H. 2004. The evolution of strong reciprocity: cooperation in heterogeneous populations. *Theoretical Population Biology* 65, 17–28.

Bowles, S. and Gintis, H. 2007. A cooperative species: human reciprocity and its evolution, in preparation.

Bowles, S., Jung-Kyoo, C. and Hopfensitz, A. 2003. The co-evolution of individual behaviors and social institutions. *Journal of Theoretical Biology* 223, 135–47.

Boyd, R., Gintis, H., Bowles, S. and Richerson, P.J. 2003. Evolution of altruistic punishment. *Proceedings of the National Academy of Sciences* 100, 3531–5.

Boyd, R. and Richerson, P.J. 1985. *Culture and the Evolutionary Process*. Chicago: University of Chicago Press.

Brandt, H. and Sigmund, K. 2004. The logic of reprobation: assessment and action rules for indirect reciprocation. *Journal of Theoretical Biology* 231, 475–86.

Brandt, H. and Sigmund, K. 2005. Indirect reciprocity, image scoring, and moral hazard. *Proceeding of the National Academy of Sciences* 102, 2666–70.

Brown, D.E. 1991. *Human Universals*. New York: McGraw-Hill.

Debreu, G. 1959. *Theory of Value*. New York: Wiley.

Durkheim, E. 1902. *De La Division du Travail Social*. Paris: Presses Universitaires de France, 1967.

Ely, J.C. and Välimäki, J. 2002. A robust folk theorem for the Prisoner's Dilemma. *Journal of Economic Theory* 102, 84–105.

Fehr, E. and Gächter, S. 2000. Cooperation and punishment. *American Economic Review* 90, 980–94.

Fudenberg, D., Levine, D.K. and Maskin, E. 1994. The Folk Theorem with imperfect public information. *Econometrica* 62, 997–1039.

Fudenberg, D. and Maskin, E. 1986. The Folk Theorem in repeated games with discounting or with incomplete information. *Econometrica* 54, 533–54.

Gintis, H. 1976. The nature of the labor exchange and the theory of capitalist production. *Review of Radical Political Economics* 8(2), 36–54.

Gintis, H. 2000. Strong reciprocity and human sociality. *Journal of Theoretical Biology* 206, 169–79.

Gintis, H. 2003. The hitchhiker's guide to altruism: genes, culture, and the internalization of norms. *Journal of Theoretical Biology* 220, 407–18.

Gintis, H. 2007. *Modeling Cooperation with Self-regarding Agents*. Santa Fe: Santa Fe Institute.

Henrich, J., Boyd, R., Bowles, S., Camerer, C. et al. 2005. Economic man in cross-cultural perspective: behavioral experiments in 15 small-scale societies. *Behavioral and Brain Sciences* 28, 795–815.

Laffont, J.J. 2000. *Incentives and Political Economy*. Oxford: Oxford University Press.

Leimar, O. and Hammerstein, P. 2001. Evolution of cooperation through indirect reciprocity. *Proceedings of the Royal Society of London, Series B* 268, 745–53.

Mailath, G.J. and Morris, S. 2006. Coordination failure in repeated games with almost-public monitoring. *Theoretical Economics* 1, 311–40.

Nowak, M.A. and Sigmund, K. 1998. Evolution of indirect reciprocity by image scoring. *Nature* 393, 573–7.

Panchanathan, K. and Boyd, R. 2003. A tale of two defectors: the importance of standing for evolution of indirect reciprocity. *Journal of Theoretical Biology* 224, 115–26.

Panchanathan, K. and Boyd, R. 2004. Indirect reciprocity can stabilize cooperation without the second-order free rider problem. *Nature* 432, 499–502.

Parsons, T. and Shils, E. 1951. *Toward a General Theory of Action*. Cambridge, MA: Harvard University Press.
Piccione, M. 2002. The repeated Prisoner's Dilemma with imperfect private monitoring. *Journal of Economic Theory* 102, 70–83.
Salomonsson, M. and Weibull, J. 2006. Natural selection and social preferences. *Journal of Theoretical Biology* 239, 79–92.
Sekiguchi, T. 1997. Efficiency in repeated Prisoner's Dilemma with private monitoring. *Journal of Economic Theory* 76, 345–61.
Sethi, R. and Somanathan, E. 2001. Preference evolution and reciprocity. *Journal of Economic Theory* 97, 273–97.
Shubik, M. 1959. *Strategy and Market Structure: Competition, Oligopoly, and the Theory of Games*. New York: Wiley.
Stiglitz, J. 1987. The causes and consequences of the dependence of quality on price. *Journal of Economic Literature* 25, 1–48.
Sugden, R. 1986. *The Economics of Rights, Co-operation and Welfare*. Oxford: Basil Blackwell.
Taylor, M. 1976. *Anarchy and Cooperation*. London: Wiley.
Tirole, J. 1988. *The Theory of Industrial Organization*. Cambridge, MA: MIT Press.

deterministic evolutionary dynamics

1. Introduction

Deterministic evolutionary dynamics for games first appeared in the mathematical biology literature, where Taylor and Jonker (1978) introduced the *replicator dynamic* to provide an explicitly dynamic foundation for the static evolutionary stability concept of Maynard Smith and Price (1973). But one can find precursors to this approach in the beginnings of game theory: Brown and von Neumann (1950) introduced differential equations as a tool for computing equilibria of zero-sum games. In fact, the replicator dynamic appeared in the mathematical biology literature long before game theory itself: while Maynard Smith and Price (1973) and Taylor and Jonker (1978) studied game theoretic models of animal conflict, the replicator equation is equivalent to much older models from population ecology and population genetics. These connections are explained by Schuster and Sigmund (1983), who also coined the name 'replicator dynamic', borrowing the word 'replicator' from Dawkins (1982).

In economics, the initial phase of research on deterministic evolutionary dynamics in the late 1980s and early 1990s focused on populations of agents who are randomly matched to play normal form games, with evolution described by the replicator dynamic or other closely related dynamics. The motivation behind the dynamics continued to be essentially biological: individual agents are preprogrammed to play specific strategies, and the dynamics themselves are driven by differences in birth and death rates. Since that time the purview of the literature has broadened considerably, allowing more general sorts of large population interactions, and admitting dynamics derived from explicit models of active myopic decision making.

This article provides a brief overview of deterministic evolutionary dynamics in game theory. More detailed treatments of topics introduced here can be found in the recent survey article by Hofbauer and Sigmund (2003), and in books by Maynard Smith (1982), Hofbauer and Sigmund (1988; 1998), Weibull (1995), Vega-Redondo (1996), Samuelson (1997), Fudenberg and Levine (1998), Cressman (2003), and Sandholm (2007).

2. Population games

Population games provide a general model of strategic interactions among large numbers of anonymous agents. For simplicity, we focus on games played by a single population, in which agents are not differentiated by roles; allowing for multiple populations is mostly a matter of introducing more elaborate notation.

In a single-population game, each agent from a unit-mass population chooses a strategy from the finite set $S = \{1, \ldots, n\}$, with typical elements i and j. The distribution of strategy choices at a given moment in time is described by a *population state* $x \in X = \{x \in \mathbf{R}^n_+ : \sum_{i \in S} x_i = 1\}$. The *payoff* to strategy i, denoted $F_i : X \to \mathbf{R}$, is a continuous function of the population state; we use the notation $F : X \to \mathbf{R}^n$ to refer to all strategies' payoffs at once. By taking the set of strategies S as fixed, we can refer to F itself as a *population game.*

The simplest example of a population game is the most commonly studied one: random matching to play a symmetric normal form game $A \in \mathbf{R}^{n \times n}$, where A_{ij} is the payoff obtained by an agent choosing strategy i when his opponent chooses strategy j. When the population state is $x \in X$, the expected payoff to strategy i is simply the weighted average of the elements of the ith row of the payoff matrix: $F_i(x) = \sum_{j \in S} A_{ij} x_j = (Ax)_i$. Thus, the population game generated by random matching in A is the linear population game $F(x) = Ax$.

Many models of strategic interactions in large populations that arise in applications do not take this simple linear form. For example, in models of highway congestion, payoff functions are convex: increases in traffic when traffic levels are low have virtually no effect on delays, while increases in traffic when traffic levels are high increase delays substantially (see Beckmann, McGuire and Winsten, 1956; Sandholm, 2001). Happily, allowing nonlinear payoffs extends the range of possible applications of population games without making evolutionary dynamics especially more difficult to analyse, since the dynamics themselves are nonlinear even when the underlying payoffs are not.

3. Foundations of evolutionary dynamics

Formally, an *evolutionary dynamic* is a map that assigns to each population game F a differential equation $\dot{x} = V^F(x)$ on the state space X. While one can define evolutionary dynamics directly, it is preferable to derive them from explicit models of myopic individual choice.

We can accomplish this by introducing the notion of a *revision protocol* $\rho : \mathbf{R}^n \times X \to \mathbf{R}^{n \times n}_+$. Given a payoff vector $F(x)$ and a population state x, a revision protocol specifies for each pair of strategies i and j a non-negative number $\rho_{ij}(F(x), x)$, representing the rate at which strategy i players who are considering switching strategies switch to strategy j. Revision protocols that are most consistent with the evolutionary paradigm require agents to possess only limited information: for example, a revising agent might know only the current payoffs of his own strategy i and his candidate strategy j.

A given revision protocol can admit a variety of interpretations. For one all-purpose interpretation, suppose each agent is equipped with an exponential alarm clock. When the clock belonging to an agent playing strategy i rings, he selects a strategy $j \in S$ at random, and then switches to this strategy with probability proportional to $\rho_{ij}(F(x), x)$. While this interpretation is always available, others may be simpler in certain instances. For example, if the revision protocol is of the imitative form $\rho_{ij} = x_j \times \hat{\rho}_{ij}$, we can incorporate the x_j term into our story by supposing that the revising agent selects his candidate strategy j not by drawing a strategy at random, but by drawing an opponent at random and observing this opponent's strategy.

A population game F and a revision protocol ρ together generate an ordinary differential equation $\dot{x} = V^F(x)$ on the state space X. This equation, which captures the population's *expected* motion under F and ρ, is known as the *mean dynamic* or *mean field* for F and ρ:

$$\dot{x} = V_i^F(x) = \sum_{j \in S} x_j \rho_{ji}(F(x), x) - x_i \sum_{j \in S} \rho_{ij}(F(x), x). \tag{M}$$

The form of the mean dynamic is easy to explain. The first term describes the 'inflow' into strategy i from other strategies; it is obtained by multiplying the mass of agents playing each strategy j by the rate at which such agents switch to strategy i, and then summing over j. Similarly, the second term describes the 'outflow' from strategy i to other strategies. The difference between these terms is the net rate of change in the use of strategy i.

To obtain a formal link between the mean dynamic (M) and our model of individual choice, imagine that the population game F is played not by a continuous mass of agents but rather by a large, finite population with N members. Then the model described above defines a Markov process $\{X_t^N\}$ on a fine but discrete grid in the state space X. The foundations for deterministic evolutionary dynamics are provided by the following finite horizon deterministic approximation theorem: Fix a time horizon $T < \infty$. Then the behaviour of the stochastic process $\{X_t^N\}$ through time T is approximated by a solution of the mean dynamic (M); the approximation is uniformly good with probability close to 1 once the population size N is large enough. (For a formal statement of this result, see Benaïm and Weibull, 2003.)

In cases where one is interested in phenomena that occur over very long time horizons, it may be more appropriate to consider the infinite horizon behaviour of the stochastic process $\{X_t^N\}$. Over this infinite time horizon, the deterministic approximation fails, as a correct analysis must explicitly account for the stochastic nature of the evolutionary process. For more on the distinction between the two time scales, see Benaïm and Weibull (2003).

4. Examples and families of evolutionary dynamics

We now describe revision protocols that generate some of the most commonly studied evolutionary dynamics. In the table below, $\bar{F}(x) = \sum_{i \in S} x_i F_i(x)$ represents the population's average payoff at state x, and $B^F(x) = \operatorname{argmax}_{y \in X} y' F(x)$ is the best response correspondence for the game F.

A common critique of evolutionary analysis of games is that the choice of a specific revision protocol, and hence the evolutionary analysis that follows, is necessarily arbitrary. There is surely some truth to this criticism: to the extent that one's analysis is sensitive to the fine details of the choice of protocol, the conclusions of the analysis are cast into doubt. But much of the force of this critique is dispelled by this important observation: *evolutionary dynamics based on qualitatively similar revision protocols lead to qualitatively similar aggregate behaviour.* We call a collection of dynamics generated by similar revision protocols a 'family' of evolutionary dynamics.

To take one example, many properties that hold for the replicator dynamic also hold for dynamics based on revision protocols of the form $\rho_{ij} = x_j \hat{\rho}_{ij}$ where $\hat{\rho}_{ij}$ satisfies

$$\text{sgn}((\hat{\rho}_{ki} - \hat{\rho}_{ik}) - (\hat{\rho}_{kj} - \hat{\rho}_{jk})) = \text{sgn}(F_i - F_j) \quad \text{for all } k \in S.$$

(In words: if i earns a higher payoff than j, then the net conditional switch rate from k to i is higher than that from k to j for all $k \in S$.) For reasons described in Section 3, dynamics generated in this way are called 'imitative dynamics'. (See Björnerstedt and Weibull, 1996, for a related formulation.) For another example, most properties of the pairwise difference dynamic remain true for dynamics based on protocols of the form $\rho_{ij} = \varphi(F_i - F_j)$, where $\varphi : \mathbf{R} \to \mathbf{R}_+$ satisfies sign-preservation:

$$\text{sgn}(\varphi(d)) = \text{sgn}([d]_+).$$

Dynamics in this family are called 'pairwise comparison dynamics'. For more on these and other families of dynamics, see Sandholm (2007, ch. 5).

5. Rest points and local stability

Having introduced families of evolutionary dynamics, we now turn to questions of prediction: if agents playing game F follow the revision protocol ρ (or, more broadly, a revision protocol from a given family), what predictions can we make about how they will play the game? To what extent do these predictions accord with those provided by traditional game theory?

A natural first question to ask concerns the relationship between the rest points of an evolutionary dynamic V^F and the Nash equilibria of the underlying game F. In fact, one can prove for a very wide range of evolutionary dynamics that if a state $x^* \in X$ is a Nash equilibrium (that is, if $x \in B(x)$), then x^* is a rest point as well.

One way to show that $NE(F) \subseteq RP(V^F)$ is to first establish a *monotonicity* property for V^F: that is, a property that relates strategies' growth rates under V^F with their payoffs in the underlying game (see, for example, Nachbar, 1990; Friedman, 1991; and Weibull, 1995). The most general such property, first studied by Friedman (1991) and Swinkels (1993), we call 'positive correlation':

$$\text{If } x \notin RP(V^F), \text{ then } F(x)'V^F(x) > 0. \tag{PC}$$

Property (PC) is equivalent to requiring a positive correlation between strategies' growth rates $V_i^F(x)$ and payoffs $F_i(x)$ (where the underlying probability measure is the uniform measure on the strategy set S). This property is satisfied by the first three dynamics in Table 1, and modifications of it hold for the remaining two as well. Moreover, it is not difficult to show that if V^F satisfies (PC), then all Nash equilibria of F are rest points of V^F: that is, $NE(F) \subseteq RP(V^F)$, as desired (see Sandholm, 2007, ch. 5).

In many cases, one can also prove that every rest point of V^F is a Nash equilibrium of F, and hence that $NE(F) = RP(V^F)$. In fact, versions of this statement are true for all of the dynamics introduced above, with the notable exception of the replicator

Table 1

Revision protocol	Evolutionary dynamic	Name	Origin
$\rho_{ij} = x_j(K - F_i)$, or $\rho_{ij} = x_j(K + F_j)$, or $\rho_{ij} = x_j[F_j - F_i]_+$	$\dot{x}_i = x_i(F_i(x) - \bar{F}(x))$	Replicator	Taylor and Jonker (1978)
$\rho_{ij} = [F_j - \bar{F}]_+$	$\dot{x}_i = [F_i(x) - \bar{F}(x)]_+ - x_i \sum_{j \in S} [F_j(x) - \bar{F}(x)]_+$	Brown–von Neumann–Nash (BNN)	Brown and von Neumann (1950)
$\rho_{ij} = [F_j - F_i]_+$	$\dot{x}_i = \sum_{j \in S} x_j [F_i(x) - F_j(x)]_+ - x_i \sum_{j \in S} [F_i(x) - F_j(x)]_+$	Pairwise difference (PD)	Smith (1984)
$\rho_{ij} = \frac{\exp(\eta^{-1} F_j)}{\sum_{k \in S} \exp(\eta^{-1} F_k)}$	$\dot{x}_i = \frac{\exp(\eta^{-1} F_i(x))}{\sum_{k \in S} \exp(\eta^{-1} F_k(x))} x_i$	Logit	Fudenberg and Levine (1998)
$\rho_{ij} = B_i^F(x)$	$\dot{x} = B_i^F(x) - x_i$	Best response	Gilboa and Matsui (1991)

dynamic and other imitative dynamics. The reason for this failure is easy to see: when revisions are based on imitation, unused strategies, even ones that are optimal, are never chosen. On the other hand, if we introduce a small number of agents playing an unused optimal strategy, then these agents will be imitated. Developing this logic, Bomze (1986) and Nachbar (1990) show that, under many imitative dynamics, every Lyapunov stable rest point is a Nash equilibrium.

As we noted at the onset, the original motivation for the replicator dynamic was to provide a foundation for Maynard Smith and Price's (1973) notion of an evolutionarily stable strategy (ESS). Hofbauer, Schuster and Sigmund (1979) and Zeeman (1980) show that an ESS is asymptotically stable under the replicator dynamic, but that an asymptotically state need not be an ESS.

More generally, when is a Nash equilibrium a dynamically stable rest point, and under which dynamics? Under differentiable dynamics, stability of isolated equilibria can often be determined by linearizing the dynamic around the equilibrium. In many cases, the question of the stability of the rest point x^* reduces to a question of the negativity of certain eigenvalues of the Jacobian matrix $DF(x^*)$ of the payoff vector field. In non-differentiable cases, and in cases where the equilibria in question form a connected component, stability can sometimes be established by using another standard approach: the construction of suitable Lyapunov functions. For an overview of work in these directions, see Sandholm (2007, ch. 6).

In the context of random matching in normal form games, it is natural to ask whether an equilibrium that is stable under an evolutionary dynamic also satisfies the restrictions proposed in the equilibrium refinements literature. Swinkels (1993) and Demichelis and Ritzberger (2003) show that this is true in great generality under even the most demanding refinements: in particular, any component of rest points that is asymptotically stable under a dynamic that respects condition (PC) contains a strategically stable set in the sense of Kohlberg and Mertens (1986). While proving this result is difficult, the idea behind the result is simple. If a component is asymptotically stable under an evolutionary dynamic, then this dynamic stability ought not to be affected by slight perturbations of the payoffs of the game. *A fortiori*, the *existence* of the component ought not to be affected by the payoff perturbations either. But this preservation of existence is precisely what strategic stability demands.

This argument also shows that asymptotic stability under evolutionary dynamics is a qualitatively stronger requirement than strategic stability: while strategic stability requires equilibria not to vanish after payoff perturbations, it does not demand that they be attracting under a disequilibrium adjustment process. For example, while all Nash equilibria of simple coordination games are strategically stable, only the pure Nash equilibria are stable under evolutionary dynamics.

Demichelis and Ritzberger (2003) establish their results using tools from index theory. Given an evolutionary dynamic V^F for a game F, one can assign each component of rest points an integer, called the *index*, that is determined by the behaviour of the dynamic in a neighbourhood of the rest point; for instance, regular, stable rest points are assigned an index of 1. The set of all indices for the dynamic V^F is constrained by the *Poincaré–Hopf theorem*, which tells us that the sum of the indices of the equilibrium components of V^F must equal 1. As a consequence of this deep topological result, one can sometimes determine the local stability of one component of rest points by evaluating the local stability of the others.

6. Global convergence: positive and negative results

To provide the most satisfying evolutionary justification for the prediction of Nash equilibrium play, it is not enough to link the rest points of a dynamic and the Nash equilibria of the underlying game, or to prove local stability results. Rather, one must establish convergence to Nash equilibrium from *arbitrary* initial conditions.

One way to proceed is to focus on a class of games defined by some noteworthy payoff structure, and then to ask whether global convergence can be established for games in this class under certain families of evolutionary dynamics. As it turns out, general global convergence results can be proved for a number of classes of games. Among these classes are *potential games*, which include common interest games, congestion games, and games generated by externality pricing schemes; *stable games*, which include zero-sum games, games with an interior ESS, and (perturbed) concave potential games; and *supermodular games*, which include models of Bertrand oligopoly, arms races, and macroeconomic search. A fundamental paper on global convergence of evolutionary dynamics is Hofbauer (2000); for a full treatment of these results, see Sandholm (2007).

Once we move beyond specific classes of games, global convergence to Nash equilibrium cannot be guaranteed; cycling and chaotic behaviour become possible. Indeed, Hofbauer and Swinkels (1996) and Hart and Mas-Colell (2003) construct examples of games in which all reasonable deterministic evolutionary dynamics fail to converge to Nash equilibrium from most initial conditions. These results tell us that general guarantees of convergence to Nash equilibrium are impossible to obtain.

In light of this fact, we might instead consider the extent to which solution concepts simpler than Nash equilibrium are supported by evolutionary dynamics. Cressman and Schlag (1998) and Cressman (2003) investigate whether imitative dynamics lead to subgame perfect equilibria in reduced normal forms of extensive form games – in particular, generic games of perfect information. In these games, interior solution trajectories do converge to Nash equilibrium components, and only subgame perfect components can be interior asymptotically stable. But even in very simple games interior asymptotically stable components need not exist, so the dynamic analysis may fail to select subgame perfect equilibria. For a full treatment of these issues, see Cressman (2003).

What about games with strictly dominated strategies? Early results on this question were positive: Akin (1980), Nachbar (1990), Samuelson and Zhang (1992), and Hofbauer and Weibull (1996) prove that dominated strategies are eliminated under certain classes of imitative dynamics. However, Berger and Hofbauer (2006) show that dominated strategies need not be eliminated under the BNN dynamic. Pushing this argument further, Hofbauer and Sandholm (2006) find that dominated strategies can survive under any continuous evolutionary dynamic that satisfies positive correlation and *innovation*; the latter condition requires that agents choose unused best responses with positive probability. Thus, whenever there is some probability that agents base their choices on direct evaluation of payoffs rather than imitation of successful opponents, evolutionary dynamics may violate even the mildest rationality criteria.

7. Conclusion

Because the literature on evolutionary dynamics came to prominence shortly after the literature on equilibrium refinements, it is tempting to view the former literature as a branch of the latter. But, while it is certainly true that evolutionary models have something to say about selection among multiple equilibria, viewing them simply as equilibrium selection devices can be misleading. As we have seen, evolutionary dynamics capture the behaviour of large numbers of myopic, imperfectly informed decision makers. Using evolutionary models to predict behaviour in interactions between, say, two well-informed players is daring at best.

The negative results described in Section 6 should be understood in this light. If we view evolutionary dynamics as an equilibrium selection device, the fact that they need not eliminate strictly dominated strategies might be viewed with disappointment. But, if we take the result at face value, it becomes far less surprising: if agents switch to strategies that perform reasonably well at the moment of choice, that a strategy is never optimal need not deter agents from choosing it.

A similar point can be made about failures of convergence to equilibrium. From a traditional point of view, persistence of disequilibrium behaviour might seem to pundermine the very possibility of a satisfactory economic analysis. But the work described in this entry suggests that in large populations, this possibility is not only real but is also one that game theorists are well equipped to analyse.

WILLIAM H. SANDHOLM

See also **learning and evolution in games: adaptive heuristics; learning and evolution in games: an overview; learning and evolution in games: ESS; Nash equilibrium, refinements of; stochastic adaptive dynamics.**

I thank John Nachbar for helpful comments. Financial support from NSF Grant SES-0092145 is gratefully acknowledged.

Bibliography

Akin, E. 1980. Domination or equilibrium. *Mathematical Biosciences* 50, 239–50.
Beckmann, M., McGuire, C. and Winsten, C. 1956. *Studies in the Economics of Transportation*. New Haven, CT: Yale University Press.
Benaïm, M. and Weibull, J. 2003. Deterministic approximation of stochastic evolution in games. *Econometrica* 71, 873–903.
Berger, U. and Hofbauer, J. 2006. Irrational behavior in the Brown–von Neumann–Nash dynamics. *Games and Economic Behavior* 56, 1–6.
Björnerstedt, J. and Weibull, J. 1996. Nash equilibrium and evolution by imitation. In *The Rational Foundations of Economic Behavior*, ed. K. Arrow et al. New York: St Martin's Press.
Bomze, I. 1986. Non-cooperative two-person games in biology: a classification. *International Journal of Game Theory* 15, 31–57.
Brown, G. and von Neumann, J. 1950. Solutions of games by differential equations. In *Contributions to the Theory of Games I*, ed. H. Kuhn and A. Tucker. Annals of Mathematics Studies 24. Princeton: Princeton University Press.
Cressman, R. 2003. *Evolutionary Dynamics and Extensive Form Games*. Cambridge, MA: MIT Press.
Cressman, R. and Schlag, K. 1998. The dynamic (in)stability of backwards induction. *Journal of Economic Theory* 83, 260–85.
Dawkins, R. 1982. *The Extended Phenotype*. San Francisco: Freeman.
Demichelis, S. and Ritzberger, K. 2003. From evolutionary to strategic stability. *Journal of Economic Theory* 113, 51–75.
Friedman, D. 1991. Evolutionary games in economics. *Econometrica* 59, 637–66.
Fudenberg, D. and Levine, D. 1998. *Theory of Learning in Games*. Cambridge, MA: MIT Press.
Gilboa, I. and Matsui, A. 1991. Social stability and equilibrium. *Econometrica* 59, 859–67.
Hart, S. and Mas-Colell, A. 2003. Uncoupled dynamics do not lead to Nash equilibrium. *American Economic Review* 93, 1830–6.
Hofbauer, J. 2000. From Nash and Brown to Maynard Smith: equilibria, dynamics, and ESS. *Selection* 1, 81–8.
Hofbauer, J. and Sandholm, W. 2006. Survival of dominated strategies under evolutionary dynamics. Mimeo, University College London and University of Wisconsin.
Hofbauer, J., Schuster, P. and Sigmund, K. 1979. A note on evolutionarily stable strategies and game dynamics. *Journal of Theoretical Biology* 27, 537–48.
Hofbauer, J. and Sigmund, K. 1988. *Theory of Evolution and Dynamical Systems*. Cambridge: Cambridge University Press.

Hofbauer, J. and Sigmund, K. 1998. *Evolutionary Games and Population Dynamics*. Cambridge: Cambridge University Press.

Hofbauer, J. and Sigmund, K. 2003. Evolutionary game dynamics. *Bulletin of the American Mathematical Society N.S.* 40, 479–519.

Hofbauer, J. and Swinkels, J. 1996. A universal Shapley example. Mimeo, University of Vienna and Northwestern University.

Hofbauer, J. and Weibull, J. 1996. Evolutionary selection against dominated strategies. *Journal of Economic Theory* 71, 558–73.

Kohlberg, E. and Mertens, J.-F. 1986. On the strategic stability of equilibria. *Econometrica* 54, 1003–38.

Maynard Smith, J. 1982. *Evolution and the Theory of Games*. Cambridge: Cambridge University Press.

Maynard Smith, J. and Price, G. 1973. The logic of animal conflict. *Nature* 246, 15–18.

Nachbar, J. 1990. 'Evolutionary' selection dynamics in games: convergence and limit properties. *International Journal of Game Theory* 19, 59–89.

Samuelson, L. 1997. *Evolutionary Games and Equilibrium Selection*. Cambridge, MA: MIT Press.

Samuelson, L. and Zhang, J. 1992. Evolutionary stability in asymmetric games. *Journal of Economic Theory* 57, 363–91.

Sandholm, W. 2001. Potential games with continuous player sets. *Journal of Economic Theory* 97, 81–108.

Sandholm, W. 2007. *Population Games and Evolutionary Dynamics*. Cambridge, MA: MIT Press.

Schuster, P. and Sigmund, K. 1983. Replicator dynamics. *Journal of Theoretical Biology* 100, 533–8.

Smith, M. 1984. The stability of a dynamic model of traffic assignment – An application of a method of Lyapunov. *Transportation Science* 18, 245–52.

Swinkels, J. 1993. Adjustment dynamics and rational play in games. *Games and Economic Behavior* 5, 455–84.

Taylor, P. and Jonker, L. 1978. Evolutionarily stable strategies and game dynamics. *Mathematical Biosciences* 40, 145–56.

Vega-Redondo, F. 1996. *Evolution, Games, and Economic Behavior*. Oxford: Oxford University Press.

Weibull, J. 1995. *Evolutionary Game Theory*. Cambridge, MA: MIT Press.

Zeeman, E. 1980. Population dynamics from game theory. In *Global Theory of Dynamical Systems*, ed. Z. Nitecki and C. Robinson. Berlin: Springer.

epistemic game theory: an overview

The following three articles survey some aspects of the foundations of non-cooperative game theory. The goal of work in foundations is to examine in detail the basic ingredients of game analysis.

The starting point for most of game theory is a 'solution concept' – such as Nash equilibrium or one of its many variants, backward induction, or iterated dominance of various kinds. These are usually thought of as the embodiment of 'rational behaviour' in some way and used to analyse game situations.

One could say that the starting point for most game theory is more of an endpoint of work in foundations. Here, the primitives are more basic. The very idea of rational – or irrational – behaviour needs to be formalized. So does what each player might know or believe about the game – including about the rationality or irrationality of other players. Foundational work shows that even what each player knows or believes about what other players know or believe, and so on, can matter.

Investigating the basis of existing solution concepts is one part of work in foundations. Other work in foundations has uncovered new solution concepts with useful properties. Still other work considers changes even to the basic model of decision making by players – such as departures from the expected utility model or reasoning in various formal logics.

The first article, EPISTEMIC GAME THEORY: BELIEFS AND TYPES, by Marciano Siniscalchi, describes the formalism used in most work on foundations. This is the 'types' formalism going back to Harsanyi (1967–8). Originally proposed to describe the players' beliefs about the structure of the game (such as the payoff functions), the types approach is equally suited to describing beliefs about the play of the game or beliefs about both what the game is and how it will be played. Indeed, in its most general form, the formalism is simply a way to describe any multi-person uncertainty. Harsanyi's conception of a 'type' was a crucial breakthrough in game theory. Still, his work left many fundamental questions about multi-person uncertainty unanswered. Siniscalchi's article surveys these later developments.

The second and third articles apply these tools to the two kinds of uncertainty mentioned. The second article, EPISTEMIC GAME THEORY: COMPLETE INFORMATION, concerns the case where the matrix or tree itself is 'transparent' to the players, and what is uncertain are the actual strategies chosen by the players. The third article, EPISTEMIC GAME THEORY: INCOMPLETE INFORMATION, by Aviad Heifetz, has the opposite focus: it covers the case of uncertainty about the game itself. (Following Harsanyi, the third article focuses on uncertainty about the payoffs, in particular.)

Both cases are important to the foundations programme. Because Nash equilibrium is 'as if' each player is certain (and correct) about the strategies chosen by the other players (Aumann and Brandenburger, 1995, Section 7h), uncertainty of the first kind has played a small role in game theory to date. Uncertainty of the second kind is the topic of the large literatures on information asymmetries, incentives, and so on.

Interestingly, though, von Neumann and Morgenstern (1944) already appreciated the significance of both complete and incomplete information environments. Indeed, they asserted that phenomena often thought to be characteristic of incomplete-information settings could, in fact, arise in complete-information settings (1944, p. 31):

> Actually, we think that our investigations – although they assume 'complete information' without any further discussion – do make a contribution to the study of this subject. It will be seen that many economic and social phenomena which are usually ascribed to the individual's state of 'incomplete information' make their appearance in our theory and can be satisfactorily interpreted with its help.

This is indeed true, as work in the modern foundations programme shows. (Some instances are mentioned in what follows.) Overall, the foundations programme aims at a 'neutral' and comprehensive treatment of all ingredients of a game.

ADAM BRANDENBURGER

See also **epistemic game theory: beliefs and types; epistemic game theory: complete information; epistemic game theory: incomplete information; Nash equilibrium, refinements of.**

My thanks to Rena Henderson and Michael James. The Stern School of Business provided financial support.

Bibliography

Aumann, R. and Brandenburger, A. 1995. Epistemic conditions for Nash equilibrium. *Econometrica* 63, 1161–80.

Harsanyi, J. 1967–8. Games with incomplete information played by 'Bayesian' players, I–III. *Management Science* 14, 159–82, 320–34, 486–502.

Von Neumann, J. and Morgenstern, O. 1944. *Theory of Games and Economic Behavior.* Princeton: Princeton University Press. 60th anniversary edn, 2004.

epistemic game theory: beliefs and types

John Harsanyi (1967–8) introduced the formalism of type spaces to provide a simple and parsimonious representation of belief hierarchies. He explicitly noted that his formalism was not limited to modelling a player's beliefs about payoff-relevant variables: rather, its strength was precisely the ease with which Ann's beliefs about Bob's beliefs about payoff variables, Ann's beliefs about Bob's beliefs about Ann's beliefs about payoff variables, and so on, could be represented.

This feature plays a prominent role in the epistemic analysis of solution concepts (see EPISTEMIC GAME THEORY: COMPLETE INFORMATION), as well as in the literature on global games (Morris and Shin, 2003) and on robust mechanism design (Bergemann and Morris, 2005). All these applications place particular emphasis on the expressiveness of the type-space formalism. Thus, a natural question arises: just how expressive is Harsanyi's approach?

For instance, solution concepts such as Nash equilibrium or rationalizability can be characterized by means of restrictions on the players' mutual beliefs. In principle, these assumptions could be formulated directly as restrictions on players' hierarchies of beliefs; but in practice the analysis is mostly carried out in the context of a type space à la Harsanyi. This is without loss of generality only if Harsanyi type spaces do not themselves impose restrictions on the belief hierarchies that can be represented. Similar considerations apply in the context of robust mechanism design.

A rich literature addresses this issue from different angles, and for a variety of basic representations of beliefs. This article focuses on hierarchies of probabilistic beliefs; however, some extensions are also mentioned. For simplicity, attention is restricted to two players, denoted '1' and '2' or 'i' and '$-i$.'

Probabilistic type spaces and belief hierarchies

Begin with some mathematical preliminaries. A topology on a space X is deemed Polish if it is separable and completely metrizable; in this case, X is itself deemed a Polish space. Examples include finite sets, Euclidean space $\mathbb{R}^n$ and closed subsets thereof. A countable product of Polish spaces, endowed with the product topology, is itself Polish. For any topological space X, the notation $\Delta(X)$ indicates the set of Borel probability measures on X. If the topology on X is Polish, then the weak* topology on $\Delta(X)$ is also Polish (for example, Aliprantis and Border, 1999, Theorem 14.15). A sequence $\{\mu^k\}_{k\geq 1}$ in $\Delta(X)$ converges in the weak* sense to a measure $\mu \in \Delta(X)$, written $\mu^k \xrightarrow{w^*} \mu$, if and only if, for every bounded, continuous function $\psi : X \to \mathbb{R}$, $\int_X \psi d\mu^k \to \int_X \psi d\mu$. The weak* topology on $\Delta(X)$ is especially meaningful and convenient when X is a Polish space: see Aliprantis and Border (1999, ch. 14) for an overview of its properties. Finally, if μ is a measure on some product space $X \times Y$, the marginal of μ on X is denoted $\mathrm{marg}_x\mu$.

The basic ingredient of the players' hierarchical beliefs is a description of payoff-relevant or fundamental uncertainty. Fix two sets S_1 and S_2, hereinafter called the *uncertainty domains*; the intended interpretation is that S_{-i} describes aspects of the strategic situation that Player i is uncertain about. For example, in an independent private-values auction, each set S_i could represent bidder i's possible valuations of the object being sold, which is not known to bidder $-i$. In the context of interactive epistemology, S_i is usually taken to be Player i's strategy space. It is sometimes convenient to let $S_1 = S_2 \equiv S$; in this case, the formalism introduced below enables one to formalize the assumption that each player observes different aspects of the common uncertainty domain S (for instance, different signals correlated with the common, unknown value of an object offered for sale).

An (S_1, S_2)-*based type space* is a tuple $\mathcal{T} = (T_i, g_i)_{i=1,2}$ such that, for each $i = 1, 2$, T_i is a Polish space and $g_i : T_i \to \Delta(S_{-I} \times T_{-i})$ is continuous. As noted above, type spaces can represent hierarchies of beliefs; it is useful to begin with an example. Let $S_1 = S_2 = \{a, b\}$ and consider the type space defined in Table 1. To interpret, for every $i = 1, 2$, the entry in the row corresponding to t_i and (s_{-i}, t_{-i}) is $g_i(t_i)(\{(s_{-i}, t_{-i})\})$. Thus, for instance, $g_1(t_1)(\{(a, t_2')\}) = 0$; $g_2(t_2)(\{b\} \times T_1) = 0.5$.

Consider type t_1 of Player 1. She is certain that $s_2 = a$; furthermore, she is certain that Player 2 believes that $s_1 = a$ and $s_1 = b$ are equally likely. Taking this one step further, type t_1 is certain that Player 2 assigns probability 0.5 to the event that Player 1 believes that $s_2 = b$ with probability 0.7.

These intuitive calculations can be formalized as follows. Fix an (S_1, S_2)-based type space $\mathcal{T} = (T_i, g_i)_{i=1,2}$; for every $i = 1, 2$ define the set X^0_{-i} and the function $h_i^1 : T_i \to \Delta(X_0^{-i})$ by

$$X^0_{-i} = S_{-i} \quad \text{and} \quad \forall t_i \in T_i,\ h_i^1(t_i) = \text{marg}_{S_{-i}} g_i(t_i). \tag{1}$$

Thus, $h_i^1(t_i)$ represents the *first-order beliefs* of type t_i in type space T – her beliefs about the uncertainty domain S_{-i}. Note that each $X^0_{-i} = S_{-i}$ is Polish. Proceeding inductively, assuming that $X^0_{-i}, \ldots, X^{k-1}_{-i}$ and $h_i^1, \ldots, h_i^k$ have been defined up to some $k>0$ for $i = 1, 2$, and that all sets X^ℓ_{-i}, $\ell = 0, \ldots, k-1$ are Polish, define the set X^k_{-i} and the functions $h_i^{k+1} : T_i \to \Delta(X^k_{-i})$ for $i = 1, 2$ by

$$X^k_{-i} = X^{k-1}_{-i} \times \Delta(X^{k-1}_i) \quad \text{and} \quad \forall t_i \in T_i,\ h_i^{k+1}(t_i)(E) = g_i(t_i)(\{(s_{-i}, t_{-i}) \in S_{-i} \times T_{-i} : (s_{-i}, h^k_{-i}(t_{-i})) \in E\}) \tag{2}$$

for every Borel subset E of X^k_{-i}. Thus, $h_1^2(t_1)$ represents the *second-order beliefs* of type t_1 – her beliefs about *both* the uncertainty domain $S_2 = X_2^0$ and Player 2's beliefs about S_1, which by definition belong to the set $\Delta(X_1^0) = \Delta(S_1)$. Similarly, $h_i^{k+1}(t_i)$ represents type t_i's $(k+1)$-th order beliefs.

Observe that type t_i's second-order beliefs are defined over $X_2^0 \times \Delta(X_1^0) = S_2 \times \Delta(S_1)$, rather than just over $\Delta(X_1^0) = \Delta(S_1)$; a similar statement holds for her $(k+1)$-th order beliefs. This is crucial in many applications. For instance, a typical assumption in the literature on epistemic foundations of solution concepts is that

Table 1 *A type space*

T_1	a, t_2	a, t'_2	b, t_2	b, t'_2
t_1	1	0	0	0
t'_1	0	0.3	0	0.7

T_2	a, t_1	a, t'_1	b, t_1	b, t'_1
t_2	0	0.5	0.5	0
t'_2	0	0	0	1

Player 1 believes that Player 2 is rational. Letting S_i be the set of actions or strategies of Player i in the game under consideration, this can be modelled by assuming that the support of $h_1^2(t_1)$ consists of pairs $(s_2, \mu_1) \in S_2 \times \Delta(S_1)$ wherein s_2 is a best response to μ_1. Clearly, such an assumption could not be formalized if $h_1^2(t_1)$ only conveyed information about type t_1's beliefs on Player 2's first-order beliefs: even though type t_1's beliefs about the action played by Player 2 could be retrieved from $h_1^1(t_1)$, it would be impossible to tell whether each action that type t_1 expects to be played is matched with a belief that rationalizes it.

Note that, since X_i^{k-1} and X_{-i}^{k-1} are assumed Polish, so are $\Delta(X_i^{k-1})$ and X_{-i}^k. Also, each function h_i^k is continuous.

Finally, it is convenient to define a function that associates to each type $t_i \in T_i$ an entire *belief hierarchy*: to do so, define the set H_i and, for $i = 1, 2$, the function $h_i : T_i \to H_i$ by

$$H_i = \prod_{k \geq 0} \Delta(X_{-i}^k) \text{ and } \forall t_i \in T_i,\ h_i(t_i) = (h_i^1(t_i), \ldots, h_i^{k+1}(t_i), \ldots). \tag{3}$$

Thus, H_i is the set of all hierarchies of beliefs; notice that, since each X_{-i}^k is Polish, so is H_i.

Rich type spaces

The preceding construction suggests a rather direct way to ask how expressive Harsanyi's notion of a type space is: can one construct a type space that generates *all* hierarchies in H_i?

A moment's reflection shows that this question must be refined. Fix a type space $(T_i, g_i)_{i=1,2}$ and a type $t_i \in T_i$; recall that, for reasons described above, the first- and second-order beliefs of type t_i satisfy $h_i^1(t_i) \in \Delta(S_{-i})$ and $h_i^2(t_i) \in \Delta(X_{-i}^0 \times \Delta(X_i^0)) = \Delta(S_{-i} \times \Delta(S_i))$ respectively. This, however, creates the potential for redundancy or even contradiction, because both $h_i^1(t_i)$ and $\text{marg}_{S_{-i}} h_i^2(t_i)$ can be viewed as 'type t_i's beliefs about S_{-i}'. A similar observation applies to higher-order beliefs. Fortunately, it is easy to verify that, for every type space $(T_i, g_i)_{i=1,2}$ and type $t_i \in T_i$, the following *coherency* condition holds:

$$\forall k > 1, \text{marg}_{X_{-i}^{k-2}} h_i^k(t_i) = h_i^{k-1}(t_i); \tag{4}$$

To interpret, recall that $h_i^k(t_i) \in \Delta(X_{-i}^{k-1}) = \Delta(X_{-i}^{k-2} \times \Delta(X_{-i}^{k-2}))$. Thus, in particular, $\text{marg}_{S_{-i}} h_i^2(t_i) = h_i^1(t_i)$.

Since H_i is defined as the set of *all* hierarchies of beliefs for Player i, some (in fact, 'most') of its elements are not coherent. As noted above, no type space can generate incoherent hierarchies; more importantly, coherency can be viewed as an integral part of the interpretation of interactive beliefs. How could an individual simultaneously hold (infinitely) many distinct first-order beliefs? Which of these should be used, say, to verify whether she is rational? This motivates restricting attention to coherent hierarchies, defined as follows:

$$H_i^c = \{(\mu_i^1, \mu_i^2, \ldots) \in H_i \; : \; \forall k > 1, \text{marg}_{X_{-i}^{k-2}} \mu_i^k = \mu_i^{k-1}\}. \tag{5}$$

Since $\text{marg}_{X_{-i}^{k-2}} : \Delta(X_{-i}^{k-1}) \to \Delta(X_{-i}^{k-2})$ is continuous, H_i^c is a closed, hence Polish subspace of H_i.

Brandenburger and Dekel (1993, Proposition 1) show that there exist homeomorphisms $g_i^c : H_i^c \to \Delta(S_{-i} \times H_{-i})$: that is, every coherent hierarchy corresponds to a distinct belief over the uncertainty domain and the hierarchies of the opponent, and conversely. Furthermore, this homeomorphism is canonical, in the following sense. Note that $S_{-i} \times H_{-i} = S_{-i} \times \prod_{k \geq 0} \Delta(X_i^k) = X_{-i}^k \times \prod_{\ell > k} \Delta(X_i^\ell)$. Then it can be shown that, if $\mu_i = (\mu_i^1, \mu_i^2, \ldots) \in H_i^c$, then $\text{marg}_{X_{-i}^k} g_i^c(\mu_i) = \mu_i^{k+1}$. Intuitively, the marginal belief associated with μ_i over the first k orders of the opponent's beliefs is precisely what it should be, namely μ_i^{k+1}. The proof of these results builds upon Kolmogorov's extension theorem, as may be suggested by the similarity of the coherency condition in eq. (5) with the notion of Kolmogorov consistency: cf. for example Aliprantis and Border (1999, theorem 14.26).

This result does not quite imply that all coherent hierarchies can be generated in a suitable type space; however, it suggests a way to obtain this result. Notice that the belief on $S_{-i} \times H_{-i}$ associated by the homeomorphism g_i^c to a coherent hierarchy μ_i may include *in*coherent hierarchies $\nu_{-i} \in H_{-i} \backslash H_{-i}^c$ in its support. This can be interpreted in the following terms: if Player i's hierarchical beliefs are given by μ_i, then she is coherent, but she is not certain that her opponent is. On the other hand, consider a type space $(T_i, g_i)_{i=1,2}$; as noted above, for every player i, each type $t_i \in T_i$ generates a coherent hierarchy $h_i(t_i) \in H_i^c$. So, for instance, if (s_1, t_1) is in the support of $g_2(t_2)$ then t_1 also generates a coherent hierarchy. Thus, not only is type t_2 of Player 2 coherent: he is also certain (believes with probability one) that Player 1 is coherent. Iterating this argument suggests that *hierarchies of beliefs generated by type spaces display common certainty of coherency.*

Motivated by these considerations, let

$$H_i^0 = H_i^c \quad \text{and} \quad \forall k > 0, H_i^k = \{\mu_i \in H_i^{k-1} \; : \; g_i^c(\mu_i)(S_{-i} \times H_{-i}^{k-1}) = 1\}. \tag{6}$$

Thus, H_i^0 is the set of coherent hierarchies for Player i; H_i^1 is the set of hierarchies that are coherent and correspond to beliefs that display certainty of the opponent's coherency; and so on. Finally, let $H_i^* = \cap_{k \geq 0} H_i^k$. Each element of H_i^* is intuitively consistent with coherency and common certainty of coherency.

Brandenburger and Dekel (1993, Proposition 2) show that the restriction g_i^* of g_i^c to H_i^* is a homeomorphism between H_i^* and $\Delta(S_{-i} \times H_{-i}^*)$; furthermore, it is canonical in the sense described above. This implies that the tuple $(H_i^*, g_i^*)_{i=1,2}$ is a type space in its own right – the (S_1, S_2)-based *universal type space.*

The existence of a universal type space fully addresses the issue of richness. Since the homeomorphism g_i^* is canonical, it is easy to see that the hierarchy generated as per eqs (1) and (2) by any 'type' $t_i = (\mu^1, \mu^2, s) \in H_i^*$ in the universal type space $(H_i^*, g_i^*)_{i=1,2}$ is t_i itself; thus, since H_i^* *consists* of all hierarchies that are coherent and display common certainty of consistency, the universal type space also *generates* all such hierarchies.

The type space $(H_i^*, g_i^*)_{i=1,2}$ is rich in two additional, related senses. First, as may be expected, every belief hierarchy for Player i generated by an arbitrary type space is an element of H_i^*; this implies that every type space $(T_i, g_i)_{i=1,2}$ can be uniquely embedded in $(H_i^*, g_i^*)_{i=1,2}$ as a 'belief-closed' subset: see Battigalli and Siniscalchi (1999, Proposition 8.8). Call a type space *terminal* if, like $(H_i^*, g_i^*)_{i=1,2}$, it embeds all other type spaces as belief-closed subsets.

Second, since each function g_i^* is a homeomorphism, in particular it is a surjection (that is, onto). Call a type space $(T_i, g_i)_{i=1,2}$ *complete* if every map g_i is onto. (This should not be confused with the topological notion of completeness.) Thus, the universal type space $(H_i^*, g_i^*)_{i=1,2}$ is complete. It is often the case that, when a universal type space is employed in the epistemic analysis of solution concepts, the objective is precisely to exploit its completeness. Furthermore, for certain representations of beliefs, it is not known whether universal type spaces can be constructed; however, the existence of complete type spaces can be established, and is sufficient for the purposes of epistemic analysis. The next section provides examples.

Alternative constructions and extensions

The preceding discussion adopts the approach proposed by Brandenburger and Dekel (1993), which has the virtue of relying on familiar ideas from the theory of stochastic processes. However, the first constructions of universal type spaces consisting of hierarchies of beliefs are due to Armbruster and Böge (1979), Böge and Eisele (1979) and Mertens and Zamir (1985).

From a technical point of view, Mertens and Zamir (1985) assume that the state space S is compact Hausdorff and beliefs are regular probability measures. Heifetz and Samet (1998b) instead drop topological assumptions altogether: in their approach, both the underlying set of states and the sets of types of each player are modelled as measurable spaces. They show that a terminal type space can be explicitly constructed in this environment.

In all the contributions mentioned so far, beliefs are modelled as countably additive probabilities. The literature has also examined other representations of beliefs, broadly defined.

A *partitional structure* (Aumann, 1976) is a tuple $(\Omega,(\sigma_i, P_i)_{i=1,2})$, where Ω is a (typically finite) space of 'possible worlds', every $\sigma_i : \Omega \to S_i$ indicates the realization of the basic uncertainty corresponding to each element of Ω, and every P_i is a partition of Ω. The interpretation is that, at any world $\omega \in \Omega$, Player i is only informed that the true world lies in the cell of the partition P_i containing ω, denoted $P_i(\omega)$. The *knowledge operator* for Player i can then be defined as

$$\forall E \subset \Omega, K_i(E) = \{\omega \in \Omega : P_i(\omega) \subseteq E\}.$$

Notice that no probabilistic information is provided in this environment (although it can be easily added).

Heifetz and Samet (1998a) show that a terminal partitional structure does not exist. This result was extended to more general 'possibility' structures by Meier (2005). Brandenburger and Keisler (2006) establish related non-existence results for complete structures. However, recent contributions show that topological assumptions, which play a key role in the constructions of Mertens and Zamir (1985) and Brandenburger and Dekel (1993), can also deliver existence results in non-probabilistic settings. For instance, Mariotti, Meier and Piccione (2005) construct a structure that is universal, complete and terminal for possibility structures.

Other authors investigate richer probabilistic representations of beliefs. Battigalli and Siniscalchi (1999) construct a universal, terminal, and complete type space for *conditional probability system*, or collections of probability measures indexed by relevant conditioning events (such as histories in an extensive game) and related by a version of Bayes's rule. This type space is used in (2002) to provide an epistemic analysis of forward induction. Brandenburger, Friedenberg and Keisler (2006) construct a complete type space for *lexicographic sequences*, which may be thought of as an extension of lexicographic probability systems (Blume, Brandenburger and Dekel, 1991) for infinite domains. They then use it to provide an epistemic characterization of iterated admissibility.

Non-probabilistic representations of beliefs that reflect a concern for ambiguity (Ellsberg, 1961) have also been considered. Heifetz and Samet (1998b) observe that their measure-theoretic construction extends to beliefs represented by continuous *capacities*, that is non-additive set functions that preserve monotonicity with respect to set inclusion. Motivated by the multiple-priors model of Gilboa and Schmeidler (1989), Ahn (2006) constructs a universal type space for sets of probabilities.

Epstein and Wang (1996) approach the richness issue taking *preferences*, rather than beliefs, as primitive objects. In their setting, an S-based type space is a tuple $(T_i, g_i)_i = 1, 2$, where, for every type t_i, $g_i(t_i)$ is a suitably regular preference over *acts* defined on the set $S \times T_{-i}$. The analysis in the preceding section can be viewed as a special case of Epstein and Wang (1996), where preferences conform to expected-utility theory. Epstein and Wang construct a universal type space in this framework (see also Di Tillio, 2006).

Finally, constructions analogous to that of a universal type space appear in other, unrelated contexts. For instance, Epstein and Zin (1989) develop a class of recursive preferences over infinite-horizon temporal lotteries; to construct the domain of such preferences, they employ arguments related to Mertens and Zamir's. Gul and

Pesendorfer (2004) employ analogous techniques to analyse self-control preferences over infinite-horizon consumption problems.

MARCIANO SINISCALCHI

See also **epistemic game theory: an overview; epistemic game theory: complete information; epistemic game theory: incomplete information.**

Bibliography

Ahn, D. 2007. Hierarchies of ambiguous beliefs. *Journal of Economic Theory* 136, 286–301.

Aliprantis, C. and Border, K. 1999. *Infinite Dimensional Analysis*, 2nd edn. Berlin: Springer.

Armbruster, W. and Böge, W. 1979. Bayesian game theory. In *Game Theory and Related Topics*, ed. O. Moeschlin and D. Pallaschke. Amsterdam: North-Holland.

Aumann, R. 1976. Agreeing to disagree. *Annals of Statistics* 4, 1236–9.

Battigalli, P. and Siniscalchi, M. 1999. Hierarchies of conditional beliefs and interactive epistemology in dynamic games. *Journal of Economic Theory* 88, 188–230.

Battigalli, P. and Siniscalchi, M. 2002. Strong belief and forward induction reasoning. *Journal of Economic Theory* 106, 356–91.

Bergemann, D. and Morris, S. 2005. Robust mechanism design. *Econometrica* 73, 1521–34.

Blume, L., Brandenburger, A. and Dekel, E. 1991. Lexicographic probabilities and choice under uncertainty. *Econometrica* 59, 61–79.

Böge, W. and Eisele, T. 1979. On solutions of Bayesian games. *International Journal of Game Theory* 8, 193–215.

Brandenburger, A. and Dekel, E. 1993. Hierarchies of beliefs and common knowledge. *Journal of Economic Theory* 59, 189–98.

Brandenburger, A., Friedenberg, A. and Keisler, H.J. 2006. Admissibility in games. Unpublished, Stern School of Business, New York University.

Brandenburger, A. and Keisler, J. 2006. An impossibility theorem on beliefs in games. *Studia Logica* 84, 211–40.

Di Tillio, A. 2006. Subjective expected utility in games. Working Paper No. 311, IGIER, Università Bocconi.

Ellsberg, D. 1961. Risk, ambiguity, and the Savage axioms. *Quarterly Journal of Economics* 75, 643–69.

Epstein, L. and Wang, T. 1996. Beliefs about beliefs without probabilities. *Econometrica* 64, 1343–73.

Epstein, L. and Zin, S. 1989. Substitution, risk aversion, and the temporal behavior of consumption and asset returns: a theoretical framework. *Econometrica* 57, 937–69.

Gilboa, I. and Schmeidler, D. 1989. Maxmin-expected utility with a non-unique prior. *Journal of Mathematical Economics* 18, 141–53.

Gul, F. and Pesendorfer, W. 2004. Self-control and the theory of consumption. *Econometrica* 72, 119–58.

Harsanyi, J. 1967–8. Games of incomplete information played by Bayesian players. Parts I, II, III. *Management Science* 14, 159–82, 320–34, 486–502.

Heifetz, A. and Samet, D. 1998a. Knowledge spaces with arbitrarily high rank. *Games and Economic Behavior* 22, 260–73.

Heifetz, A. and Samet, D. 1998b. Topology-free typology of beliefs. *Journal of Economic Theory* 82, 324–81.

Mariotti, T., Meier, M. and Piccione, M. 2005. Hierarchies of beliefs for compact possibility models. *Journal of Mathematical Economics* 41, 303–24.

Meier, M. 2005. On the nonexistence of universal information structures. *Journal of Economic Theory* 122, 132–9.

Mertens, J.F. and Zamir, S. 1985. Formulation of Bayesian analysis for games with incomplete information. *International Journal of Game Theory* 14, 1–29.

Morris, S. and Shin, H. 2003. Global games: theory and applications. In *Advances in Economics and Econometrics (Proceedings of the Eight World Congress of the Econometric Society)*, ed. M. Dewatripont, L. Hansen and S. Turnovsky. Cambridge: Cambridge University Press.

epistemic game theory: complete information

1. Epistemic analysis

Under the epistemic approach, the traditional description of a game is augmented by a mathematical framework for talking about the rationality or irrationality of the players, their beliefs and knowledge, and related ideas.

The first step is to add sets of *types* for each of the players. The apparatus of types goes back to Harsanyi (1967–8), who introduced it as a way to talk formally about the players' beliefs about the payoffs in a game, their beliefs about other players' beliefs about the payoffs, and so on. (See EPISTEMIC GAME THEORY: INCOMPLETE INFORMATION.) But the technique is equally useful for talking about uncertainty about the actual play of the game – that is, about the players' beliefs about the strategies chosen in the game, their beliefs about other players' beliefs about the strategies, and so on. This survey focuses on this second source of uncertainty. It is also possible to treat both kinds of uncertainty together, using the same technique.

We give a definition of a type structure as commonly used in the epistemic literature, and an example of its use.

Fix an n-player finite strategic-form game $\langle S^1, \ldots, S^n, \pi^1, \ldots, \pi^n \rangle$. Some notation: given sets X^1, …, X^n, let $X = \times_{i=1}^n X^i$ and $X^{-i} = \times_{j \neq i} X^j$. Also, given a finite set Ω, write $\mathcal{M}(\Omega)$ for set of all probability measures on Ω.

Definition 1.1 *An* $(S^1, \ldots, S^n)$*-based* (***finite***) ***type structure*** *is a structure*

$$\langle S^1, \ldots, S^n; T^1, \ldots, T^n; \lambda^1, \ldots, \lambda^n \rangle,$$

where each T^i *is a finite set, and each* $\lambda^i : T^i \rightarrow \mathcal{M}(S^{-i} \times T^{-i})$. *Members of* T^i *are called* ***types*** *for player i. Members of* $S \times T$ *are called* ***states*** *(**of the world**).*

For some purposes – see, for example, Sections 4 and 6 – it is important to consider infinite type structures. Topological assumptions are then made on the type spaces T_i.

A particular state $(s^1, t^1, \ldots, s^n, t^n)$ describes the strategy chosen by each player, and also each player's type. Moreover, a type t^i for player i induces, via a natural induction, an entire hierarchy of beliefs – about the strategies chosen by the players $j \neq i$, about the beliefs of the players $j \neq i$, and so on. (See EPISTEMIC GAME THEORY: BELIEFS AND TYPES.)

The following example is similar to one in Aumann and Brandenburger (1995, pp. 1166–7).

Example 1.1 (**A coordination game**). *Consider the coordination game in* Figure 1.1 *(where Ann chooses the row and Bob the column), and the associated type structure in* Figure 1.2.

	L	R
U	2, 2	0, 0
D	0, 0	1, 1

Figure 1.1

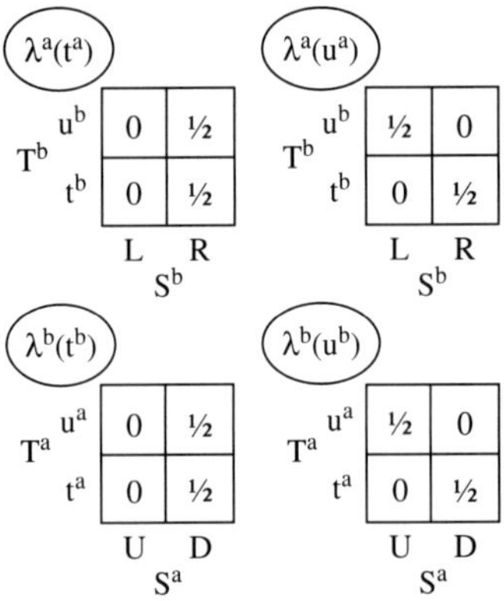

Figure 1.2

There are two types t^a, u^a for Ann, and two types t^b, u^b for Bob. The measure associated with each type is as shown. Fix the state (D, t^a, R, t^b). At this state, Ann plays D and Bob plays R. Ann is 'correct' about Bob's strategy. (Her type t^a assigns probability 1 to Bob's playing R.) Likewise, Bob is correct about Ann's strategy. Ann, though, thinks it possible Bob is wrong about her strategy. (Her type assigns probability 1/2 to type u^b for Bob, which assigns probability 1/2 to Ann's playing U, not D.) Again, likewise with Bob.

What about the rationality or irrationality of the players? At state (D, t^a, R, t^b), Ann is rational. Her strategy maximizes her expected payoff, given her first-order belief (which assigns probability 1 to R). Likewise, Bob is rational. Ann, though, thinks it possible Bob is irrational. (She assigns probability 1/2 to (R, u^b). With type u^b, Bob gets a higher expected payoff from L than R.) The situation with Bob is again symmetric.

Summing up, the example is just a description of a game situation, not a prediction. A type structure is a descriptive tool. Note, too, that the example includes both rationality and irrationality, and also allows for incorrect as well as correct beliefs (for example, Ann thinks it possible Bob is irrational, though in fact he isn't). These are typical features of the epistemic approach.

Two comments on type structures. First, we can ask whether Definition 1.1 above is to be taken as primitive or derived. Arguably, hierarchies of beliefs are the primitive, and types are simply a convenient tool for the analyst. See EPISTEMIC GAME THEORY: BELIEFS AND TYPES for further discussion.

Second, note that Definition 1.1 applies to finite games. These will be the focus of this survey. There is nothing yet approaching a developed literature on epistemic analysis of infinite games.

2. Early results

A major use of type structures is to identify conditions on the players' rationality, beliefs, and so on, that yield various solution concepts.

A very basic solution concept is iterated dominance. This involves deleting from the matrix all strongly dominated strategies, then deleting all strategies that become strongly dominated in the resulting submatrix, and so on until no further deletion is possible. (It is easy to check that in finite games – as considered in this survey – the residual set will always be non-empty.) Call the remaining strategies the *iteratively undominated* (*IU*) strategies. There is a basic equivalence: a strategy is not strongly dominated if and only if there is a probability measure on the product of the other players' strategy sets under which it is optimal. Using this, IU can also be defined as follows: delete from the matrix any strategy that isn't optimal under some measure on the product of the other players' strategy sets. Consider the resulting sub-matrix and delete strategies that don't pass this test on the sub-matrix, and so on.

The second definition suggests what a formal epistemic treatment of IU should look like. A rational player will choose a strategy which is optimal under some measure. This is the first round of deletion. A player who is rational and believes the other players are rational will choose a strategy which is optimal under a measure that assigns probability 1 to the strategies remaining after the first round of deletion. This gives the second round of deletion. And so on.

Type structures allow a formal treatment of this idea. First the formal definition of rationality. This is a property of strategy-type pairs. Say (s^i, t^i) is **rational** if s^i maximizes player i's expected payoff under the marginal on S^{-i} of the measure $\lambda^i(t^i)$.

Say type t^i of player i **believes** an event $E \subseteq S^{-i} \times T^{-i}$ if $\lambda^i(t^i)(E) = 1$, and write

$$B^i(E) = \{t^i \in T^i : t^i \text{ believes } E\}.$$

Now, for each player i, let R_1^i be the set of all rational pairs (s^i, t^i), and for $m>0$ define R_m^i inductively by

$$R_{m+1}^i = R_m^i \cap [S^i \times B^i(R_m^{-i})].$$

Definition 2.1 *If* $(s^1, t^1, \ldots, s^n, t^n) \in R_{m+1}$, *say there is* ***rationality and mth-order belief of rationality (RmBR)*** *at this state. If* $(s^1, t^1, \ldots, s^n, t^n) \in \bigcap_{m=1}^{\infty} R_m$, *say there is* ***rationality and common belief of rationality (RCBR)*** *at this state.*

These definitions yield an epistemic characterization of IU: *Fix a type structure and a state* $(s^1, t^1, \ldots, s^n, t^n)$ *at which there is RCBR. Then the strategy profile* $(s^1, \ldots, s^n)$ *is IU. Conversely, fix an IU profile* $(s^1, \ldots, s^n)$. *There is a type structure and a state* $(s^1, t^1, \ldots, s^n, t^n)$ *at which there is RCBR.* Results like this can be found in the early literature – see, among others, Brandenburger and Dekel (1987) and Tan and Werlang (1988).

An important stimulus to the early literature was the pair of papers by Bernheim (1984) and Pearce (1984), which introduced the solution concept of *rationalizability.* This differs from IU by requiring on each round that a player's probability measure on the product of the other players' (remaining) strategy sets be a product measure – that is, be independent. Thus the set of rationalizable strategy profiles is contained in the IU set. It is well known that there are games (with three or more players) in which inclusion is strict.

The argument for the independence assumption is that in non-cooperative game theory it is supposed that players do not coordinate their strategy choices. Interestingly though, correlation is consistent with the non-cooperative approach. This view is put forward in Aumann (1987). (Aumann, 1974, introduced the study of correlation into non-cooperative theory.) Consider an analogy to coin tossing. A correlated assessment over coin tosses is possible, if there is uncertainty over the coin's parameter or 'bias'. (The assessment is usually required to be conditionally i.i.d., given the parameter.) Likewise, in a game, Charlie might have a correlated assessment over Ann's and Bob's strategy choices, because, say, he thinks Ann and Bob have observed similar signals before the game (but is uncertain what the signal was).

The same epistemic tools used to understand IU can be used to characterize other solution concepts on the matrix. Aumann and Brandenburger (1995, Preliminary Observation) point out that pure-strategy Nash equilibrium is characterized by the simple condition that each player is rational and assigns probability 1 to the actual strategies chosen by the other players. (Thus, in Example 1.1 above, these conditions hold at the state (D, t^a, R, t^b), and (D, R) is indeed a Nash equilibrium.) As far as mixed strategies are concerned, in the epistemic approach to games these don't play the central role that they do under equilibrium analysis. Built into the set-up of Section 1 is that each player makes a definite choice of (pure) strategy. (If a player does have the option of making a randomized choice, this can be added to the – pure – strategy set. Indeed, in a finite game, a finite number of such choices can be added.) It is the other players who are uncertain about this choice. Harsanyi (1973) originally proposed this shift in thinking about randomization. Aumann and Brandenburger (1995) give an epistemic treatment of mixed-strategy Nash equilibrium along these lines.

Aumann (1987) asks a question about an outside observer of a game. He provides conditions under which the observer's assessment of the strategies chosen will be the distribution of a correlated equilibrium (as defined in his 1974 paper). The distinctive condition in (1987) is the so-called Common Prior Assumption, which says that the probability assessment associated with each player's type is the same as the observer's assessment, except for being conditioned on what the type in question knows. A number of papers have investigated foundations for this assumption – see, among others, Morris (1994), Samet (1998), Bonanno and Nehring (1999), Feinberg (2000), Halpern (2002), and also the exchange between Gul (1998) and Aumann (1998).

2. Next steps: the tree

An important next step in the epistemic programme was extending the analysis to game trees. A big motivation for this was to understand the logical foundation of *backward induction* (*BI*). At first sight, BI is one of the easiest ideas in game theory. If Ann, the last player to move, is rational, she will make the BI choice. If Bob, the second-to-last player to move, is rational and thinks Ann is rational, he will make the choice that is maximal given that Ann makes the BI choice – that is, he too will make the BI choice. And so on back in the tree, until the BI path is a identified (Aumann, 1995).

For example, Figure 3.1 is three-legged centipede (Rosenthal, 1981). (The top payoffs are Ann's, and the bottom payoffs are Bob's.) BI says Ann plays *Out* at her first node. But what if she doesn't? How will Bob react? Perhaps Bob will conclude that Ann is an irrational player, who plays *Across*. That is, Bob might play *In*, hoping to get a payoff of 6 (better than 4 from *Out*). Perhaps, anticipating this, Ann will in fact play *Down*, hoping to get 4 (better than 2 from playing *Out*).

Many papers have examined this conceptual puzzle with BI – see, among others, Binmore (1987), Bicchieri (1988, 1989), Basu (1990), Bonanno (1991), and Reny (1992).

A key step in resolving the puzzle is extending the epistemic tools of Section 1, to be able to talk formally about rationality, beliefs and so on in the tree.

Example 3.1 (**three-Legged centipede**). Figure 3.2 *is a type structure for three-legged Centipede.*

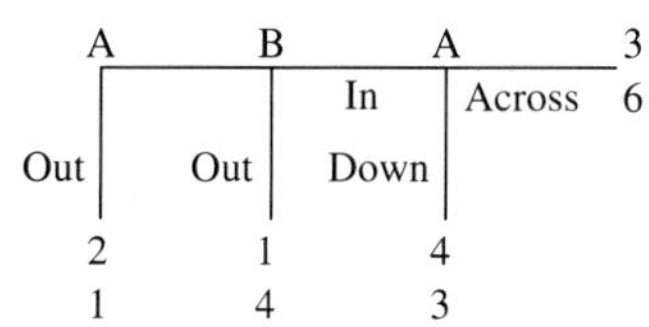

Figure 3.1

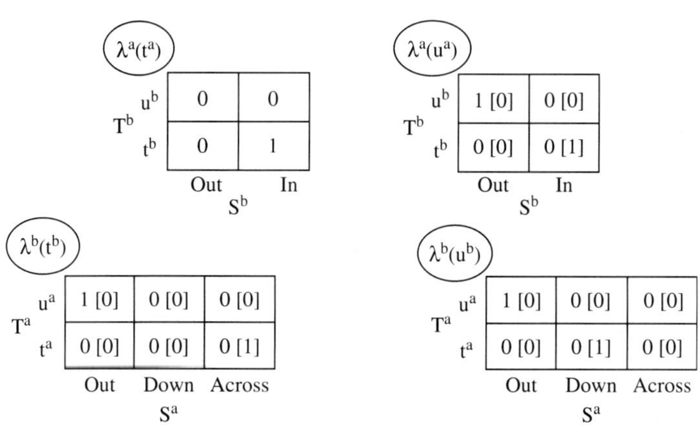

Figure 3.2

There are two types t^a, u^a for Ann. Type t^a for Ann has the measure shown in the top-left matrix. It assigns probability 1 to (In, t^b) for Bob. Type u^a has two associated measures – shown in the top-right matrix. The first measure (the numbers without parentheses) assigns probability 1 to (Out, u^b) for Bob. In this case, we also specify a second measure for Ann, because we want to specify what Ann thinks at her second node, too. Reaching this node is assigned positive probability (in fact, probability 1) under Ann's type t^a, but probability 0 under her type u^a. So, for type u^a, there isn't a well-defined conditional probability measure at Ann's second node. This is why we (separately) specify a second measure for Ann's type u^a: it is the measure in square brackets. If type u^a, Ann assigns probability 1 to (In, t^b) at her second node.

There are also two types t^b, u^b for Bob. Both types initially assign probability 1 to Ann's playing Out. For both of Bob's types, there isn't a well-defined conditional probability measure at his node. At his node, Bob's type t^b assigns probability 1 to {(Across, t^a)}, while his type u^b assigns probability 1 to {(Down, t^a)}.

This is a simple illustration of the concept of a *conditional probability system* (*CPS*), due to Rényi (1955). A CPS specifies a family of conditioning events E and a measure p_E for each such event, together with certain restrictions on these measures. The interpretation is that p_E is what the player believes, after observing E. Even if $p_\Omega(E) = 0$ (where Ω is the entire space), the measure p_E is still specified. That is, even if E is 'unexpected', the player has a measure if E nevertheless happens. This is why CPS's are well-suited to epistemic analysis of game trees – where we need to be able to describe how players react to the unexpected.

Myerson (1991, ch. 1) provided a preference-based axiomatization of a class of CPS's. Battigalli and Siniscalchi (1999; 2002) further developed both the pure theory and the game-theoretic application of CPS's (see below).

Suppose the true state in Figure 3.2 is (*Down*, t^a, *In*, t^b). In particular, Ann plays *Down*, expecting Bob to play *In*. Bob plays *In*, expecting (at his node) Ann to play *Across*. Ann expects a payoff of 4 (and gets this). Bob expects a payoff of 6 (but gets only 3). In everyday language, we can say that Ann successfully bluffs Bob. (At the state (*Down*, t^a, *In*, t^b), the bluff works. By contrast, at the state (*Down*, t^a, *Out*, u^b), Ann attempts the bluff and it fails.)

But what about epistemic conditions? Are the players rational in this situation? Does each think the other is rational? And so on.

To answer, we need a definition of rationality with CPS's. Fix a strategy-type pair (s^i, t^i), where t^i is associated with a CPS. Call this pair **rational (in the tree)** if the following holds: fix any information set H for i allowed by s^i, and look at the measure on the other players' strategies, given H. (This means given the event that the other players' strategies allow H.) Require that s^i maximizes i's expected payoff under this measure, among all strategies r^i of i that allow H.

With this definition, the rational strategy-type pairs in Figure 3.2 are (*Down*, t^a), (*Out*, u^a), (*In*, t^b), and (*Out*, u^b).

Next, what does Ann think about Bob's rationality? To answer, we need a CPS-analogue to belief (as defined in Section 2). Ben Porath (1997) proposed the following (we have taken the liberty of changing terminology, for consistency with 'strong belief' below): Say player i **initially believes** event E if, under i's CPS, E gets probability 1 at the root of the tree. (Formally, the conditioning event consists of all strategy profiles of the other players.) Battigalli and Siniscalchi (2002) strengthened this definition to: Say player i **strongly believes** event E if, under i's CPS, E gets probability 1 at every information set at which E is possible. Under initial belief, E also gets probability 1 at any information set H that gets positive probability under i's initial measure (that is, i's measure given the root). This is just standard conditioning on non-null events. But under strong belief, this conclusion holds for any information set H which has a non-empty intersection with E – even if H is null under i's initial measure. This is why strong belief is stronger than initial belief.

Let us apply these definitions to Figure 3.2. Does Ann initially believe that Bob is rational? Yes. Both of Ann's types initially believe Bob is rational. Type t^a initially assigns probability 1 to the rational pair (In, t^b). Type u^a initially assigns probability 1 to the rational pair (Out, u^b). In fact, both types strongly believe Bob is rational. Since, under type t^a, Ann's second node gets positive probability (in fact, probability 1) under her initial measure, we need only check this for type u^a. But at Ann's second node, type u^a assigns probability 1 to the rational pair (In, t^b).

Turning to Bob, both of his types initially believe that Ann is rational. Type u^b even strongly believes Ann is rational; but type t^b doesn't. This is because, at Bob's node, type t^b assigns positive probability (in fact, probability 1) to the irrational pair ($Across$, t^a).

Staying with initial belief (we come back to strong belief below), we can parallel Definition 2.1 and define inductively **rationality and *m*th-order initial belief of rationality (R*m*IBR)** at a state of a type structure, and **rationality and common initial belief of rationality (RCIBR)** (see Ben Porath, 1997). In Figure 3.2, since all four types initially believe the other player is rational, a simple induction gives that at the state ($Down$, t^a, In, t^b) for instance, RCIBR holds.

In words, Ann plays across at her first node, believing (initially) that Bob will play *In*, so she can get a payoff of 4. Why would Bob play In? Because he initially believes that Ann plays *Out*. But in the probability-0 event that Ann plays across at her first node, Bob then assigns probability 1 to Ann's playing across at her second node – that is, to Ann's being irrational. He therefore (rationally) plays *In*. All this is consistent with RCIBR.

4. Conditions for backward induction

Interestingly, this is exactly the line of reasoning which, as we said, was the original stimulus for investigating the foundations of BI. So, there is no difficulty with it – we've just seen a formal set-up in which it holds. The resolution of the BI puzzle is simply to accept that the BI path may not result.

But one can also argue that RCIBR is not the right condition: it is too weak. In the above example, Bob realizes that he might be 'surprised' in the play of the game – that's why he has a CPS, not just an ordinary probability measure. If he realizes he might be surprised, should he abandon his (initial) belief that Ann is rational when he is surprised? Bob's type t^b does so. This is the step taken by Battigalli and Siniscalchi (2002) with their concept of strong belief. The argument says that we want t^b to strongly believe, not just initially believe, that Ann is rational. Type t^b will strongly believe Ann is rational if we move the probability-1 weight (in square brackets) on (*Across*, t^a) to (*Down*, t^a). But now (*In*, t^b) isn't rational for Bob, so Ann doesn't (even initially) believe Bob is rational. It looks as if the example unravels.

We can again parallel Definition 2.1 and define inductively **rationality and *m*th-order strong belief of rationality** (**R*m*SBR**), and **rationality and common strong belief of rationality (RCSBR)** (see Battigalli and Siniscalchi, 2002). The question is then: does RCSBR yield BI?

The answer is yes. *Fix a CPS-based type structure for n-legged Centipede* (Figure 4.1), *and a state at which there is RCSBR. Then Ann plays Out.* The result follows from Friedenberg (2002), who shows that in a PI game (satisfying certain payoff restrictions), RCSBR yields a Nash-equilibrium outcome. In Centipede, there is a unique Nash path and it coincides with the BI path. Of course, this isn't true in general.

Example 4.1 (A second coordination game) *Consider the coordination game in* Figure 4.2 *and the associated CPS-based type structure in* Figure 4.3.

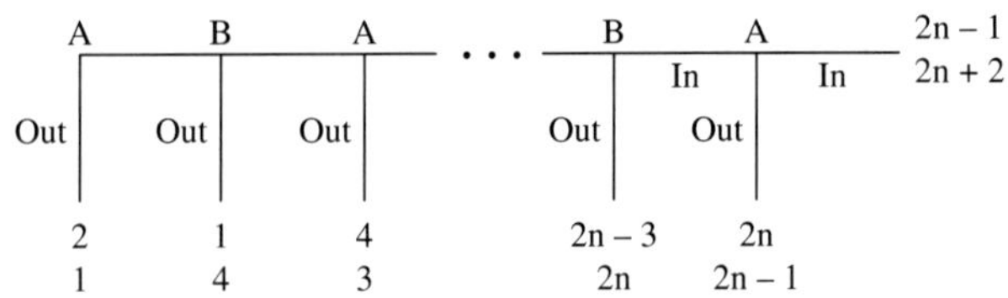

Figure 4.1

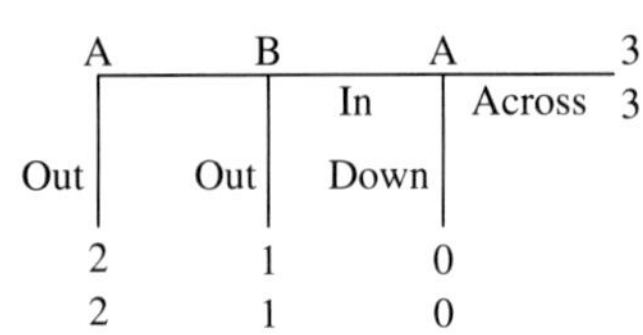

Figure 4.2

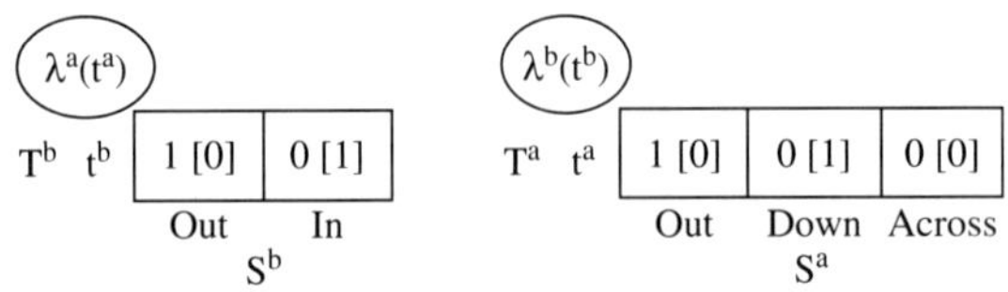

Figure 4.3

The rational strategy-type pairs are (Out, t^a) and (Out, t^b) for Ann and Bob respectively. Ann's type t^a strongly believes {(*Out*, t^b)}, *and Bob's type t^b strongly believes* {(*Out*, t^a)}. *By induction, RCSBR holds at the state (Out, t^a, Out, t^b).*

Here, the BI path need not be played under RCSBR. The key is to see that both (*Down*, t^a) and (*Across*, t^a) are irrational for Ann, since she (strongly) believes Bob plays *Out*. So at his node, Bob can't believe Ann is rational. If he considers it sufficiently more likely Ann will play *Down* rather than *Across*, he will rationally play *Out* (as happens). In short, if Ann doesn't play *Out*, she is irrational and so 'all bets are off' as to what she will do. She could play *Down*.

This situation may be surprising, at least at first blush, but there does not appear to be anything conceptually wrong with it. Indeed, it points to an interesting way in which the players in a game can literally be trapped by their beliefs – which here prevent them from getting their mutually preferred (3, 3) outcome.

But one can also argue differently. If Ann forgoes the payoff of 2 she can get by playing *Out* at the first node, then surely she must be playing *Across* to get 3. Playing *Down* to get 0 makes little sense since this is lower than the payoff she gave up at the first node. (This is forward-induction reasoning à la Kohlberg and Mertens, 1986, Section 2.3, introduced in the context of non-PI games. Interestingly, epistemic analysis makes clear that the issue already arises in PI games, such as Figure 4.2.) But if Bob considers *Across* (sufficiently) more likely than *Down*, he will play *In*. Presumably then, Ann will indeed play *Across*, and the BI path results.

There is no contradiction with the previous analysis because in Figure 4.3 Ann is irrational once she doesn't play *Out*, so we can't say Ann should then rationally play *Across* not *Down*. To make *Across* rational for Ann, we have to add more types to the structure – specifically, we would want to add a second type for Ann that assigns (initial) probability 1 to Bob's playing *In* not *Out*. This key insight is due to Stalnaker (1998) and Battigalli and Siniscalchi (2002).

Battigalli and Siniscalchi formulate a general result of this kind. They consider a **complete** CPS-based type structure, which contains, in a certain sense, every possible type for each player (a complete type structure will be uncountably infinite), and prove: *Fix a complete CPS-based type structure. If there is RCSBR at the state* (s^1, t^1, …, s^n, t^n), *then the strategy profile* (s^1, …, s^n) *is extensive-form rationalizable. Conversely, if the profile* (s^1, …, s^n) *is extensive-form rationalizable, then there is a state* (s^1, t^1, …, s^n, t^n) *at which there is RCBR.*

The extensive-form rationalizability strategies (Pearce, 1984) yield the BI outcome in a PI game (under an assumption ruling out certain payoff ties; Battigalli, 1997), so the Battigalli and Siniscalchi analysis gives epistemic conditions for BI.

There are other routes to getting BI in PI games. Asheim (2001) develops an epistemic analysis using the properness concept (Myerson, 1978). Go back to Example 4.1. The properness idea says that Bob's type t^b should view (*Across*, t^a) as infinitely more likely than (*Down*, t^a) since *Across* is the less costly 'mistake' for Ann, given her type t^a. Unlike the completeness route taken above, the irrationality of both *Down* and *Across* (given Ann's type t^a) is accepted. But the relative ranking of these 'mistakes' must be in the right order. With this ranking, Bob is irrational to play *Out* rather than *In*. Ann presumably will play *Across*, and we get BI again. Asheim (2001) formulates a general such result.

Another strand of the literature on BI employs knowledge models rather than belief models. As pointed out in Example 1.1, players' beliefs don't have to be correct in any sense. For example, a type might even assign probability 1 to a strategy-type pair for another player different from the actual one. Knowledge as usually formalized is different, in that if a player knows an event E, then E indeed happens.

Aumann (1995) formulates a knowledge-based epistemic model for PI trees. In his set-up, the condition of common knowledge of rationality implies that the players choose their BI strategies. Stalnaker (1996) finds that non-BI outcomes are possible, under a different formulation of the same condition. The explanation lies in differences in how counterfactuals are treated. These play an important role in a knowledge-based analysis, when we talk about what a player thinks at an information set that cannot be reached given what he knows. Halpern (2001) provides a synthesis in which these differences can be understood. See also the exchange between Binmore (1996) and Aumann (1996), and the analyses by Samet (1996), Balkenborg and Winter (1997), and Halpern (1999).

Aumann (1998) provides knowledge-based epistemic conditions under which Ann plays *Out* in Centipede. The conditions are weaker than in his (1995) paper, and the conclusion weaker (about outcomes not strategies). There is an obvious parallel between this result and the belief-based result on Centipede we stated above (also about outcomes). More generally, there may be an analogy between counterfactuals in knowledge models and extended probabilities in belief models. But, for one thing, completeness is crucial to the belief-based approach, as we have seen, and an analogous concept does not appear to be present in the knowledge-based approach. As yet, there does not appear to be any formal treatment of the relationship between the two approaches.

5. Next steps: weak dominance

Extending the epistemic analysis of games from the matrix to the tree has been the focus of much recent work in the literature. Another area has been extending the analysis on the matrix from strong dominance (described in Section 2) to weak dominance.

Weak dominance (admissibility) says that a player considers as possible (even if unlikely) any of the strategies for the other players. In the game context, we are naturally led to consider *iterated admissibility* (*IA*) – the weak-dominance analogue to IU. This is an old concept in game theory, going back at least to Gale (1953). Like BI, it is a powerful solution concept, delivering sharp answers in many games – Bertrand, auctions, voting games, and others. (Mertens, 1989, p. 582, and Marx and Swinkels, 1997, pp. 224–5, list various games involving weak dominance.)

But, also like BI, there is a conceptual puzzle. Suppose Ann conforms to the admissibility requirement, so that she considers possible any of Bob's strategies. Suppose Bob also conforms to the requirement, and this leads him not to play a strategy, say *L*. If Ann thinks Bob adheres to the requirement (as he does), then she can rule out Bob's playing *L*. But this conflicts with the requirement that she not rule anything out (see Samuelson, 1992).

Can a sound argument be made for IA? To investigate this, the epistemic tools of Section 1 have to be extended again.

Example 5.1 (**Bertrand**) Figure 5.1 *is a Bertrand pricing game, where each firm chooses a price in {0,1, 2, 3}. (Ken Corts kindly provided this example.) The left payoff is to A, the right payoff to B. Each firm has capacity of two units and zero cost. Two units are demanded. If the firms charge the same price, they each sell one unit.* Figure 5.2 *is an associated type structure (with one type for each player).*

The rational strategy-type pairs are $R_1^a = \{0, 1, 2, 3\} \times \{t^a\}$ *and* $R_1^b = \{0, 1, 2, 3\} \times \{t^b\}$. *Since both types assign positive probability only to a rational strategy-type pair for the other player, we get* $R_m^a = R_1^a$ *and* $R_m^b = R_1^b$ *for all m. In particular, there is RCBR at the state* $(3, t^a, 3, t^b)$.

But a price of 3 is inadmissible (as is a price of 0). The IA set is just {(1,1)}, where each firm charges the lowest price above cost. (This is a plausible scenario: while pricing at cost is inadmissible, competition forces price down to the first price above cost.)

A \ B	3	2	1	0
3	3, 3	0, 4	0, 2	0, 0
2	4, 0	2, 2	0, 2	0, 0
1	2, 0	2, 0	1, 1	0, 0
0	0, 0	0, 0	0, 0	0, 0

Figure 5.1

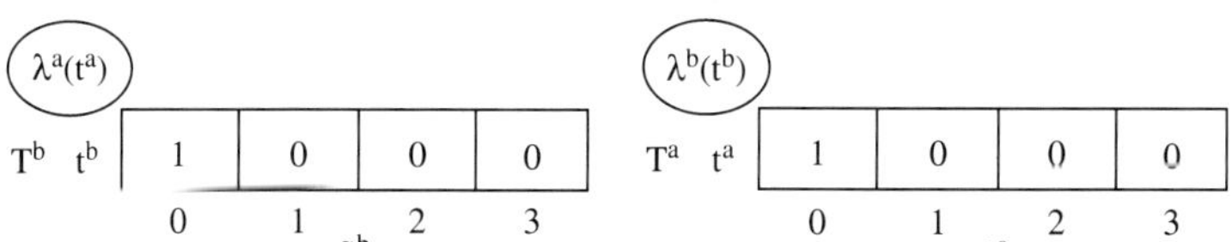

Figure 5.2

A tool to incorporate admissibility is *lexicographic probability systems* (*LPS*'s), introduced and axiomatized by Blume, Brandenburger and Dekel (1991a; 1991b). An LPS specifies a sequence of probability measures. The interpretation is that the first measure is the player's primary hypothesis about the true state. But the player recognizes that his primary hypothesis might be mistaken, and so also forms a secondary hypothesis. This is his second measure. Then his tertiary hypothesis, and so on. The primary states can be thought of as infinitely more likely than the secondary states, which are infinitely more likely than the tertiary states, and so on. Stahl (1995), Stalnaker (1998), Asheim (2001), Brandenburger, Friedenberg and Keisler (2006), and Asheim and Perea (2005), among other papers, use LPS's.

Example 5.2 (**Bertrand contd.**) Figure 5.3 *is a type structure for Bertrand* (Figure 5.1) *that now specifies LPS's.*

Each player has a primary hypothesis which assigns probability 1 to the other player's charging a price of 0. But each player also has a secondary hypothesis that assigns equal probability to each of the three remaining choices for the other player. This measure is shown in parentheses. Note that every state (that is, strategy-type pair) gets positive probability under some measure. But states can also be ruled out, in the sense that they can be give infinitely less weight than other states.

What about epistemic conditions? Are the players rational in this situation? Does each think the other is rational? And so on.

To answer, we need a definition of rationality with LPS's. Fix strategy-type pairs (s^i, t^i) and (r^i, t^i) for player i, where t^i is now associated with an LPS. Calculate the tuple of expected payoffs to i from s^i, using first the primary measure associated with t^i, then the secondary measure associated with t^i, and so on. Calculate the corresponding tuple for r^i. If the first tuple lexicographically exceeds the second, then s^i is preferred to r^i. (If $x = (x_1, \ldots, x_n)$ and $y = (y_1, \ldots, y_n)$, then x lexicographically exceeds y if $y_j > x_j$ implies $x_k > y_k$ for some $k < j$.) A strategy-type pair (s^i, t^i) is **rational** (**in the lexicographic sense**) if s^i is maximal under this ranking.

So $(3, t^a)$ and $(3, t^b)$ are irrational. All choices give each player an expected payoff of 0 under the primary measure. But a price of 2 gives each player an expected payoff of 2 under the secondary measure, as opposed to an expected payoff of 1 from a price of 3. Conceptually, we want $(3, t^a)$ and $(3, t^b)$ to be irrational (because a price of 3 is inadmissible).

What does each player think about the other's rationality? For this, we again need an LPS-based definition. An early candidate in the literature was: Say player i **believes**

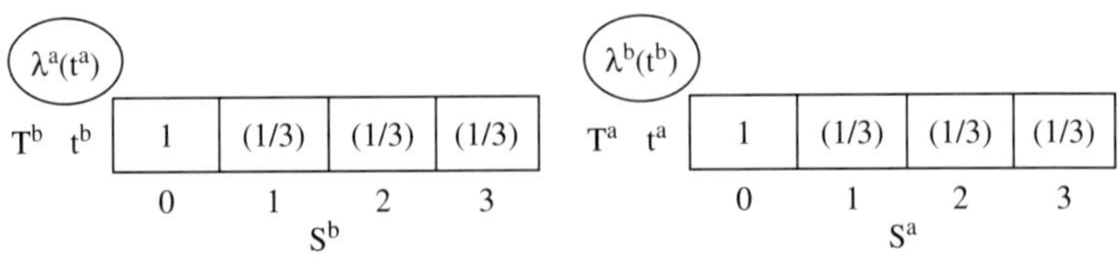

Figure 5.3

event E **at the 1st level** if E gets primary probability 1 under i's LPS (Börgers, 1994; Brandenburger, 1992). A stronger concept is: Say i **assumes** E if all states not in E are infinitely less likely than all states in E, under i's LPS (Brandenburger, Friedenberg and Keisler, 2006). In other words, a player who assumes E recognizes E may not happen, but is prepared to 'count on' E versus not-E.

In Figure 5.3, type t^a doesn't 1st-level believe (so certainly doesn't assume) the other player is rational. Likewise with t^b. Again, this is right conceptually.

6. Conditions for iterated admissibility

Once again we can parallel Definition 2.1 and define inductively **rationality and *m*th-order 1st-level belief of rationality (R*m*1BR)** at a state of a type structure, and **rationality and common 1st-level belief of rationality (RC1BR)**. Likewise, one can define **rationality and *m*th-order assumption of rationality (R*m*AR), and rationality and common assumption of rationality (RCAR)**. What do these conditions yield?

In fact, just as we saw in Sections 3 and 4 that neither RCIBR not RCSBR yields BI, so neither RC1BR nor RCAR yields IA. RC1BR is characterized by the $S^{\infty}W$ concept (Dekel and Fudenberg, 1990), that is , the set of strategies that remain after one round of deletion of inadmissible strategies followed by iterated deletion of strongly dominated strategies. RCAR is characterized by the self-admissible set concept (Brandenburger, Friedenberg and Keisler, 2006). Self-admissible sets may be viewed as the weak-dominance analogue to Pearce (1984) best-response sets.

But while the IA set is one self-admissible set in a game, there may well be others. To select the IA set, a completeness assumption is needed, similar to Section 4: *Fix a complete LPS-based type structure. If there is RmAR at the state* $(s^1, t^1, \ldots, s^n, t^n)$, *then the strategy profile* $(s^1, \ldots, s^n)$ *survives* $(m+1)$ *rounds of iterated admissibility. Conversely, if the profile* $(s^1, \ldots, s^n)$ *survives* $(m+1)$ *rounds of iterated admissibility, then there is a state* $(s^1, t^1, \ldots, s^n, t^n)$ *at which there is RmAR* (Brandenburger, Friedenberg and Keisler, 2006).

This result is stated for R*m*AR and not RCAR. See the next section for the reason. Of course, for a given game, there is an m such that IA stabilizes after m rounds.

IA yields the BI outcome in a PI game (again ruling out certain payoff ties; Marx and Swinkels, 1997), so, understanding IA gives, in particular, another analysis of BI.

Related analyses of IA include Stahl (1995) and Ewerhart (2002). Stahl uses LPS's and directly assumes that Ann considers one of Bob's strategies infinitely less likely than another if the first is eliminated on an earlier round of IA than the second. Ewerhart gives an analysis of IA couched in terms of provability (from mathematical logic).

7. Strategic versus extensive analysis

Kohlberg and Mertens (1986, Section 2.4) argued that a 'fully rational' analysis of games should be invariant – that is , should depend only on the fully reduced strategic form of a game. (This is the strategic form after elimination of any – pure – strategies that are duplicates or convex combinations of other strategies.) In this, they appealed to early results in game theory (Dalkey, 1953; Thompson, 1952) which established that two trees sharing the same reduced strategic form differ from each other by a (finite) sequence of elementary transformations of the tree, each of which can be argued to be 'strategically inessential'. Kohlberg and Mertens added a fourth transformation involving convex combinations, to get to the fully reduced strategic form.

In decision theory, invariance is implied by (and implies) admissibility. (Kohlberg and Mertens, 1986, Section 2.7, gave the essential idea. See Brandenburger, 2007, for the decision-theory argument.) If we build up our game analysis using a decision theory that satisfies admissibility, we can hope to get invariance at this level too. LPS-based decision theory satisfies admissibility. Indeed, IA, and also the $S^{\infty}W$ and self-admissible set concepts, are invariant in the Kohlberg–Mertens sense. The extensive-form rationalizability concept (Section 4) is not.

There does appear to be a price paid for invariance, however. The extensive-form conditions of RCSBR and (CPS-based) completeness are consistent (in any tree). That is, for any tree, we can build a complete type structure and find a state at which RCSBR holds. But Brandenburger, Friedenberg and Keisler (2006) show the strategic-form conditions of RCAR and (LPS-based) completeness are inconsistent (in any matrix satisfying a non-triviality condition).

A possible interpretation is that rationality, even as a theoretical concept, appears to be inherently limited. There are purely theoretical limits to the Kohlberg-Mertens notion of a 'fully rational' analysis of games.

The epistemic programme has uncovered a number of impossibility results (see EPISTEMIC GAME THEORY: BELIEFS AND TYPES for some others). We don't see this as a deficiency of the programme, but rather as a sign it has reached a certain depth and maturity. Also, central to the programme is the analysis of scenarios (we have seen several in this survey) that are 'a long way from' these theoretical limits. Under the epistemic approach to game theory there is not one right set of assumptions to make about a game.

ADAM BRANDENBURGER

See also **epistemic game theory: an overview; epistemic game theory: beliefs and types; epistemic game theory: incomplete information; Nash equilibrium, refinements of.**

This survey is based on Brandenburger (2007). I am grateful to Springer for permission to use this material. I owe a great deal to joint work and many conversations with Robert Aumann, Eddie Dekel, Amanda Friedenberg, Jerry Keisler and Harborne Stuart. My thanks to Konrad Grabiszewski for important input, John Nachbar for very important editorial advice, and Michael James for valuable assistance. The Stern School of Business provided financial support.

Bibliography

Asheim, G. 2001. Proper rationalizability in lexicographic beliefs. *International Journal of Game Theory* 30, 453–78.

Asheim, G. and Perea, A. 2005. Sequential and quasi-perfect rationalizability in extensive games. *Games and Economic Behavior* 53, 15–42.

Aumann, R. 1974. Subjectivity and correlation in randomized strategies. *Journal of Mathematical Economics* 1, 67–96.

Aumann, R. 1987. Correlated equilibrium as an expression of Bayesian rationality. *Econometrica* 55, 1–18.

Aumann, R. 1995. Backward induction and common knowledge of rationality. *Games and Economic Behavior* 8, 6–19.

Aumann, R. 1996. Reply to Binmore. *Games and Economic Behavior* 17, 138–46.

Aumann, R. 1998. On the centipede game. *Games and Economic Behavior* 23, 97–105.

Aumann, R. 1998. Common priors: a reply to Gul. *Econometrica* 66, 929–38.

Aumann, R. and Brandenburger, A. 1995. Epistemic conditions for Nash equilibrium. *Econometrica* 63, 1161–80.

Balkenborg, D. and Winter, E. 1997. A necessary and sufficient epistemic condition for playing backward induction. *Journal of Mathematical Economics* 27, 325–45.

Basu, K. 1990. On the existence of a rationality definition for extensive games. *International Journal of Game Theory* 19, 33–44.

Battigalli, P. 1997. On rationalizability in extensive games. *Journal of Economic Theory* 74, 40–61.

Battigalli, P. and Siniscalchi, M. 1999. Hierarchies of conditional beliefs and interactive epistemology in dynamic games. *Journal of Economic Theory* 88, 188–230.

Battigalli, P. and Siniscalchi, M. 2002. Strong belief and forward-induction reasoning. *Journal of Economic Theory* 106, 356–91.

Ben Porath, E. 1997. Rationality, Nash equilibrium, and backward induction in perfect information games. *Review of Economic Studies* 64, 23–46.

Bernheim, D. 1984. Rationalizable strategic behavior. *Econometrica* 52, 1007–28.

Bicchieri, C. 1988. Strategic behavior and counterfactuals. *Synthese* 76, 135–69.

Bicchieri, C. 1989. Self-refuting theories of strategic interaction: a paradox of common knowledge. *Erkenntnis* 30, 69–85.

Binmore, K. 1987. Modelling rational players I. *Economics and Philosophy* 3, 179–214.

Binmore, K. 1996. A note on backward induction. *Games and Economic Behavior* 17, 135–7.

Blume, L., Brandenburger, A. and Dekel, E. 1991a. Lexicographic probabilities and choice under uncertainty. *Econometrica* 59, 61–79.

Blume, L., Brandenburger, A. and Dekel, E. 1991b. Lexicographic probabilities and equilibrium refinements. *Econometrica* 59, 81–98.

Bonanno, G. 1991. The logic of rational play in games of perfect information. *Economics and Philosophy* 7, 37–65.

Bonanno, G. and Nehring, K. 1999. How to make sense of the common prior assumption under incomplete information. *International Journal of Game Theory* 28, 409–34.

Börgers, T. 1994. Weak dominance and approximate common knowledge. *Journal of Economic Theory* 64, 265–76.

Brandenburger, A. 1992. Lexicographic probabilities and iterated admissibility. In *Economic Analysis of Markets and Games*, ed. P. Dasgupta, D. Gale, O. Hart and E. Maskin. Cambridge, MA: MIT Press.

Brandenburger, A. 2007. The power of paradox: some recent results in interactive epistemology. *International Journal of Game Theory* 35, 465–92.

Brandenburger, A. and Dekel, E. 1987. Rationalizability and correlated equilibria. *Econometrica* 55, 1391–402.

Brandenburger, A., Friedenberg, A. and Keisler, H.J. 2006. Admissibility in games. Unpublished, Stern School of Business, New York University.

Dalkey, N. 1953. Equivalence of information patterns and essentially determinate games. In *Contributions to the Theory of Games*, vol. 2, ed. H. Kuhn and A. Tucker. Princeton: Princeton University Press.

Dekel, E. and Fudenberg, D. 1990. Rational behavior with payoff uncertainty. *Journal of Economic Theory* 52, 243–67.

Ewerhart, C. 2002. Ex-ante justifiable behavior, common knowledge, and iterated admissibility. Unpublished, Department of Economics, University of Bonn.

Feinberg, Y. 2000. Characterizing common priors in terms of posteriors. *Journal of Economic Theory* 91, 127–79.

Friedenberg, A. 2002. When common belief is correct belief. Unpublished, Olin School of Business, Washington University.

Gale, D. 1953. A theory of *n*-person games with perfect information. *Proceedings of the National Academy of Sciences* 39, 496–501.

Gul, F. 1998. A comment on Aumann's Bayesian view. *Econometrica* 66, 923–7.

Halpern, J. 1999. Hypothetical knowledge and counterfactual reasoning. *International Journal of Game Theory* 28, 315–30.

Halpern, J. 2001. Substantive rationality and backward induction. *Games and Economic Behavior* 37, 425–35.

Halpern, J. 2002. Characterizing the common prior assumption. *Journal of Economic Theory* 106, 316–55.

Harsanyi, J. 1967–8. Games with incomplete information played by 'Bayesian' players, I–III. *Management Science* 14, 159–82, 320–34, 486–502.

Harsanyi, J. 1973. Games with randomly disturbed payoffs: a new rationale for mixed strategy equilibrium points. *International Journal of Game Theory* 2, 1–23.

Kohlberg, E. and Mertens, J.-F. 1986. On the strategic stability of equilibria. *Econometrica* 54, 1003–37.

Marx, L. and Swinkels, J. 1997. Order independence for iterated weak dominance. *Games and Economic Behavior* 18, 219–45.

Mertens, J.-F. 1989. Stable equilibria – a reformulation. *Mathematics of Operations Research* 14, 575–625.

Morris, S. 1994. Trade with heterogeneous prior beliefs and asymmetric information. *Econometrica* 62, 1327–47.

Myerson, R. 1978. Refinements of the Nash equilibrium concept. *International Journal of Game Theory* 1, 73–80.

Myerson, R. 1991. *Game Theory*. Cambridge, MA: Harvard University Press.

Pearce, D. 1984. Rational strategic behavior and the problem of perfection. *Econometrica* 52, 1029–50.

Reny, P. 1992. Rationality in extensive form games. *Journal of Economic Perspectives* 6(4), 103–18.

Rényi, A. 1955. On a new axiomatic theory of probability. *Acta Mathematica Academiae Scientiarum Hungaricae* 6, 285–335.

Rosenthal, R. 1981. Games of perfect information, predatory pricing and the chain-store paradox. *Journal of Economic Theory* 25, 92–100.

Samet, D. 1996. Hypothetical knowledge and games with perfect information. *Games and Economic Behavior* 17, 230–51.

Samet, D. 1998. Common priors and the separation of convex sets. *Games and Economic Behavior* 24, 172–4.

Samet, D. 1998. Iterated expectations and common priors. *Games and Economic Behavior* 24, 131–41.

Samuelson, L. 1992. Dominated strategies and common knowledge. *Games and Economic Behavior* 4, 284–313.
Stahl, D. 1995. Lexicographic rationalizability and iterated admissibility. *Economic Letters* 47, 155–9.
Stalnaker, R. 1996. Knowledge, belief and counterfactual reasoning in games. *Economics and Philosophy* 12, 133–63.
Stalnaker, R. 1998. Belief revision in games: forward and backward induction. *Mathematical Social Sciences* 36, 31–56.
Tan, T. and Werlang, S. 1988. The Bayesian foundations of solution concepts of games. *Journal of Economic Theory* 45, 370–91.
Thompson, F. 1952. Equivalence of games in extensive form. Research Memorandum RM-759. The RAND Corporation.

epistemic game theory: incomplete information

A game of incomplete information is a game in which at least some of the players possess private information which may be relevant to the strategic interaction. The private information of a player may be about the payoff functions in the game, as well as about some exogenous, payoff-irrelevant events. The player may also form beliefs about other players' beliefs about payoffs and exogenous events, about their beliefs about the beliefs of others, and so forth.

Harsanyi (1967–8) introduced the idea that such a state of affairs can be succinctly described by a *type space*. With this formulation, T_i denotes the set of player i's *types*. Each type $t_i \in T_i$ is associated with a belief $\lambda_i(t_i) \in \Delta(K \times T_{-i})$ about some basic space of uncertainty, K, and the combination T_{-i} of the other players' types. The basic space of uncertainty K is called the space of *states of nature*, and $\Omega = K \times \Pi_{i \in I} T_i$, where I is the set of players, is called the space of *states of the world*.

A type space models a game of incomplete information once each state of nature $k \in K$ is associated with a payoff matrix of the game, or, more generally, with a payoff function u_i^k for each player $i \in I$. This payoff function specifies the player's payoff $u_i^k(s)$ for each combination of strategies $s = (s_i)_{i \in I} \in S = \Pi_{i \in I} S_i$ of the players. (In the particular case in which k is associated with a payoff matrix, that is, the game is such that each player has finitely many strategies, the payoffs $u_i^k(s)$ to the players $i \in I$ appear in the entry of the matrix corresponding to the combination of strategies $s = (s_i)_{i \in I}$.) As usual, the set of strategies S_i of player $i \in I$ may be a complex object by itself. For instance, it may be the set of mixed strategies over some set of pure strategies S_i^0. The payoff function of player i in the state of nature k is $u_i^k : S \to \mathbb{R}$.

Obviously, different types of a player may want to choose different strategies. Thus, a *Bayesian strategy* of player i in a game of incomplete information specifies the strategy $\sigma_i(t_i) \in S_i$ that the player chooses given each one of her types $t_i \in T_i$.

Given a profile of Bayesian strategies $\sigma = (\sigma_j : T_j \to S_j)_{j \in I}$ of the players, the expected payoff of player i of type t_i is

$$U_i(\sigma, t_i) = \sum_{(k, t_{-i}) \in K \times T_{-i}} u_i^k(\sigma_i(t_i), \sigma_{-i}(t_{-i})) \times \lambda_i(t_i)(k, t_{-i})$$

where $\sigma_{-i}(t_{-i}) = (\sigma_j(t_j))_{j \neq i}$. If there is a continuum of states of nature and types, the sum becomes an integral:

$$U_i(\sigma, t_i) = \int_{K \times T_{-i}} u_i^k(\sigma_i(t_i), \sigma_{-i}(t_{-i})) d\lambda_i(t_i)(k, t_{-i})$$

(In this case, the expected payoff function $U_i(\sigma, t_i)$ is well defined if the Bayesian strategies $\sigma_j : T_j \to S_j$ are measurable functions and if the payoff function $u_i : K \times S \to R$ is measurable as well; we omit the details of this technical requirement).

We assume that the players are expected payoff maximizers. Thus, player i prefers the Bayesian strategy σ over σ' if and only if $U_i(\sigma, t_i) \geq U_i(\sigma', t_i)$ for each of her types $t_i \in T_i$. It follows that given a Bayesian strategy profile σ_{-i} of the other players, the Bayesian strategy σ_i is a *best reply* of player i if for any other strategy σ_i' of hers, $U_i((\sigma_i, \sigma_{-i}), t_i) \geq U_i((\sigma', \sigma_{-i}), t_i)$ for each of her types $t_i \in T_i$. A *Bayes–Nash equilibrium* or a *Bayesian equilibrium* is a profile of Bayesian strategies $\sigma^* = (\sigma_i^*)_{i \in I}$ such that σ_i^* is a best reply against σ_{-i}^* for every player $i \in I$.

A simple, discrete variant of an example by Gale (1996) may clarify these abstract definitions. There are two investors $i = 1, 2$ and three possible states of nature $k \in K = \{-1, 0, 1\}$. Each investor i only knows her own type

$$t_i \in T_i = \{-10, -6, -2, 2, 6, 10\}.$$

Every type t_i of investor i believes that all of the other investor's types $t_j \in T_j$, $j \neq i$, are equally likely, so that each of them has probability $\frac{1}{6}$. Moreover, every type t_i believes that the state of nature is $k = 1$ when $t_i + t_j > 0$; that the state of nature is $k = 0$ when $t_i + t_j = 0$; and that the state of nature is $k = -1$ when $t_i + t_j < 0$. Formally, the belief $\lambda_i(t_i)$ of type $t_i \in T_i$ is defined by

$$\lambda_i(t_i)(k, t_j) = \begin{cases} \frac{1}{6} & k \text{ has the same sign as } t_i + t_j \\ 0 & \text{otherwise.} \end{cases}$$

The investors cannot communicate their types to one another. They can invest in at most one of two available investment periods. Each investor has three relevant strategies: invest *immediately*, in the first period; *wait* to the second period and invest only if the other investor has invested in the first period; or *never* invest. The payoff of each of the investors depends on the state of nature $k \in K = \{-1, 0, 1\}$ and on her own investment strategy, but not on the investment strategy of the other investor. The payoffs are as follows:

- Investing *immediately* when the state of nature is k yields investor i a payoff of k

$$u_i^k(\textit{immediately}', \cdot) = k$$

(The $\cdot$ stands for the investment decision of the other investor $j \neq i$, which, as we said, does not effect the payoff of investor i.)

- If investor i chooses to *wait* to the second period and invest only if the other investor has invested in the first period, investor i's payoff in the state of nature k is

$$u_i^k(\textit{wait}', \cdot) = \frac{3}{4}k.$$

- If the investor *never* invests, her payoff is 0 irrespective of the state of nature:

$$u_i^k(\textit{never}', \cdot) = 0.$$

How will the different types behave at a Bayesian equilibrium?

The type $t_i = 10$ assesses that by investing immediately her expected payoff is

$$U_i(immediately', 10) = \frac{1}{6} \times 0 + \frac{5}{6} \times 1 = \frac{5}{6}$$

(immediate investment yields 0 in case $t_j = -10$, and yields 1 in case $t_j = -6, -2, 2, 6, 10$). This is higher than $\frac{3}{4}$, the maximum payoff she could possibly get by waiting for the second period, and higher than the payoff 0 of never investing. So at a Bayesian equilibrium

$$\sigma_i^*(10) = immediately', \quad i = 1, 2.$$

Next, the expected payoff to the type $t_i = 6$ from immediate investment is

$$U_i(immediately', 6) = \frac{1}{6} \times (-1) + \frac{1}{6} \times 0 + \frac{4}{6} \times 1 = \frac{1}{2}$$

(immediate investment yields 1 unless $t_j = -10$, in which case the payoff is -1, or $t_j = -6$, in which case the payoff is 0). So investing immediately is preferred for her over never investing. But how about waiting until the second period? That's an inferior option as well, since the types $t_j = -10, -6, -2$ will never invest in the first period (this would yield them a negative expected payoff). So only the positive types $t_j = 2$, 6, 10 could *conceivably* invest immediately, with overall probability reaching at most $\frac{3}{6}$. So waiting to see if they invest yields to the type $t_i = 6$ an expected payoff not higher than $\frac{3}{6} \times \frac{3}{4} = \frac{3}{8}$, which is smaller than $\frac{1}{2}$. We conclude that the preferable strategy of $t_i = 6$ at equilibrium is

$$\sigma_i^*(6) = immediately', \quad i = 1, 2.$$

What about $t_i = 2$? Immediate investment yields her

$$U_i(immediately', 2) = \frac{2}{6} \times (-1) + \frac{1}{6} \times 0 + \frac{3}{6} \times 1 = \frac{1}{6}$$

(-1 is the payoff when $t_j = -10, -6$; 0 is the payoff when $t_j = -2$; the payoff is 1 otherwise). However, given that the types $t_j = 6{,}10$ invest immediately at equilibrium, and that the negative types $t_j = -10, -6, -2$ do not invest immediately, the type $t_i = 2$ figures out that by waiting and investing only if the other investor has invested first would yield her an expected payoff

$$U_i(wait', 2) \geq \frac{2}{6} \times \frac{3}{4} = \frac{1}{4} > \frac{1}{6}$$

($\frac{2}{6}$ is the probability assigned by $t_i = 2$ to the event that $t_j \in \{6, 10\}$ and hence j invests immediately, and $\frac{3}{4}$ is the payoff from the second period investment). The preferred strategy of $t_i = 2$ at equilibrium is therefore

$$\sigma_i^*(2) = wait', \quad i = 1, 2.$$

We can now compute inductively, in a similar way, that also

$$\sigma_i^*(-2) = wait', \quad i = 1, \quad 2\sigma_i^*(-6) = wait', \quad i = 1, 2$$

and that

$$\sigma_i^*(-10) = \text{'}never\text{'}, \quad i = 1, 2.$$

Notice that the equilibrium in the example is inefficient. For instance, when the pair of types is $(t_1,t_2) = (2,2)$ the investment is profitable, but both investors wait to see if the other one invests, and thus end up not investing at all. In this case, behaviour would become efficient if the investors could communicate their types to each other. Indeed, they would have been happy to do so, because their interests are aligned.

Obviously, there are other strategic situations with incomplete information in which the interests of the players are not completely aligned. For example, a potential seller of an object would like to strike a deal with a potential buyer at a price which is as high as possible, while the potential buyer would like the price to be as low as possible. That's why the traders might not volunteer to communicate honestly their private valuations of the object, even if they are technically able to do so. Still, in case the buyer values the object more than the seller, they would both prefer to trade at some price in-between their valuations rather than forgoing trade altogether. Therefore, the traders would nevertheless like to avoid a complete lack of communication. Myerson and Satterthwaite (1983) phrase general conditions under which no Bayesian equilibrium of any trade mechanism is ever fully efficient due to this tension between interests alignment and interests mismatch. Under these conditions, even if the traders are able to communicate their private information, at no Bayesian equilibrium does trade take place in all instances in which there exist gains from trade.

In the above variant of Gale's example we were able to find the unique Bayesian equilibrium using iterative dominance arguments. We have iteratively crossed out strategies that are inferior for some types, which enabled us to eliminate inferior strategies for other types, and so forth. As in games of complete information, this technique is not applicable in general, and there are games with incomplete information in which a Bayesian equilibrium is not the outcome of any process of iterative elimination of dominated strategies (Battigalli and Siniscalchi, 2003; Dekel, Fudenberg and Morris, 2007).

Games with incomplete information are discussed in many game theory textbooks (for example, Dutta, 1999; Gibbons, 1992; Myerson, 1991; Osborne, 2003; Rasmusen, 1989; Watson, 2002). Aumann and Heifetz (2002), Battigalli and Bonanno (1999) and Dekel and Gul (1997) are advanced surveys.

AVIAD HEIFETZ

See also **epistemic game theory: an overview; epistemic game theory: beliefs and types; epistemic game theory: complete information.**

Bibliography

Aumann, R.J. and Heifetz, A. 2002. Incomplete information. In *Handbook of Game Theory*, vol. 3, ed. R.J. Aumann and S. Hart. Amsterdam: North-Holland.

Battigalli, P. and Bonanno, G. 1999. Recent results on belief, knowledge and the epistemic foundations of game theory. *Research in Economics* 53, 149–225.

Battigalli, P. and Siniscalchi, M. 2003. Rationalization with incomplete information. *Advances in Theoretical Economics* 3(1), article 3. Online. Available at http://www.bepress.com/bejte/advances/vol3/iss1/art3, accessed 25 April 2007.

Dekel, E., Fudenberg, D. and Morris, S. 2007. Interim correlated rationalizability. *Theoretical Economics* 2, 15–40.

Dekel, E. and Gul, F. 1997. Rationality and knowledge in game theory. In *Advances in Economics and Econometrics*, ed. D. Kreps and K. Wallis. Cambridge, UK: Cambridge University Press.

Dutta, P.K. 1999. *Strategies and Games: Theory and Practice.* Cambridge, MA: MIT Press.

Gale, D. 1996. What have we learned from social learning. *European Economic Review* 40, 617–28.

Gibbons, R. 1992. *Game Theory for Applied Economists.* Princeton: Princeton University Press.

Harsanyi, J.C. 1967–8. Games with incomplete information played by Bayesian players, parts I–III. *Management Science* 14, 159–82, 320–34, 486–502.

Myerson, R. 1991. *Game Theory: Analysis of Conflict.* Cambridge, MA: Harvard University Press.

Myerson, R. and Satterthwaite, M. 1983. Efficient mechanisms for bilateral trading. *Journal of Economic Theory* 29, 265–81.

Osborne, M. 2003. *Introduction to Game Theory.* Oxford: Oxford University Press.

Rasmusen, E. 1989. *Games and Information: An Introduction to Game Theory.* Oxford: Basil Blackwell.

Watson, J. 2002. *Strategy: An Introduction to Game Theory.* New York: W.W. Norton.

game theory and biology

In biology, game theory is a branch of evolutionary theory that is particularly suited to the study of interactions between individuals. The evolution of animal fighting behaviour was among the first applications and it was in this context that Maynard Smith and Price (1973) developed the concept of an evolutionarily stable strategy (ESS) (see LEARNING AND EVOLUTION IN GAMES: ESS). Cooperative interactions (Trivers, 1971) and signalling interactions (Grafen, 1991), such as when males signal their quality to females, are examples of other important areas of application. There is an overlap of ideas between economics and biology, which has been quite noticeable since the 1970s and, in a few instances, earlier (Sigmund, 2005). In the early 21st century, the interchange takes the form of a joint exploration of theoretical and empirical issues by biologists and economists (Hammerstein and Hagen, 2005).

Strategies in games inspired by biology can represent particular behaviour patterns, including rules about which behaviour to perform in which circumstance. Other aspects of an individual's phenotype can also be viewed as the result of strategic choice. A life-history strategy specifies choices that have major impact on an individual's course of life, for instance, whether to become a male or a female or, for certain insects, whether or not to develop wings. Interactions between individuals are modelled as games where the payoffs represent Darwinian fitness. Random matching of players drawn from a large population is one common game model, which was used to study fighting behaviour (Maynard Smith and Price, 1973; Maynard Smith and Parker, 1976). 'Playing the field' (Maynard Smith, 1982) is a more general modelling approach, where the payoff to an individual adopting a particular strategy depends on some average property of the population (cf. population games in DETERMINISTIC EVOLUTIONARY DYNAMICS).

Game theory is needed for situations where payoffs to strategies depend on the state of a population, and this state in turn depends on the strategies that are present. For matching of players drawn from a population, the distribution of opposing strategies is of course given by the population distribution, but there are other reasons why the distribution of strategies influences expected payoffs. A 'playing-the-field' example is the choice by an individual, or by its mother, to develop into a male or a female. The two sexes occur in roughly equal proportions in many species. This observation intrigued Darwin, who was unable to provide a satisfactory explanation, writing that 'I formerly thought that when a tendency to produce the two sexes in equal numbers was advantageous to the species, it would follow from natural selection, but I now see that the whole problem is so intricate that it is safer to leave its solution to the future' (Darwin, 1874, pp 399). The solution to the problem was found by Düsing (1884; see also Edwards, 2000, and Fisher, 1930), and rested on the principle that, in diploid sexual organisms, the total genetic contribution to offspring by all males in a generation must equal the contribution by all females in the same generation. This gives a reproductive advantage to the rarer sex in the passing of genes to future generations. The payoffs to a mother from producing a son or a daughter must then depend on the population sex ratio, and this dependence can result in an evolutionary equilibrium at an even sex ratio (see below). The idea arose before the development of the concept of an ESS by Maynard Smith and Price (1973), but it can be regarded as the first instance of game-theoretical reasoning in biology.

Payoffs, reproductive value, and evolutionary dynamics

Class-structured populations in discrete time (Caswell, 2001) are often used as settings for evolutionary analysis. The classes or states are properties like female and male, and time might be measured in years. Let $n_i(t)$ denote the number of individuals in state i at time t. We can write a deterministic population dynamics as $n(t+1) = An(t)$, where n is the vector of the n_i and A is the so-called population projection matrix. The elements a_{ij} of A can depend on n and on the strategies that are present in the population. They represent per capita genetic contributions of individuals in state j to state i, in terms of offspring or individual survival. A common evolutionary analysis is to determine a stationary n for the case where all individuals use a strategy x and to examine whether rare mutants with strategy x' would increase in number in such a population.

Let us apply this scheme to the just mentioned sex ratio problem. Suppose that a mother can determine the sex of her offspring and that females in the population produce a son with probability x and a daughter with probability $1 - x$. For non-overlapping generations, the dynamics $n(t+1) = An(t)$ can be written as

$$\begin{pmatrix} n_f(t+1) \\ n_m(t+1) \end{pmatrix} = \begin{pmatrix} 0.5(1-x)b & 0.5(1-x)bq \\ 0.5xb & 0.5xbq \end{pmatrix} \begin{pmatrix} n_f(t) \\ n_m(t) \end{pmatrix},$$

where b is the reproductive output (number of offspring) of a female, bq is the reproductive output of a male, and the factor 0.5 accounts for the genetic shares of the parents in their offspring. Because the reproductive output of all males must equal that of all females, it follows that $q = n_f(t)/n_m(t)$ and thus that $q = (1-x)/x$. In a stationary population, $b = 1/(1 - x)$ must hold, which could come about through a dependence of b on the total population size. Introducing the matrix

$$B(x', x) = \frac{1}{2} \begin{pmatrix} (1-x')/(1-x) & (1-x)/x \\ x'/(1-x) & 1 \end{pmatrix},$$

the population projection matrix for a stationary population is $A = B(x, x)$ and the stationary $n = (n_f, n_m)$ is proportional to the leading eigenvector, $w = (1-x, x)$, of this A. Suppose a mutant gene causes a female to adopt the sex ratio strategy x', but has no effect in a male. As long as the mutant gene is rare, only the strategy of heterozygous mutant females needs to be taken into account, and the dynamics of the mutant sub-population can be written as $n'(t+1) = A'n'(t)$ with $A' = B(x', x)$. The mutant can invade if $\lambda(x', x) > 1$ holds for the leading eigenvalue $\lambda(x', x)$ of $B(x', x)$. Direct computation of the leading eigenvalue shows that a mutant with $x' > x$ can invade if $x < 0.5$ and one with $x' < x$ can invade if $x' > 0.5$, resulting in an evolutionary equilibrium at $x = 0.5$.

The reproductive value of state i is defined as the ith component of the leading 'left eigenvector' v of the stationary population projection matrix $A = B(x, x)$, that is, v is the leading eigenvector of the transpose of A. It is convenient to normalize v so that its scalar product $v \cdot w$ with the leading eigenvector w equals 1. For our sex ratio problem we have $v = \frac{1}{2}(1/(1-x), 1/x)$. The reproductive value of state i can be interpreted
as being proportional to the expected genetic contribution to future generations of an individual in state i. The eigenvectors v and w can be used to investigate how the leading eigenvalue depends on x' near $x' = x$. It is easy to show that $\partial\lambda(x', x)/\partial x' = \partial(v \cdot B(x', x)w)/\partial x'$ holds at $x' = x$ (for example, Caswell, 2001), and this result can be used to identify evolutionary equilibria. If a mutation has an effect in only one of the states, like the females in our example, there is further simplification in that only one column of $B(x', x)$ depends on x'. It follows that evolutionary change through small mutational steps in the sex ratio example can be described as if females were selected to maximize the expected reproductive value per offspring, given by $V(x', x) = \frac{1}{2}(1 - x')/(1 - x) + \frac{1}{2}x'/x$. Payoff functions having this form were introduced by Shaw and Mohler (1953), in what may have been the first worked out game-theoretical argument in biology. As we have seen, analysis of such payoff functions corresponds to an analysis of mutant invasion in a stationary population.

The concept of reproductive value was introduced by Fisher (1930) and plays an important role in the very successful field of sex ratio theory (Charnov, 1982; Pen and Weissing, 2002), as well as in evolutionary theory in general (McNamara and Houston, 1996; Houston and McNamara, 1999; Grafen, 2006). The concept is useful to represent payoffs in games played in populations in stationary environments. Reproductive value can be regarded as a Darwinian representation of the concept of utility in economics. For populations exposed to large-scale environmental fluctuations, as well as for those with limit-cycle or chaotic attractors of the population dynamics, concepts similar to reproductive value have proven less useful. In such situations, one needs the more general approach of explicitly considering evolutionary dynamics for populations of players of strategies. There are several influential approaches to the study of evolutionary dynamics in biology (Nowak and Sigmund, 2004), ranging from replicator dynamics (see DETERMINISTIC EVOLUTIONARY DYNAMICS) and adaptive dynamics (Metz, Nisbet and Geritz, 1992; Metz et al., 1996; Hofbauer and Sigmund, 1998) to the traditional modelling styles of population genetics and quantitative genetics (Rice, 2004). These approaches make different assumptions about such things as the underlying genetics and the rate and distribution of mutation. Recent years have seen an increasing emphasis on explicitly dynamical treatments in evolutionary theory.

Are there mixed strategies in nature?

Biologists have wondered how individuals, as players of a game, come to play one strategy or another. For life-history strategies, involving choices between alternative phenotypes, a population containing a mixture of phenotypes could be the result of randomization at the level of an individual, which corresponds to a mixed strategy, or there could be a genetic polymorphism of pure strategies (Maynard Smith, 1982). These two possibilities can be contrasted with a third, where individuals (or their parents) use information about themselves or their local environment to make life-history choices, which could correspond to a conditional strategy in a Bayesian game. The general question is related to the issue of purification of mixed strategy equilibria in game theory (see PURIFICATION). When observing populations that are mixtures of discrete phenotypes, biologists have tried to establish if one of the above three possibilities applies and, if so, what the evolutionary explanation might be. This question has been asked, for instance, about the phenomenon of alternative reproductive strategies (Gross, 1996; Shuster and Wade, 2003), like the jack and hooknose males in coho salmon (Gross, 1985) or the winged and wingless males in fig wasps (Hamilton, 1979). Since there are likely to be a number of factors that influence the relative success of reproductive alternatives and could be known to a developing individual – for instance, its juvenile growth rate and thus its potential adult size – one might expect some form of conditional strategy to evolve. This expectation agrees with the observation that conditional determination is common (Gross, 1996). There are also instances of genetic determination of reproductive alternatives (Shuster and Wade, 2003) but, somewhat surprisingly, there is as yet no empirically confirmed case of a mixed strategy of this kind. Perhaps it has been difficult for evolution to construct a well-functioning randomization device, leaving genetic polymorphism as a more easily achieved evolutionary outcome.

Evolution of cooperation

Among the various applications of game theory in biology, the evolution of cooperation is by far the most studied issue. This great interest is based on the belief that cooperation has played a crucial role in the evolution of biological organization, from the structure of chromosomes, cells and organisms to the level of animal societies. An extreme form of cooperation is that of the genes operating in an organism. Several thousand genes coordinate and direct cellular activities that in the main serve the well-being of their organism. Kin selection (Hamilton, 1964), which predicts that agents have an evolutionary interest in assisting their genetic relatives, cannot be the main explanation for this cooperation, since the different genes in an organism are typically not closely related by descent (except for a given gene in one cell and its copies in other cells). It is instead division of labour that is the principle that unites the parts of an organism into a common interest, of sufficient strength to make it evolutionarily unprofitable for any one gene to abandon its role in the organism for its own advantage. There are of course exceptions, in the form of selfish genetic elements, but these represent a minority of cases (Burt and Trivers, 2006).

Trivers (1971) and Axelrod and Hamilton (1981) promoted the idea that many of the features of the interactions between organisms would find an explanation in the give and take of direct reciprocation. In particular, the strategy of tit for tat (Axelrod and Hamilton, 1981) for the repeated Prisoner's Dilemma game was thought to represent a general mechanism for reciprocity in cooperative interactions and received much attention from biologists. On the whole, this form of direct reciprocity has subsequently failed to be supported by empirical observation. Two reasons for this failure have been proposed (Hammerstein, 2003). One is that the structure of real biological interactions differs in important ways from the original theoretical assumptions of a repeated game. The other is that the proposed strategies, like tit for tat, are unlikely to be reached by evolutionary change in real organisms, because they correspond to unlikely behavioural mechanisms. In contrast to reciprocity, both the influence of genetic relatedness through kin selection (Hamilton, 1964) and the presence of direct fitness benefits to cooperating individuals have relatively strong empirical support. Division of labour and the direct advantages of the trading of benefits between agents are likely to be crucial ingredients in the explanation of cooperation between independent organisms. The idea of a market, where exchanges take place, is thus relevant in both biology and economics (Noe and Hammerstein, 1994).

Evolution of signalling

Signals are found in a wide variety of biological contexts, for instance in aggressive interactions, parent–offspring interactions, and in connection with mate choice. There is now a fairly well developed set of theories about biological signals (Maynard Smith and Harper, 2003). One of the most influential ideas in the field is Zahavi's handicap principle (Zahavi, 1975). It states that a signal can reliably indicate high quality of the signaller only if the signal is costly, to the extent that it does not pay low-quality individuals to display the signal. The idea can be seen as a non-mathematical version of Spence's signalling theory (Spence, 1973; 1974), but, because biologists, including Zahavi, were unaware of Spence's work in economics, Zahavi's principle remained controversial in biology until Grafen (1991) provided a game-theoretical justification. The turn of events illustrates that biologists might have benefited from being more aware of theoretical developments in economics.

An example where Zahavi's handicap principle could apply is female mate choice in stalk-eyed flies (David et al., 2000). Males of stalk-eyed flies have long eye stalks, increasing the distance between their eyes, which is likely to be an encumbrance in their day-to-day life. A high level of nutrition, but also the possession of genes for high phenotypic quality, cause males to develop longer eye stalks. Female stalk-eyed flies prefer to mate with males with eyes that are far apart, and in this way their male offspring have a greater chance of receiving genes for long eye stalks. Female choice will act to reduce genetic variation in males, but if a sufficiently broad range of genetic loci can influence eye-stalk length, because they have a general effect on the phenotypic quality of a male, processes like deleterious mutation could maintain a substantial amount of genetic variation. In this way, signalling theory can explain the evolution of elaborate male ornaments, together with a mating preference for these ornaments in females, illustrating the power of game-theoretical arguments to increase our understanding of biological phenomena.

Learning

Viewing strategies as genetically coded entities on which natural selection operates, with evolutionarily stable strategies as endpoints of evolutionary change, is not the only game-theoretical perspective that is of relevance in biology. For many categories of behaviour, learning or similar adjustment processes are important in shaping the distribution of strategies in a population. For instance, when animals search for food or locate suitable living quarters, they may have the opportunity to evaluate the relative success of different options and to adjust their behaviour accordingly. A well-studied example is the so-called producer–scrounger game, for which there are experiments with birds that forage in groups on the ground (Barnard and Sibley, 1981; Giraldeau and Caraco, 2000). The game is played by a group of foragers and consists of a number of rounds. In each round an individual can choose between two behavioural options. Producers search for and utilize new food sources, and scroungers exploit food found by producers. The game presupposes that the activities of producing and scrounging are incompatible and cannot be performed simultaneously, which is experimentally supported (Coolen, Giraldeau and Lavoie, 2001). The payoffs to the options, measured as the expected food intake per round, depend on the frequencies of the options in the group. For instance, scrounging is most profitable to an individual if no one else scrounges and yields a lower payoff with more scrounging in the group. By specifying the details of the model, one can compute an equilibrium probability of scrounging at which the payoffs to producing and scrounging are equal. This equilibrium is influenced by parameters like expected search times and the amount of food found in a new location. It has been experimentally verified that groups of spice finches converge on such an equilibrium over a period of a few days of foraging (Mottley and Giraldeau, 2000; Giraldeau and Caraco, 2000). It is not known precisely which rules are used by individuals in these experiments to modify their behaviour, but such rules are likely to play an important role in shaping behaviour in many animals, including humans (see LEARNING AND EVOLUTION IN GAMES: ADAPTIVE HEURISTICS). The study of these kinds of adjustments of behaviour could therefore represent an important area of overlap between biology and economics.

OLOF LEIMAR

See also **deterministic evolutionary dynamics; learning and evolution in games: adaptive heuristics; learning and evolution in games: ESS; mixed strategy equilibrium; purification.**

Bibliography

Axelrod, R. and Hamilton, W.D. 1981. The evolution of cooperation. *Science* 211, 1390–6.

Barnard, C.J. and Sibley, R.M. 1981. Producers and scroungers: a general model and its application to captive flocks of house sparrows. *Animal Behaviour* 29, 543–550.

Burt, A. and Trivers, R. 2006. *Genes in Conflict: The Biology of Selfish Genetic Elements*. Cambridge: Harvard University Press.

Caswell, H. 2001. *Matrix Population Models*, 2nd edn. Sunderland: Sinauer.

Charnov, E.L. 1982. *The Theory of Sex Allocation*. Princeton: Princeton University Press.

Coolen, I., Giraldeau, L.-A. and Lavoie, M. 2001. Head position as an indicator of producer and scrounger tactics in a ground-feeding bird. *Animal Behaviour* 61, 895–903.

Darwin, C. 1874. *The Descent of Man and Selection in Relation to Sex*, 2nd edn. London: Murray.

David, P., Bjorksten, T., Fowler, K. and Pomiankowski, A. 2000. Condition-dependent signalling of genetic variation in stalk-eyed flies. *Nature* 406, 186–8.

Düsing, C. 1884. *Die Regulierung des Geschlechtsverhältnisses*. Jena: Fischer.

Edwards, A.W.F. 2000. Carl Düsing 1884 on the regulation of the sex-ratio. *Theoretical Population Biology* 58, 255–7.

Fisher, R.A. 1930. *The Genetical Theory of Natural Selection*. Oxford: Clarendon Press.

Giraldeau, L.-A. and Caraco, T. 2000. *Social Foraging Theory*. Princeton: Princeton University Press.

Grafen, A. 1991. Biological signals as handicaps. *Journal of Theoretical Biology* 144, 517–46.

Grafen, A. 2006. A theory of Fisher's reproductive value. *Journal of Mathematical Biology* 53, 15–60.

Gross, M.R. 1985. Disruptive selection for alternative life histories in salmon. *Nature* 313, 47–8.

Gross, M.R. 1996. Alternative reproductive strategies and tactics: diversity within sexes. *Trends in Ecology & Evolution* 11, 92–8.

Hamilton, W.D. 1964. The genetical evolution of social behaviour, I, II. *Journal of Theoretical Biology* 7, 1–52.

Hamilton, W.D. 1979. Wingless and fighting males in fig wasps and other insects. In *Reproductive Competition, Mate Coice and Sexual Selection in Insects*, ed. M.S. Blum, and N.A. Blum. New York: Academic Press.

Hammerstein, P. 2003. Why is reciprocity so rare in social animals? A protestant appeal. In *Genetic and Cultural Evolution of Cooperation*, ed. P. Hammerstein. Cambridge: MIT Press.

Hammerstein, P. and Hagen, E.H. 2005. The second wave of evolutionary economics in biology. *Trends in Ecology & Evolution* 20, 604–9.

Hofbauer, J. and Sigmund, K. 1998. *Evolutionary Games and Population Dynamics*. Cambridge: Cambridge University Press.

Houston, A.I. and McNamara, J.M. 1999. *Models of Adaptive Behaviour: An Approach Based on State*. Cambridge: Cambridge University Press.

Maynard Smith, J. 1982. *Evolution and the Theory of Games*. Cambridge: Cambridge University Press.

Maynard Smith, J. and Harper, D. 2003. *Animal Signals*. Oxford: Oxford University Press.

Maynard Smith, J. and Parker, G.R. 1976. The logic of asymmetric contests. *Animal Behaviour* 24, 159–75.

Maynard Smith, J. and Price, G.R. 1973. The logic of animal conflict. *Nature* 246, 15–8.

McNamara, J.M. and Houston, A.I. 1996. State-dependent life histories. *Nature* 380, 215–21.

Metz, J.A.J., Nisbet, R.M. and Geritz, S.A.H. 1992. How should we define 'fitness' for general ecological scenarios? *Trends in Ecology & Evolution* 7, 198–202.

Metz, J.A.J., Geritz, S.A.H., Meszéna, G., Jacobs, F.J.A. and van Heerwaarden, J.S. 1996. Adaptive dynamics, a geometrical study of nearly faithful reproduction. In *Stochastic and Spatial*

Structures of Dynamical Systems. Proceedings of the Royal Dutch Academy of Science (KNAV Verhandelingen), ed. S.J. van Strien, and S.M. Verduyn Lunel. Amsterdam: North-Holland.

Mottley, K. and Giraldeau, L.-A. 2000. Experimental evidence that group foragers can converge on predicted producer-scrounger equilibria. *Animal Behaviour* 60, 341–50.

Noe, R. and Hammerstein, P. 1994. Biological markets: supply and demand determine the effect of partner choice in cooperation, mutualism and mating. *Behavioral Ecology and Sociobiology* 35, 1–11.

Nowak, M.A. and Sigmund, K. 2004. Evolutionary dynamics of biological games. *Science* 303, 793–9.

Pen, I. and Weissing, F.J. 2002. Optimal sex allocation: steps towards a mechanistic theory. In *Sex Ratios – Concepts and Research Methods*, ed. I.C.W. Hardy. Cambridge: Cambridge University Press.

Rice, S.H. 2004. *Evolutionary Theory – Mathematical and Conceptual Foundations*. Sunderland: Sinauer.

Shaw, R.F. and Mohler, J.D. 1953. The selective significance of the sex ratio. *American Naturalist* 87, 337–42.

Shuster, S.M. and Wade, M.J. 2003. *Mating Systems and Strategies*. Princeton: Princeton University Press.

Sigmund, K. 2005. John Maynard Smith and evolutionary game theory. *Theoretical Population Biology* 68, 7–10.

Spence, M. 1973. Job market signaling. *Quarterly Journal of Economics* 87, 355–74.

Spence, M. 1974. *Market Signaling*. Cambridge: Harvard University Press.

Trivers, R.L. 1971. The evolution of reciprocal altruism. *Quarterly Review of Biology* 46, 35–57.

Zahavi, A. 1975. Mate selection – a selection for a handicap. *Journal of Theoretical Biology* 53, 205–14.

games in coalitional form

1. Introduction

In their seminal book, von Neumann and Morgenstern (1944) introduced two theories of games: strategic and coalitional. Strategic game theory concentrates on the selection of strategies by payoff-maximizing players. coalitional game theory concentrates on coalition formation and the distribution of payoffs.

The next two examples illustrate situations in the domain of the coalitional approach.

1.1 Games with no strategic structure

Example 1 ***Cost allocation of a shared facility***. *Three municipalities, E, W, and S, need to construct water purification facilities. Costs of individual and joint facilities are described by the cost function c: $c(E)=20$, $c(W)=30$, and $c(S)=50$; $c(E, W)=40$, $c(E, S)=60$, and $c(W, S)=80$; $c(E, W, S)=80$. For example, a facility that serves the needs of W and S would cost \$80 million.*

The optimal solution is to build, at the cost of 80, one facility that serves all three municipalities. How should its cost be allocated?

1.2 Games with many Nash equilibria

Example 2 ***Repeated sales***. *A seller and a buyer play the following stage game on a daily basis. The seller decides on the quality level, H, M, or L, of the item sold (at a fixed price); without knowledge of the seller's selected quality, the buyer decides whether or not to buy. If she does not buy, the payoffs of both are zero; if she buys, the corresponding payoffs are (0, 3), (3, 2) or (4, 0), depending on whether the quality is H, M, or L.*

Under perfect monitoring of past choices and low discounting of future payoffs, the folk theorem of repeated games states that any pair of numbers in the convex hull of (0, 0), (0, 3), (3, 2), and (4, 0) are Nash-equilibrium average payoffs. What equilibrium and what average payoffs should they select?

We proceed with a short survey of the major models and selected solution concepts. More elaborate overviews are available in GAME THEORY, Myerson (1991), and other surveys mentioned below.

2. Types of coalitional game

In what follows, N is a fixed set of n *players*; the set of *coalitions* $\mathscr{C}$ consists of the nonempty subsets of N; $|S|$ denotes the number of players in a coalition S. The terms 'profile' and 'S-profile' denote vectors of items (payoffs, costs, commodities, and so on) indexed by the names of the players.

For every coalition S, R^S denotes the $|S|$-dimensional Euclidean space indexed by the names of the players; for single-player coalitions the symbol i replaces $\{i\}$. A profile $u^S \in R^S$ denotes *payoffs* u_i^S of the players $i \in S$.

Definition 1 *An (n person) game (also known as a game with no transferable utility, or NTU game) is a function V that assigns every coalition S a set* $V(S) \subset R^s$.

Remark 1 *The initial models of coalitional games were presented in von Neumann and Morgenstern (1944) for the special case of TU games described below, Nash (1950) for the special case of two-person games, and Aumann and Peleg (1960) for the general case.*

The interpretation is that $V(S)$ describes all the *feasible payoff profiles* that the coalition S can generate for its members. Under the assumption that the *grand coalition* N is formed, the central question is which payoff profile $u^N \in V(N)$ to select. Two major considerations come into play: the relative strength of different coalitions, and the relative strength of players within coalitions.

To separate these two issues, game theorists study the two simpler types of games defined below: TU games and bargaining games. In TU games the players in every coalition are symmetric, so only the relative strength of coalitions matters. In bargaining games only one coalition is active, so only the relative strength of players' within that coalition matters. Historically, solutions of games have been developed first for these simpler classes of games, and only then extended to general (NTU) games. For this reason, the literature on these simpler classes is substantially richer than the general theory of (NTU) games.

Definition 2 *V is a transferable-utility game (TU game) if for a real-valued function* $v = (v(S))_{S \in \mathscr{C}}$, $V(S) = \{u^S \in R^S : \sum_i u_i^S \leq v(S)\}$.

It is customary to identify a TU game by the function v instead of V.

TU games describe many interactive environments. Consider, for example, any environment with individual outcomes consisting of prizes p and monetary payoffs m, and individual utilities that are additive and separable in money ($u_i(p, m) = v_i(p) + m$). Under the assumption that the players have enough funds to make transfers, the TU formulation presents an accurate description of the situation.

Definition 3 *A Nash (1950) bargaining game is a two-person game. An n-person bargaining game is a game V in which* $V(S) = \times_{i \in S} V(i)$ *for every coalition* $S \subsetneq N$.

Remark 2 Partition games (Lucas and Thrall, 1963) *use a more sophisticated function V to describe coalitional payoffs. For every partition of the set of players* $\pi = (T_1, T_2, \ldots, T_m)$, $V_\pi(T_j)$ *is the set of* T_j*'s feasible payoff profiles, under the cooperation structure described by* π. *Thus, what is feasible for a coalition may depend on the strategic alignment of the opponents. The literature on partition games is not highly developed.*

3. Some special families of games

Coalitional game theory is useful for analysing special types of interactive environments. And conversely, such special environments serve as a laboratory to test the usefulness of game theoretic solutions. The following are a few examples.

3.1 Profit sharing and cost allocation

Consider a partnership that needs to distribute its total profits, $v(N)$, to its n individual partners. A profit-distribution formula should consider the potential profits $v(S)$ that coalitions of partners S can generate on their own. A TU game is a natural description of the situation.

A cost allocation problem, like Example 1, can be turned into a natural TU game by defining the worth of a coalition to be the savings obtained by joining forces: $v(S) = \sum_{i \in S} c(i) - c(S)$.

Examples of papers on cost allocation are Shubik (1962) and Billera, Heath and Raanan (1978). See Young (1994) for an extensive survey.

3.2 Markets and auctions

Restricting this discussion to simple exchange, consider an environment with n traders and m commodities. Each trader i starts with an initial *bundle* ω_i^0, an m-dimensional vector that describes the quantities of each commodity he owns. The utility of player i for a bundle ω_i is described by $u_i(\omega_i)$. An S-profile of bundles $\omega = (\omega_i)_{i \in S}$ is feasible for the coalition S if $\sum_{i \in S} \omega_i = \sum_{i \in S} \omega_i^0$.

Definition 4 *A game V is a market game, if for such an exchange environment (with assumed free-disposal of utility),* $V(S)=\{u^S \in R^S$: *for some S-feasible profile of bundles* ω, $u_i^S \leq u_i(\omega_i)$ *for every* $i \in S\}$.

Under the assumptions discussed earlier (additively separable utility and sufficient funds) the market game has the more compact TU description: $v(S) = max_\omega \sum_{i \in S} u_i(\omega_i)$, with the max taken over all S-feasible profiles ω.

As discussed below, market games play a central role in several areas of game theory.

Definition 5 *An auction game is a market game with a seller whose initial bundle consists of items to be sold, and bidders whose initial bundles consist of money.*

3.3 Matching games

Many theoretical and empirical studies are devoted to the subject of efficient and stable matching: husbands with wives, sellers with buyers, students with schools, donors with receivers, and more. The first of these was introduced by Gale and Shapley in their pioneering study (1962) using the following example.

Consider a matching environment with q males and q females. Payoff functions $u_m(f)$ and $u_m(none)$ describe the utilities of male m paired with female f or with no one; $u_f(m)$ and $u_f(none)$ describe the corresponding utilities of the females. A pairing P_S of a coalition S is a specification of male-female pairs from S, with the remaining S members being unpaired.

Definition 6 *A game V is a marriage game if for such an environment,* $V(S)=\{u^S \in R^S$: *for some pairing* P_S, $u_i^S \leq u_i(P_S)$ *for every* $i \in S\}$.

Solutions of marriage games that are efficient and stable (that is, no divorce) can be computed by Gale–Shapley algorithms.

3.4 Optimization games

Optimization problems from operations research have natural extensions to multiperson coalitional games, as the following examples illustrate.

3.4.1 Spanning-tree games

A cost-allocation TU spanning-tree game (Bird, 1976) is described by an undirected connected graph, with one node designated as the centre C and every other node corresponding to a player. Every arc has an associated nonnegative connectivity cost. The cost of a coalition S, $c(S)$, is defined to be the minimum sum of all the arc costs, taken over all subgraphs that connect all the members of S to C.

3.4.2 Flow games

A TU flow game (Kalai and Zemel, 1982b) is described by a directed graph, with two nodes, s and t, designated as the *source* and the *sink*, respectively. Every arc has an associated capacity and is owned by one of the n players. For every coalition S, $v(S)$ is the maximal s-to-t flow that the coalition S can generate through the arcs owned by its members.

3.4.3 Linear programming games

Finding minimal-cost spanning trees and maximum flow can be described as special types of linear programmes. Linear (and nonlinear) programming problems have been generalized to multiperson games (see Owen, 1975; Kalai and Zemel, 1982a; Dubey and Shapley, 1984). The following is a simple example.

Fix a $p \times q$ matrix A and a q-dimensional vector w, to consider standard linear programmes of the form max wx s.t. $Ax \leq b$. Endow each player i with a p-dimensional vector b_i, and define the linear-programming TU game v by $v(S) = max_x wx$ s.t. $Ax \leq \sum_{i \in S} b_i$.

3.5 Simple games and voting games

A TU game is simple if for every coalition S, $v(S)$ is either zero or 1. Simple games are useful for describing the power of coalitions in political applications. For example, if every player is a party in a certain parliament, then $v(S) = 1$ means that under the parliamentary rules the parties in the coalition S have the ability to pass legislation (or *win*) regardless of the positions of the parties not in S; $v(S) = 0$ (or S *loses*) otherwise.

In applications like the one above, just formulating the game may already offer useful insights into the power structure. For example, consider a parliament that requires 50 votes in order to pass legislation, with three parties that have 12 votes, 38 votes, and 49 votes, respectively. Even though the third party seems strongest, a simple formulation of the game yields the symmetric *three-person majority game*: any coalition with two or more parties wins; single-party coalitions lose.

Beyond the initial stage of formulation, standard solutions of game theory offer useful insights into the power structure of such institutions and other political structures (see, for example, Shapley and Shubik, 1954; Riker and Shapley, 1968; Brams, Lucas and Straffin, 1983).

4. Solution concepts

When cooperation is beneficial, which coalitions will form and how would coalitions allocate payoffs to their members? Given the breadth of situations for which this question is relevant, game theory offers several different solutions that are motivated by different criteria. In this brief survey, we concentrate on the Core and on the Shapley value.

Under the assumptions that utility functions can be rescaled, that lotteries over outcomes can be performed, and that utility can be freely disposed of, we restrict the discussion to games V with the following properties.

Every $V(S)$ is a compact convex subset of the nonnegative orthant R^S_+, and it satisfies the following property: if $w^S \in R^S_+$ with $w^S \le u^S$ for some $u^S \in V(S)$, then $w^S \in V(S)$. And for single player coalitions, assume $V(i) = \{0\}$. For TU games this means that every $v(S) \geq 0$, the corresponding $V(S) = \{u^S \in R^S_+ : \sum_{i \in S} u^S_i \leq v(S)\}$, and for each i, $v(i) = 0$.

In addition, we assume that the games are *superadditive*: for any pair of disjoint coalitions T and S, $V(T \cup S) \supseteq V(T) \times V(S)$; for TU games this translates to $v(T \cup S) \geq v(T) + v(S)$. Under superadditivity, the maximal possible payoffs are generated by the grand coalition N. Thus, the discussion turns to how the payoffs of the grand coalition should be allocated, ignoring the question of which coalitions would form.

A payoff profile $u \in R^N$ is *feasible for a coalition* S, if $u^S \in V(S)$, where u^S is the projection of u to R^S. The translation to TU games is that $u(S) \equiv \sum_{i \in S} u_i \leq v(S)$. A profile $u \in R^N$ can be *improved upon* by the coalition S if there is an S-feasible profile w with $w_i > u_i$ for all $i \in S$.

Definition 7 *An imputation of a game is a grand-coalition-feasible payoff profile that is both individually rational (that is, no individual player can improve upon it) and Pareto optimal (that is, the grand coalition cannot improve upon it).*

Given the uncontroversial nature of individual rationality and Pareto optimality, solutions of a game are restricted to the selection of imputations.

4.1 The core

Definition 8 *The* core *of a game (see Shapley, 1952, and Gillies, 1953, for TU, and Aumann, 1961, for NTU)* is the set of imputations that cannot be improved upon by any coalition.

The core turns out to be a compact set of imputations that may be empty. In the case of TU games it is a convex set, but in general games (NTU) it may even be a disconnected set. The core induces stable cooperation in the grand coalition because no sub-coalition of players can reach a consensus to break away when a payoff profile is in the core.

Remark 3 *More refined notions of stability give rise to alternative solution concepts, such as the stable sets of von Neumann and Morgenstern (1944), and the kernel and bargaining sets of Davis and Maschler (1965). The nucleolus of Schmeidler (1969), with its NTU extension in Kalai (1975), offers a 'refinement' of the core. It consists of a finite number of points (exactly one for TU games) and belongs to the core when the core is not empty. For more on these solutions, see Maschler (1992) and* GAME THEORY.

Unfortunately, games with an empty core are not unusual. Even the simple three-person majority game described in Subsection 3.5 has an empty core (since among any three numbers that sum to one there must be a pair that sums to less than one, there are always two players who can improve their payoffs).

4.1.1 TU games with nonempty cores

Given the coalitional stability obtained under payoff profiles in the core, it is desirable to know in which games the core is nonempty.

Bondareva (1963) and Shapley (1967) consider 'part-time coalitions' that meet the availability constraints of their members. In this sense, a collection of nonnegative coalitional weights $\lambda = (\lambda_S)_{S \in \mathscr{C}}$ is *balanced*, if for every player i, $\sum_{S: i \in S} \lambda_S = 1$. They show that a game has a nonempty core if and only if the game is *balanced*: for every balanced collection λ, $\sum_S \lambda_S v(S) \leq v(N)$.

As Scarf (1967) demonstrates, all market games have nonempty cores and even the stronger property of having nonempty subcores: for every coalition S, consider the subgame v_S which is restricted to the players of S and their subcoalitions. The game v *has nonempty subcores*, if all its subgames v_S have nonempty cores.

By applying the balancedness condition repeatedly, one concludes that a game has nonempty subcores if and only if the balancedness condition holds for all its subgames v_S. Games with this property are called *totally balanced.*

Since Shapley and Shubik (1969a) demonstrate the converse of Scarf's result, a game is thus totally balanced if and only if it is a market game. Interestingly, the following description offers yet a different characterization of this family of games.

A game w is *additive* if there is a profile $u \in R^N$ such that for every coalition S, $w(S) = \sum_{i \in S} u_i$. A game v is the *minimum of a finite collection of games* (w^r) if for every coalition S, $v(S) = min_r w^r(S)$.

Kalai and Zemel (1982b) show that a game has nonempty subcores if and only if it is the minimum of a finite collection of additive games. Moreover, a game is such a minimum if and only if it is a flow game (as defined in Subsection 3.4.2).

In summary, a game v in this important class of TU games can be characterized by any of the following five equivalent statements: (1) v has nonempty subcores, (2) v is totally balanced, (3) v is the minimum of additive games, (4) v is a market game, (5) v is a flow game.

Scarf (1967), Billera and Bixby (1973), and the follow-up literature extend some of the results above to general (NTU) games.

4.2 The Shapley TU value

Definition 9 *The Shapley (1953) value of a TU game v is the payoff allocation $\varphi(v)$ defined by*

$$\varphi_i(v) = \sum_{S:i\in S} \tfrac{(|S|-1)!(|N|-|S|)!}{N!}[v(S) - v(S\backslash i)].$$

This expression describes the expected marginal contribution of player i to a random coalition. To elaborate, imagine the players arriving at the game in a random order. When player i arrives and joins the coalition of earlier arrivers S, he is paid his *marginal contribution* to that coalition, that is, $v(S\cup i)-v(S)$. His Shapley value $\varphi_i(v)$ is the expected value of this marginal contribution when all orders of arrivals are equally likely.

Owen (1972) describes a parallel continuous-time process in which each player arrives at the game gradually. Owen extends the payoff function v to coalitions with 'fractionally present' players, and considers the instantaneous marginal contributions of each player i to such fractional coalitions. The Shapley value of player i is the integral of his instantaneous marginal contributions, when all the players arrive simultaneously at a constant rate over the same fixed time interval.

This continuous-time arrival model, when generalized to coalitional games with infinitely many players, leads to the definition of *Aumann–Shapley prices*. These are useful for the allocation of production costs to different goods produced in a nonseparable joint production process (see Tauman, 1988; Young, 1994).

A substantial literature is devoted to extensions and variations of the axioms that Shapley (1953) used to justify his value. These include extensions to infinitely many players and to general (NTU) games (discussed briefly below), and to nonsymmetric values (see Weber, 1988; Kalai and Samet, 1987; Levy and McLean, 1991;Monderer and Samet, 2002, among others).

Is the Shapley value in the core of the game? Not always. But as Shapley (1971) shows, if the game is *convex*, meaning that $v(S\cup T) + v(S\cap T) \geq v(S) + v(T)$ for every pair of coalitions S and T, then the Shapley value and all the $n!$ profiles of marginal contributions (obtained under different orders of arrival) are in the core. Moreover, Ichiishi (1981) shows that the converse is also true.

We will turn to notions of value for NTU games after we describe solutions to the special case of two-person NTU games, that is, the Nash bargaining problem.

4.3 Solutions to Nash bargaining games

Nash (1950) pioneered the study of NTU games when he proposed a model of a two-person bargaining game and, using a small number of appealing principles, axiomatized the solution below.

Fix a two-person game V and for every imputation u define the *payoff gain* of player i by $gain_i(u)=u_i-v(i)$, with $v(i)$ being the highest payoff that player i can obtain on his own, that is, in his $V(i)$.

Definition 10 *The Nash bargaining solution is the unique imputation u that maximizes the product of the gains of the two players,* $gain_1(u) \cdot gain_2(u)$.

Twenty-five years later, Kalai and Smorodinsky (1975) and others showed that other appealing axioms lead to alternative solutions, like the two defined below.

The *ideal gain* of player i is $I_i = max_u gain_i(u)$, the maximum taken over all imputations u.

Definition 11 *The Kalai–Smorodinsky solution is the unique imputation u with payoff gains proportional to the players' ideal gains,* $gain_1(u)/gain_2(u) = I_1/I_2$.

Definition 12 *The egalitarian solution of Kalai (1977a) is the unique imputation u that equalizes the gains of the players,* $gain_1(u) = gain_2(u)$.

For additional solutions, including these of Raiffa (1953) and Maschler and Perles (1981), see the comprehensive surveys of Lensberg and Thomson (1989) and Thomson (1994).

4.4 Values of NTU games

Three different extensions of the Shapley TU value have been proposed for NTU games: the *Shapley value* (extension), proposed by Shapley (1969) and axiomatized by Aumann (1985); the *Harsanyi value*, proposed by Harsanyi (1963) and axiomatized by Hart (1985); and the *egalitarian value*, proposed and axiomatized by Kalai and Samet (1985).

All three proposed extensions coincide with the original Shapley value on the class of TU games. For the class of NTU bargaining games, however, the (extended) Shapley value and the Harsanyi value coincide with the Nash bargaining solution, while the egalitarian value coincides with the egalitarian bargaining solution.

For additional material (beyond the brief discussion below) on these and related solutions, see McLean (2002).

4.5 Axiomatic characterizations of solutions

The imposition of general principles, or axioms, often leads to a unique determination of a solution. This approach is repeatedly used in game theory, as illustrated by the short summary below.

4.5.1 Nash's axioms

Nash (1950) characterizes his bargaining solution by the following axioms: individual rationality, symmetry, Pareto optimality, invariance to utility scale, and independence of irrelevant alternatives (IIA).

Invariance to utility scale means that changing the scale of the utility of a player does not change the solution. But this axiom goes further by disallowing all methods that use information extraneous to the game, even if such methods are invariant to scale.

Nash's *IIA* axiom requires that a solution that remains feasible when other payoff profiles are removed from the feasible set should not be altered.

4.5.2 Shapley's axioms

Shapley (1953) characterizes his TU value by the following axioms: symmetry, Pareto optimality, additivity, and dummy player.

A value is *additive* if in a game that is the sum of two games, the value of each player equals the sum of his values in the two component games.

A *dummy player*, that is, one who contributes nothing to any coalition, should be allocated no payoff.

4.5.3 Monotonicity axioms

Monotonicity axioms describe notions of fairness and induce incentives to cooperate. The following are a few examples.

Kalai and Smorodinsky (1975) characterize their bargaining solution using *individual monotonicity*: a player's payoff should not be reduced if the set of imputations is expanded to improve his possible payoffs.

Kalai (1977) and Kalai and Samet (1985) characterize their egalitarian solutions using *coalitional monotonicity*: expanding the feasible set of one coalition should not reduce the payoffs of any of its members.

Thomson (1983) uses *population monotonicity* to characterize the n-person Kalai-Smorodinsky solution: in dividing fixed resources among n players, no player should benefit if more players are added to share the same resources.

Maschler and Perles (1981) characterize their bargaining solution using *super-additivity* (used also in Myerson, 1977a): if a bargaining problem is to be randomly drawn, all the players benefit by reaching agreement prior to knowing the realized game.

Young (1985) shows that Shapley's TU additivity axiom can be replaced by *strong monotonicity*: a player's payoff can only depend on his marginal contributions to his coalitions, and it has to be monotonically nondecreasing in these.

4.5.4 Axiomatizations of NTU values

The NTU Shapley value is axiomatized in Aumann (1985) by adapting Shapley's TU axioms to the NTU setting, and combining them with Nash's IIA axiom. Different adaptations lead to an axiomatization of the Harsanyi (1963) value, as illustrated in Hart (1985). Kalai and Samet (1985) use coalitional monotonicity and a weak version of additivity to axiomatize the NTU egalitarian value.

For more information on axiomatizations of NTU values, see McLean (2002).

4.5.5 Consistency axioms

Consistency axioms relate the solution of a game to the solutions of 'subgames' obtained when some of the players leave the game with their share of the payoff. Authors who employ consistency axioms include: Davis and Maschler (1965) for the bargaining set, Peleg (1985; 1986; 1992) for the core, Lensberg (1988) for the Nash n-person bargaining solution, Kalai and Samet (1987) and Levy and McLean (1991) for TU- and NTU-weighted Shapley values, Hart and Mas-Colell (1989) for the TU Shapley value, and Bhaskar and Kar (2004) for cost allocation in spanning trees.

5. Bridging strategic and coalitional models

Several theoretical bridges connect strategic and coalitional models. Aumann (1961) offers two methods for reducing strategic games to coalitional games. Such reductions allow one to study specific strategic games, such as repeated games, from the perspectives of various coalitional solutions, such as the core.

One substantial area of research is the Nash program, designed to offer strategic foundations for various coalitional solution concepts. In Nash (1953), he began by constructing a strategic bargaining procedure, and showing that the strategic solution coincides with the coalitional Nash bargaining solution. See NASH PROGRAM for a survey of the extensive literature that followed.

Network games and coalition formation are the subjects of a growing literature. Amending a TU game with a communication graph, Myerson (1977b) develops an appropriate extension of the Shapley value. Using this extended value, Aumann and Myerson (1988) construct a dynamic strategic game of links formation that gives rise to stable communication graphs. For a survey of the large follow-up literature in this domain, see NETWORK FORMATION.

Networks also offer a tool for the study of market structures. For example, Kalai, Postlewaite and Roberts (1979) compare a market game with no restrictions to a star-shaped market, where all trade must flow through one middleman. Somewhat surprisingly, their comparisons of the cores of the corresponding games reveal the existence of economies in which becoming a middleman can only hurt a player.

Recent studies of strategic models of auctions point to interesting connections with the coalitional model. For example, empirical observations suggest that the better-performing auctions are the ones with outcomes in the core of the corresponding coalitional game. For related references, see Bikhchandani and Ostroy (2006), De Vries, Schummer and Vohra (2007), and Day and Milgrom (2007).

6. Large cooperative games

When the number of players is large, the exponential number of possible coalitions makes the coalitional analysis difficult. On the other hand, in games with many players each individual has less influence and the laws of large numbers reduce uncertainties.

Unfortunately, the substantial fascinating literature on games with many players is too large to survey here, so the reader is referred to Aumann and Shapley (1974) and Neyman (2002) for the theory of the Shapley value of large games, and to Shapley and Shubik (1969a), Wooders and Zame (1984), Anderson (1992), Kannai (1992), and CORE CONVERGENCE for the theory of cores of large games.

A surprising discovery drawn from the above literature is a phenomenon unique to large market games that has become known as the equivalence theorem: when applied to large market games, the predictions of almost all (with the notable exception of the von Neumann–Morgenstern stable sets) major solution concepts (in both coalitional and strategic game theory) coincide. Moreover, they all prescribe the economic price equilibrium as the solution for the game. This theorem presents the culmination of many papers, including Debreu and Scarf (1963), Aumann (1964), Shapley (1964), Shapley and Shubik (1969a) and Aumann (1975).

7. Directions for future work

Consider, for example, the task of constructing of a profit-sharing formula for a large consulting firm that has many partners with different expertise, located in offices around the world. While a coalitional approach should be suitable for the task, several current shortcomings limit its applicability. These include:

1. *Incomplete information.* Partners may have incomplete differential information about the feasible payoffs of different coalitions. While coalitional game theory has some literature on this subject (see Harsanyi and Selten, 1972; Myerson, 1984, and the follow-up literature), it is not nearly as developed as its strategic counterpart.
2. *Dynamics.* Although the feasible payoffs of coalitions vary with time, coalitional game theory is almost entirely static.
3. *Computation.* Even with a moderate number of players, the information needed for describing a game is very demanding. The literature on the complexity of computing solutions (as in Deng and Papadimitriou, 1994; Nisan et al., 2007) is growing. But, overall, coalitional game theory is still far from offering readily computable solution concepts for complex problems like the profit-sharing formula in the situation described above.

Further research on the topics above would be an invaluable contribution to coalitional game theory.

EHUD KALAI

See also **Shapley value.**

Bibliography

The list below includes more than the relatively small number of papers discussed in this article, but due to space limitations many important contributions do not appear here.

Anderson, R.M. 1992. The core in perfectly competitive economies. In Aumann and Hart (1992).

Aumann, R.J. 1961. The core of a cooperative game without side payments. *Transactions of the American Mathematical Society* 98, 539–52.

Aumann, R.J. 1964. Markets with a continuum of traders. *Econometrica* 32, 39–50.

Aumann, R.J. 1975. Values of markets with a continuum of traders. *Econometrica* 43, 611–46.

Aumann, R.J. 1985. An axiomatization of the non-transferable utility value. *Econometrica* 53, 599–612.

Aumann, R.J. and Hart, S. 1992; 1994; 2002. *The Handbook of Game Theory with Economic Application*, vols 1, 2 and 3. Amsterdam: North-Holland.

Aumann, R.J. and Maschler, M. 1985. Game theoretic analysis of a bankruptcy problem from the Talmud. *Journal of Economic Theory* 36, 195–213.

Aumann, R.J. and Maschler, M. 1964. The bargaining set for cooperative games. In *Advances in Game Theory*, ed. M. Dresher, L.S. Shapley and A.W. Tucker. Princeton: Princeton University Press.

Aumann, R.J. and Myerson, R.B. 1988. Endogenous formation of links between players and coalitions: an application of the Shapley value. In *The Shapley Value*, ed. A. Roth. Cambridge: Cambridge University Press.

Aumann, R.J. and Peleg, B. 1960. Von Neumann–Morgenstern solutions to cooperative games without side payments. *Bulletin of the American Mathematical Society* 66, 173–9.

Aumann, R.J. and Shapley, L.S. 1974. *Values of Non-Atomic Games*. Princeton: Princeton University Press.

Bhaskar, D. and Kar, A. 2004. Cost monotonicity, consistency and minimum cost spanning tree games. *Games and Econ Behavior* 48, 223–48.

Bikhchandani, S. and Ostroy, J.M. 2006. Ascending price Vickrey auctions. *Games and Econ Behavior* 55, 215–41.

Billera, L.J. 1970a. Existence of general bargaining sets for cooperative games without side payments. *Bulletin of the American Mathematical Society* 76, 375–9.

Billera, L.J. 1970b. Some theorems on the core of an n-person game without side payments. *SIAM Journal of Applied Mathematics* 18, 567–79.

Billera, L.J. and Bixby, R. 1973. A characterization of polyhedral market games. *International Journal of Game Theory* 2, 253–61.

Billera, L.J., Heath, D.C. and Raanan, J. 1978. Internal telephone billing rates: a novel application of non-atomic game theory. *Operations Research* 26, 956–65.

Binmore, K. 1987. Nash bargaining theory III. In *The Economics of Bargaining*, ed. K. Binmore and P. Dasgupta. Oxford: Blackwell.

Binmore, K., Rubinstein, A. and Wolinsky, A. 1986. The Nash bargaining solution in economic modelling. *Rand Journal of Economics* 17, 176–88.

Bird, C.G. 1976. On cost allocation for a spanning tree: a game theoretic approach. *Networks* 6, 335–60.

Bondareva, O.N. 1963. Some applications of linear programming methods to the theory of cooperative games. *Problemy kibernetiki* 10, 119–39 [in Russian].

Brams, S.J., Lucas, W.F. and Straffin, P.D., Jr. 1983. *Political and Related Models*. New York: Springer.

Chun, Y. and Thomson, W. 1990. Nash solution and uncertain disagreement points. *Games and Economic Behavior* 2, 213–23.

Davis, M. and Maschler, M. 1965. The kernel of a cooperative game. *Naval Research Logistics Quarterly* 12, 223–59.

Day, R. and Milgrom, P. 2007. Core selecting package auctions. *International Journal of Game Theory* (forthcoming).

Debreu, G. and Scarf, H. 1963. A limit theorem on the core of an economy. *International Economic Review* 4, 236–46.

De Clippel, G., Peters, H. and Zank, H. 2004. Axiomatizing the Harsanyi solution, the symmetric egalitarian solution, and the consistent solution for NTU-games. *International Journal of Game Theory* 33, 145–58.

Deng, X. and Papadimitriou, C. 1994. On the complexity of cooperative game solution concepts. *Mathematics of Operations Research* 19, 257–66.

De Vries, S., Schummer, J. and Vohra, R.V. 2007. On ascending Vickrey auctions for heterogeneous objects. *Journal of Economic Theory* 132, 95–118.

Dubey, P. and Shapley, L.S. 1979. Some properties of the Banzhaf power index. *Mathematics of Operations Research* 4, 99–131.

Dubey, P. and Shapley, L.S. 1984. Totally balanced games arising from controlled programming problems. *Mathematical Programming* 29, 245–67.

Gale, D. and Shapley, L.S. 1962. College admissions and the stability of marriage. *American Mathematical Monthly* 69, 9–15.

Gillies, D.B. 1953. Some theorems on n-person games. Ph.D. thesis, Department of Mathematics, Princeton University.

Granot, D. and Huberman, G. 1984. On the core and nucleolus of the minimum costs spanning tree games. *Mathematical Programming* 29, 323–47.

Harsanyi, J.C. 1956. Approaches to the bargaining problem before and after the theory of games: a critical discussion of Zeuthen's, Hicks' and Nash's theories. *Econometrica* 24, 144–57.

Harsanyi, J.C. 1959. A bargaining model for the cooperative n-person game. In *Contributions to the Theory of Games*, vol. 4, ed. A.W. Tucker and R.W. Luce. Princeton: Princeton University Press.

Harsanyi, J.C. 1963. A simplified bargaining model for the *n*-persons cooperative game. *International Economic Review* 4, 194–220.

Harsanyi, J.C. 1966. A general theory of rational behavior in game situations. *Econometrica* 34, 613–34.

Harsanyi, J.C. and Selten, R. 1972. A generalized Nash solution for two-person bargaining games with incomplete information. *Management Science* 18, 80–106.

Hart, S. 1973. Values of mixed games. *International Journal of Game Theory* 2, 69–86.

Hart, S. 1977a. Asymptotic values of games with a continuum of players. *Journal of Mathematical Economics* 4, 57–80.

Hart, S. 1977b. Values of non-differentiable markets with a continuum of traders. *Journal of Mathematical Economics* 4, 103–16.

Hart, S. 1980. Measure-based values of market games. *Mathematics of Operations Research* 5, 197–228.

Hart, S. 1985. An axiomatization of Harsanyi's nontransferable utility solution. *Econometrica* 53, 1295–314.

Hart, S. and Mas-Colell, A. 1989. The potential: a new approach to the value in multiperson allocation problems. *Econometrica* 57, 589–614.

Ichiishi, T. 1981. Supermodularity: applications to convex games and to the greedy algorithm for LP. *Journal of Economic Theory* 25, 283–6.

Jackson, M.O. and Wolinsky, A. 1996. A strategic model of social and economic networks. *Journal of Economic Theory* 71, 44–74.

Kalai, E. 1975. Excess functions for cooperative games without sidepayments. *SIAM Journal of Applied Mathematics* 29, 60–71.

Kalai, E. 1977a. Proportional solutions to bargaining situations: interpersonal utility comparisons. *Econometrica* 45, 1623–30.

Kalai, E. 1977b. Non-symmetric Nash solutions for replications of 2-person bargaining. *International Journal of Game Theory* 6, 129–33.

Kalai, E., Postlewaite, A. and Roberts, J. 1979. Barriers to trade and disadvantageous middlemen: nonmonotonicity of the core. *Journal of Economic Theory* 19, 200–9.

Kalai, E. and Rosenthal, R.W. 1978. Arbitration of two-party disputes under ignorance. *International Journal of Game Theory* 7, 65–72.

Kalai, E. and Samet, D. 1985. Monotonic solutions to general cooperative games. *Econometrica* 53, 307–27.

Kalai, E. and Samet, D. 1987. On weighted Shapley values. *International Journal of Game Theory* 16, 205–22.

Kalai, E. and Smorodinsky, M. 1975. Other solutions to Nash's bargaining problems. *Econometrica* 43, 513–8.
Kalai, E. and Zemel, E. 1982a. Generalized network problems yielding totally balanced games. *Operations Research* 5, 998–1008.
Kalai, E. and Zemel, E. 1982b. Totally balanced games and games of flow. *Mathematics of Operations Research* 7, 476–78.
Kannai, Y. 1992. The core and balancedness. In *Handbook of Game Theory with Economic Applications*, vol. 1, ed. R.J. Aumann and S. Hart. Amsterdam: North-Holland.
Kaneko, M. and Wooders, M. 1982. Cores of partitioning games. *Mathematical Social Sciences* 3, 313–27.
Kohlberg, E. 1972. The nucleolus as a solution to a minimization problem. *SIAM Journal of Applied Mathematics* 23, 34–49.
Larauelle, A. and Valenciano, F. 2001. Shapley–Shubik and Banzhaf indices revisited. *Mathematics of Operations Research* 26, 89–104.
Lehrer, E. 1988. An axiomatization of the Banzhaf value. *International Journal of Game Theory* 17(2), 89–99.
Lensberg, T. 1985. Bargaining and fair allocation. In *Cost Allocation, Principles, Applications*, ed. P. Young. Amsterdam: North-Holland.
Lensberg, T. 1988. Stability and the Nash solution. *Journal of Economic Theory* 45, 330–41.
Lensberg, T. and Thomson, W. 1989. *Axiomatic Theory of Bargaining With a Variable Population.* Cambridge: Cambridge University Press.
Levy, A. and McLean, R. 1991. An axiomatization of the non-symmetric NTU value. *International Journal of Game Theory* 19, 109–27.
Lucas, W.F. 1969. The proof that a game may not have a solution. *Transactions of the American Mathematical Society* 137, 219–29.
Lucas, W.F. and Thrall, R.M. 1963. n-person games in partition function forms. *Naval Research Logistics Quarterly* 10, 281–98.
Luce, R.D. and Raiffa, H. 1957. *Games and Decisions: An Introduction and Critical Survey.* New York: Wiley.
Maschler, M. 1992. The bargaining set, kernel and nucleolus. In Aumann and Hart (1992).
Maschler, M., Peleg, B. and Shapley, L.S. 1979. Geometric properties of the kernel, nucleolus, and related solution concepts. *Mathematics of Operations Research* 4, 303–38.
Maschler, M. and Perles, M. 1981. The superadditive solution for the Nash bargaining game. *International Journal of Game Theory* 10, 163–93.
Mas-Colell, A. 1975. A further result on the representation of games by markets. *Journal of Economic Theory* 10, 117–22.
Mas-Colell, A. 1977. Competitive and value allocations of large exchange economies. *Journal of Economic Theory* 14, 419–38.
Mas-Colell, A. 1989. An equivalence theorem for a bargaining set. *Journal of Mathematical Economics* 18, 129–39.
McLean, R.P. 2002. Values of non-transferable utility games. In Aumann and Hart (2002).
Milnor, J.W. and Shapley, L.S. 1978. Values of large games II: oceanic games. *Mathematics of Operations Research* 3, 290–307.
Monderer, D. and Samet, D. 2002. Variations on the Shapley value. In Aumann and Hart (2002).
Monderer, D., Samet, D. and Shapley, L.S. 1992. Weighted Shapley values and the core. *International Journal of Game Theory* 21, 27–39.
Moulin, H. 1988. *Axioms of Cooperative Decision Making*. Cambridge: Cambridge University Press.
Myerson, R.B. 1977a. Two-person bargaining problems and comparable utility, *Econometrica* 45, 1631–37.

Myerson, R.B. 1977b. Graphs and cooperation in games. *Mathematics of Operations Research* 2, 225–9.
Myerson, R.B. 1979. Incentive compatibility and the bargaining problem. *Econometrica* 47, 61–74.
Myerson, R.B. 1984. Cooperative games with incomplete information. *International Journal of Game Theory* 13, 69–96.
Myerson, R.B. 1991. *Game theory: Analysis of Conflict*. Cambridge: Harvard University Press.
Nash, J.F. 1950. The bargaining problem. *Econometrica* 18, 155–62.
Nash, J.F. 1953. Two person cooperative games. *Econometrica* 21, 128–40.
Neyman, A. 1985. Semivalues of political economic games. *Mathematics of Operations Research* 10, 390–402.
Neyman, A. 1987. Weighted majority games have an asymptotic value. *Mathematics of Operations Research* 13, 556–80.
Neyman, A. 2002. Values of games with infinitely many players. In Aumann and Hart (2002).
Nisan, N., Roughgarden, T., Tardos, E. and Vazirani, V. 2007. *Algorithmic Game Theory*. Cambridge: Cambridge University Press.
O'Neil, B. 1982. A problem of rights arbitration from the Talmud. *Mathematical Social Sciences* 2, 345–71.
Owen, G. 1972. Multilinear extensions of games. *Management Science* 18, 64–79.
Owen, G. 1975. On the core of linear production games. *Mathematical Programming* 9, 358–70.
Osborne, M.J. and Rubinstein, A. 1994. *A Course in Game Theory*. Cambridge, MA: MIT Press.
Peleg, B. 1963a. Solutions to cooperative games without side payments. *Transactions of the American Mathematical Society* 106, 280–92.
Peleg, B. 1963b. Bargaining sets of cooperative games without side payments. *Israel Journal of Mathematics* 1, 197–200.
Peleg, B. 1985. An axiomatization of the core of cooperative games without side payments. *Journal of Mathematical Economics* 14, 203–14.
Peleg, B. 1986. On the reduced games property and its converse. *International Journal of Game Theory* 15, 187–200.
Peleg, B. 1992. Axiomatizations of the core. In Aumann and Hart (1992).
Peleg, B. and Sudholter, P. 2003. *Introduction to the Theory of Cooperative Games*. Dordrecht: Kluwer Academic Publications.
Peters, H.J.M. 1992. *Axiomatic Bargaining Game Theory*. Dordrecht: Kluwer Academic Publishers.
Peters, H., Tijs, S. and Zarzuelo, A. 1994. A reduced game property for the Kalai Smorodinsky and Egalitarian bargaining solution. *Mathematical Social Sciences* 27, 11–18.
Potters, J., Curiel, I. and Tijs, S. 1992. Traveling salesman game. *Mathematical Programming* 53, 199–211.
Raiffa, H. 1953. Arbitration schemes for generalized two-person games. In *Contributions to the Theory of Games II*, ed. H. Kuhn and A.W. Tucker. Princeton: Princeton University Press.
Riker, W.H. and Shapley, L.S. 1968. Weighted voting: a mathematical analysis for instrumental judgements. In *Representation*, ed. J.R. Pennock and J.W. Chapman. New York: Atherton.
Roth, A.E. 1977. The Shapley value as a von Neumann–Morgenstern utility. *Econometrica* 45, 657–64.
Roth, A.E. 1979. *Axiomatic Models of Bargaining*. Berlin and New York: Springer.
Roth, A.E. 1984. The evolution of the labor market for medical interns and residents: a case study in game theory. *Journal of Political Economics* 92, 991–1016.
Roth, A.E. and Verrecchia, R.E. 1979 The Shapley value as applied to cost allocation: a reinpt. *Journal of Accounting Research* 17, 295–303.
Rubinstein, A. 1982. Perfect equilibrium in a bargaining model. *Econometrica* 50, 97–109.
Scarf, H.E. 1967. The core of an n-person game. *Econometrica* 35, 50–69.

Schmeidler, D. 1969. The nucleolus of a characteristic function game. *SIAM Journal of Applied Mathematics* 17, 1163–70.

Schmeidler, D. 1972. Cores of exact games I. *Journal of Mathematical Analysis and Application* 40, 214–25.

Shapley, L.S. 1952. *Notes on the n-person game iii: some variants of the Von Neumann–Morgenstern definition of solution*. Santa Monica: RAND Corporation RM 817.

Shapley, L.S. 1953. A value for n-person games. In *Contributions to the Theory of Games II*, ed. H. Kuhn and A.W. Tucker. Princeton: Princeton University Press.

Shapley, L.S. 1964. *Values of large games, VII: a general exchange economy with money*. Santa Monica: RAND Corporation RM 4248.

Shapley, L.S. 1967. On balanced sets and cores. *Naval Research Logistics Quarterly* 14, 453–60.

Shapley, L.S. 1969. Utility comparison and the theory of games. In *La Décision*. Paris: Edition du CNRS.

Shapley, L.S. 1971. Cores of convex games. *International Journal of Game Theory* 1, 11–26.

Shapley, L.S. 1973. Let's block 'block'. *Econometrica* 41, 1201–2.

Shapley, L.S. and Shubik, M. 1954. A method for evaluating the distribution of power in a committee system. *American Political Science Review* 48, 787–92.

Shapley, L.S. and Shubik, M. 1969a. On market games. *Journal. of Economic Theory* 1, 9–25.

Shapley, L.S. and Shubik, M. 1969b. Pure competition, coalitional power and fair division. *International Economic Review* 10, 337–62.

Shubik, M. 1959. *Strategy and Market Structure: Competition, Oligopoly, and the Theory of Games*. New York: Wiley.

Shubik, M. 1962. Incentives, decentralized control, the assignment of joint costs and internal pricing. *Management Science* 8, 325–43.

Shubik, M. 1982. *Game Theory in the Social Sciences: Concepts and Solutions*. Cambridge, MA: MIT Press.

Shubik, M. 1984. *A Game Theoretic Approach to Political Economy*. Cambridge, MA: MIT Press.

Sprumont, Y. 1998. Ordinal cost sharing. *Journal of Economic Theory* 81, 26–162.

Tauman, Y. 1981. Value on a class of non-differentiable market games. *International Journal of Game Theory* 10, 155–62.

Tauman, Y. 1988. The Aumann-Shapley prices: a survey. In *The Shapley Value*, ed. A. Roth. New York: Cambridge University Press.

Thomson, W. 1983. The fair division of a fixed supply among a growing population. *Mathematics of Operations Research* 8, 319–26.

Thomson, W. 1987. Monotonicity of bargaining solutions with respect to the disagreement point. *Journal of Economic Theory* 42, 50–8.

Thomson, W. 1994. Cooperative models of bargaining. *Aumann and Hart* 2002, 1237–84.

Valenciano, F. and Zarzuelo, J.M. 1994. On the interpretation of the nonsymmetric bargaining solutions and their extensions to nonexpected utility preferences. *Games and Economic Behavior* 7, 461–72.

Von Neumann, J. and Morgenstern, O. 1944. *Theory of Games and Economic Behavior*. Princeton: Princeton University Press.

Weber, R.J. 1988. Probabilistic values for games. In *The Shapley Value*, ed. A. Roth. Cambridge: Cambridge University Press.

Weber, R.J. 1994. Games in coalitional form. In Aumann and Hart (1994).

Wilson, R. 1978. Information, efficiency, and the core of an economy. *Econometrica* 46, 807–16.

Winter, E. 2002. The Shapley value. In Aumann and Hart (2002).

Wooders, M.H. and Zame, W.R. 1984. Approximate cores of large games. *Econometrica* 52, 1327–50.

Wooders, M.H. and Zame, W.R. 1987. Large games: fair and stable outcomes. *Journal of Economic Theory* 42, 59–63.
Young, H.P. 1985. Monotonic solutions of cooperative games. *International Journal of Game Theory* 14, 65–72.
Young, H.P. 1994. Cost allocation. In Aumann and Hart (1994).

global games

Complete information games often have multiple Nash equilibria. Game theorists have long been interested in finding a way of removing or reducing that multiplicity. Carlsson and van Damme (1993) (CvD) introduced an original and attractive approach to doing so. A complete information model entails the implicit assumption that there is common knowledge among the players of the payoffs of the game. In practice, such common knowledge will often be lacking. CvD suggested a convenient and intuitive way of relaxing that common knowledge assumption: suppose that, instead of observing payoffs exactly, payoffs are observed with a small amount of continuous noise; and suppose that – before observing their signals of payoffs – there was an *ex ante* stage where any payoffs were possible. Based on the latter feature, CvD dubbed such games 'global games'. It turns out that there is a unique equilibrium in the game with a small amount of noise. This uniqueness remains no matter how small the noise is and is independent of the distribution of the noise. Since complete information, or common knowledge of payoffs, is surely always an idealization anyway, the play selected in the global game with small noise can be seen as a prediction for play in the underlying complete information game.

The following example illustrates the main idea. There are two players each of whom must decide whether to invest or not invest. Action 'not invest' always gives a payoff of 0. Action 'invest' always gives a payoff of θ; but there are strategic complementarities, and if the other player does not invest, then the player loses 1. Thus the payoff matrix is:

	Invest	Not invest
Invest	θ, θ	$\theta-1, 0$
Not invest	$0, \theta-1$	$0, 0$

Let us first examine the Nash equlibria of this game when θ is common knowledge. If $\theta < 0$, then 'invest' is a strictly dominated action for each player, and thus 'not invest, not invest' is the unique Nash (and dominant strategies) equilibrium. If $\theta > 1$, then 'not invest' is a strictly dominated action for each player, and 'invest, invest' is the unique Nash (and dominant strategies) equilibrium. The multiplicity case arises if $0 < \theta < 1$. In this case, there are two strict Nash equilibria (both not invest and both invest) and there is also a strictly mixed Nash equilibrium.

But suppose the players do not exactly observe θ. Suppose for convenience that each player believes that θ is uniformly distributed on the real line (thus there is an 'improper' prior with infinite mass: this does not cause any technical or conceptually difficulties as players will always condition on signals that generate 'proper' posteriors). Suppose that each player observes a signal $x_i = \theta + \sigma\varepsilon_i$, where each ε_i is independently normally distributed with mean 0 and standard deviation 1.

In this game of incomplete information, a pure strategy for player i is a mapping $s_i : \mathbb{R} \to \{\text{Invest}, \ \text{Not invest}\}$. Suppose player 1 was sure that player 2 was going to follow a 'threshold' strategy where she invested only if her signal were above k, so

$$s_2(x_2) = \begin{cases} \text{Invest}, & \text{if } x_2 > k \\ \text{Not invest}, & \text{if } x_2 \leq k \end{cases}$$

What is player 1's best response? First, observe that his expectation of θ is x_1. Second, note that (under the uniform prior assumption) his posterior on θ is normal with mean x_1 and variance σ^2, and thus his posterior on x_2 is normal with mean x_1 and variance $2\sigma^2$. Thus his expectation that player j will not invest is $\Phi(\frac{1}{\sqrt{2}\sigma}(k - x_1))$, where Φ is cumulative distribution of the standard normal. Thus his expected payoff is

$$x_1 - \Phi\left(\frac{1}{\sqrt{2}\sigma}(k - x_1)\right), \tag{1}$$

and player 1 will invest if and only if (1) is positive. Now if we write $b(k)$ for the unique value of x_1 setting (1) equal to 0 (this is well defined since (1) is strictly increasing in x_1), the best response of player 1 is then to follow a cut-off strategy with threshold equal to $b(k)$. Observe that as $k \to -\infty$ (player 2 always invests), (1) tends to x_1, so $b(k) \to 0$. As $k \to \infty$ (player 2 never invests), (1) tends to $x_1 - 1$, so $b(k) \to 1$. Also observe that if $k = \frac{1}{2}$, then $b(k) = \frac{1}{2}$, since if player 1 observes signal $\frac{1}{2}$, his expectation of θ is $\frac{1}{2}$ but he assigns probability $\frac{1}{2}$ to player 2 not investing. Finally, observe that (by total differentiation)

$$b'(k) = \frac{1}{1 + \frac{\sqrt{2\pi}}{\phi\left(\frac{1}{\sqrt{2}\sigma}(k - x_1)\right)}} \in (0, 1),$$

so $b(k)$ is strictly increasing in k and we can immediately conclude that there is a unique 'threshold' equilibrium where each player uses a threshold of $\frac{1}{2}$.

The strategy with threshold $\frac{1}{2}$ is in fact the unique strategy surviving iterated deletion of (interim) strictly dominated strategies. In fact, a strategy s survives n rounds of iterated deletion of strict dominated strategies if and only if

$$s(x) = \begin{cases} \text{Invest}, & \text{if } x > b^n(1) \\ \text{Not invest}, & \text{if } x < b^n(0) \end{cases}$$

where $b^n(k) = \overbrace{b(b(\ldots b}^{n \text{ times}}(k)))$.

The key intuition for this example is that the uniform prior assumption ensures that each player, whatever his signal, attaches probability $\frac{1}{2}$ to his opponent having a higher signal and probability $\frac{1}{2}$ to him having a lower signal. This property remains true no matter how small the noise is, but breaks discontinuously in the limit: when noise is zero, he attaches probability 1 to his opponent having the same signal.

In this article, I will first report how Carlsson and van Damme's (1993) analysis can be used to give a complete general analysis of two player two action games. I will then report in turn theoretical extensions of their work and a literature that has used insights from global games in economic applications. This dichotomy is somewhat arbitrary (many 'applied' papers have significant theoretical contributions) but convenient.

1. Two-player, two-action games

Let the payoffs of a two-player, two-action game be given by the following matrix:

	A	B
A	θ_1, θ_2	θ_3, θ_4
B	θ_5, θ_6	θ_7, θ_8

Thus a vector $\theta \in \mathbb{R}^8$ describes the payoffs of the game and is drawn from some distribution. For a generic choice of θ, there are three possible configurations of Nash equilibria.

1. There is a unique Nash equilibrium with both players using strictly mixed strategies.
2. There is a unique strict Nash equilibrium with both players using pure strategies.
3. There are two pure strategy strict Nash equilibria and one strictly mixed strategy Nash equilibrium.

In the last case, Harsanyi and Selten (1988) proposed the criterion of risk dominance to select among the multiple Nash equilibria. Suppose that (A, A) and (B, B) are strict Nash equilibria of the above game (that is, $\theta_1 > \theta_5$, $\theta_7 > \theta_3$, $\theta_2 > \theta_4$ and $\theta_8 > \theta_6$). Then (A, A) is a risk dominant equilibrium if

$$(\theta_1 - \theta_5)(\theta_2 - \theta_4) > (\theta_7 - \theta_3)(\theta_8 - \theta_6).$$

Generically, exactly one of the two pure Nash equilibria will be risk dominant.

Now consider the following incomplete information game $G(\sigma)$. Each player i observes a signal $x_i = \theta + \sigma\varepsilon_i$, where the ε_i are eight-dimensional noise terms. Thus we have an incomplete information game parameterized by $\sigma \geq 0$. A strategy for a player is a function from possible signals $\mathbb{R}^8$ to the action set $\{A, B\}$. For any given strategy profile of players in the game $G(\sigma)$ and any actual realization of the payoffs θ, we can ask what is the distribution over action profiles in the game (averaging across signal realizations).

Theorem For any sequence of games $G(\sigma^k)$ where $\sigma^k \to 0$ and any sequence of equilibria of those games, average play converges at almost all payoff realizations to the unique Nash equilibrium (if there is one) and to the risk dominant Nash equilibrium (if there are multiple Nash equilibria).

This is shown by the main result of Carlsson and van Damme (1993) in cases (2) and (3) above. They generalize the argument from the example described above to show that, if an action is part of a risk dominant equilibrium or a unique strict Nash equilibrium of the complete information game θ, then – for sufficiently small σ – that action is the unique action surviving iterated deletion strictly dominated strategies. Kajii and Morris (1997) show that, if a game has a unique correlated equilibrium, then that equilibrium is 'robust to incomplete information', that is, will continue to be played in some equilibrium if we change payoffs with small probability. This argument can be extended to show the theorem for case (1) (the extension is discussed in Morris and Shin, 2003).

2. Theoretical extensions; many players and many actions

Carlsson and van Damme (1993) dubbed their perturbed games for the two player, two action case 'global games' because all possible payoff profiles were possible. They showed that there was a general way of adding noise to the payoff structure such that, as the noise went to zero, there was a unique action surviving iterated deletion of (interim) dominated strategies (a 'limit uniqueness' result). And they showed that the action that got played in the limit was independent of the distribution of noise added (a 'noise independent selection' result). Their result does not extend in general to many player many action games. In discussing known extensions, we must carefully distinguish which of their results extend.

Frankel, Morris and Pauzner (2003) consider games with strategic complementarities (that is, supermodular payoffs). Rather than allow for all possible payoff profiles, they restrict attention to a one-dimensional set of possible payoff functions, or states, which are ordered so that higher states lead to higher actions. The idea of 'global' games is captured by a 'limit dominance' property: for sufficiently low values of θ, each player has a dominant strategy to choose his lowest action, and that for sufficiently high values of θ, each player has a dominant strategy to choose his highest action. Under these restrictions, they are able to present a complete analysis of the case with many players, asymmetric payoffs and many actions. In particular, a limit uniqueness result holds: if each player observes the state with noise, and the size of noise goes to zero, then in the limit there is a unique strategy profile surviving iterated deletion of strictly dominated strategies. Note that while Carlsson and van Damme required no strategic complementarity and other monotonicity properties, when there are multiple equilibria in a two-player, two-action game – the interesting case for Carlsson and van Damme's analysis – there are automatically strategic complementarities.

Within this class of monotonic global games where limit uniqueness holds, Frankel, Morris and Pauzner (2003) also provide sufficient conditions for 'noise independent selection'. That is, for some complete information games, which action gets played in the limit as noise goes to zero does not depend on the shape of the noise. Frankel, Morris and Pauzner (2003) show that a generalization of the potential maximizing action profile is sufficient for noise independent selection. This sufficient condition encompasses the risk dominant selection in two player binary action games; the selection of the 'Laplacian' action (a best response to a uniform distribution over others' actions) in many player, binary action games (Morris and Shin, 2003). It also yields unique predictions in the continuum player currency crisis of Guimaeres and Morris (2004) and in two-player, three-action games with symmetric payoffs. Morris and Ui (2005) give further sufficient conditions for equilibria to be 'robust to incomplete information' in the sense of Kajii and Morris (1997), which will also ensure noise independent section.

However, Frankel, Morris and Pauzner (2003) also provide an example of a two-player, four-action, symmetric payoff game where noise independent selection fails. Thus there is a unique limit as the noise goes to zero, but the nature of the limit depends on the exact distribution of the noise. Carlsson (1989) gave a three-player, two-action example in which noise independent selection failed. Corsetti et al. (2004) describe a global games model of currency crises, where there is a continuum of small traders and a single large trader. This is thus a many-player, two-action game with *asymmetric* payoffs. The equilibrium selected as noise goes to zero depends on the relative informativeness of the large and small traders' signals. This is thus an application where noise-independent selection fails.

More limited results are available on global games without supermodular payoffs. In many applications – such as bank runs – there are some strategic complementarities but payoffs are not supermodular everywhere: conditional on enough people running on the bank to cause collapse, I am better off if I run if few people run and share in the liquidation of the bank's assets. An important paper of Goldstein and Pauzner (2005) has shown equilibrium uniqueness for 'bank run payoffs' – satisfying a single crossing property – with uniform prior and uniform noise. This analysis has been followed in a number of applications. They establish that there is a unique equilibrium in threshold strategies and there are no non-threshold equilibria. However, their analysis does not address the question of which strategies survive iterated deletion of strictly dominated strategies. Morris and Shin (2003) discuss how the existence of a unique threshold equilibrium can be established more generally under a signal crossing property on payoffs and a monotone likelihood ratio property on signals (not required for global games analysis with supermodular payoffs); however, these arguments do not rule out the existence of non-monotonic equilibria. Results of van Zandt and Vives (2007) can be used more generally to establish the existence of a unique monotone equilibrium under weaker conditions than supermodularity.

The original analysis of Carlsson and van Damme (2003) relaxed the assumption of common knowledge of payoffs in a particular way: they assumed that there was a common prior on payoffs and that each player observes a small conditionally independent signal of payoffs. This is an intuitively small perturbation of the game and this is the perturbation that has been the focus of study in the global games literature. However, when the noise is small one can show that types in the perturbed game are close to common knowledge types in the product topology on the universal type space: that is, for each type t in the perturbed game, there is a common knowledge type t' such that type t and t' almost agree in their beliefs about payoffs, they almost agree about their beliefs about the opponents' beliefs, and so on up to any finite level. Thus the 'discontinuity' in equilibrium outcomes in global games when noise goes to zero is illustrating the same sensitivity to higher order beliefs of the famous example of Rubinstein (1989). Now we can ask: how general is the phenomenon that Rubinstein (1989) and Carlsson and van Damme (1993) identified? That is, for which games and actions is it the case that, under common knowledge, the action is part of an equilibrium (and thus survives iterated deletion of strictly dominated strategies) but for a type 'close' to common knowledge of that game, that action is the unique action surviving iterated deletion of strictly dominated strategies. Weinstein and Yildiz (2007) shows that this is true for every action surviving iterated deletion of strictly dominated strategies in the original game. This observation highlights the fact that the selections that arise in standard global games arise not just because one relaxes common knowledge, but because it is relaxed in a particular way: the common prior assumption is maintained and outcomes are analysed under that common prior, and the noisy signal technology ensures particular properties of higher-order beliefs, that is, that each player's beliefs about how other players' beliefs differ from his is not too dependent on the level of his beliefs.

3. Applications; public signals and dynamic games

Complete information models are often used in applied economic analysis for tractability: the complete information game payoffs capture the essence of the economic problem. Presumably there is not in fact common knowledge of payoffs, but if asymmetries of information are not the focus of the economic analysis, this assumption seems harmless. But complete information games often have multiple equilibria, and policy analysis – and comparative statics more generally – are hard to carry out in multiple equilibrium models. The global games analysis surveyed above has highlighted how natural relaxations of the common knowledge assumptions often lead to intuitive selections of a unique equilibrium. This suggests these ideas might be useful in applications. Fukao (1994) and Morris and Shin (1995) were two early papers that pursued this agenda. The latter paper – published as Morris and Shin (1998) – was an application to currency crises, where the existing literature builds on a dichotomy between 'fundamentals-driven' models and multiple equilibrium or 'sunspot' equilibria views of currency crises. This dichotomy does not make sense in a global games model of currency crises: currency attacks are 'self-fulfilling' – in the sense that speculators are attacking only because they expect others to do so – but their expectations of others' behaviour may nonetheless be pinned down by higher order beliefs (see Heinemann, 2000, for an important correction of the equilibrium characterization in Morris and Shin, 1998). Morris and Shin (2000) laid out the methodological case for using global games as a framework for economic applications. Morris and Shin (2003) surveys many early applications to currency crises, bank runs, the design of international institutions and asset pricing, and there have been many more since. Rather than attempt to survey these applications, I will highlight two important methodological issues – public signals and dynamics – that have played an important role in the developing applied literature.

To do this, it is useful to consider an example that has become a workhorse of the applied literature, dubbed the 'regime change' game in a recent paper of Angeletos, Hellwig and Pavan (2007). The example comes from a 1999 working paper on 'Coordination Risk and the Price of Debt' presented as a plenary talk at the 1999 European meetings of the Econometric Society, eventually published as Morris and Shin (2004). A continuum of players must decide whether to invest or not invest. The cost of investing is c. The payoff to investing is one if the proportion investing is at least $1 - \theta$, 0 otherwise. If there is common knowledge of θ and $\theta \in (0, 1)$, there are multiple Nash equilibria of this continuum player complete information game: 'all invest' and 'all not invest'. But now suppose that θ is normally distributed with mean y and standard deviation τ. Each player in the continuum population observes the mean y (which is thus a public signal of θ). But in addition, each player i observes a private signal x_i, where the private signals are distributed in the continuum population with mean θ and standard deviation σ (that is, as in the example at the beginning of this article). Morris and Shin (2004) show that the resulting game of incomplete information has a unique equilibrium if and only if $\sigma \leq \sqrt{2\pi}\tau^2$, that is, if private signals are sufficiently accurate relative to the accuracy of public signals. This result is intuitive: we know that if there is common knowledge of θ, there are multiple equilibria. A very small value of τ means that the public signal is very accurate and there is 'almost' common knowledge.

This result makes it possible to conduct comparative statics within a unique equilibrium not only in the uniform prior, no 'public' information, limit but also with non-trivial public information. A distinctive comparative static that arises is that the unique equilibrium is very sensitive to the public signal y, even conditioning on the true state θ (see Morris and Shin, 2003; 2004; Angeletos and Werning, 2006). This is because, for each player, the public signal y becomes a more accurate prediction of others' behaviour than his private signal, even if they are of equal precision.

But the sensitivity of the uniqueness result to public signals also raises a robustness question. Public information is endogenously generated in economic settings, and thus a qst that arises in many dynamic applications of global games in general and the regime change game in particular is when endogenous information generates enough public information to get back multiplicity (Tarashev, 2003; Dasgupta, 2007; Angeletos, Hellwig and Pavan, 2006; 2007; Angeletos and Werning, 2006; Hellwig, Mukherji and Tsyvinski, 2006). This literature has highlighted the importance of endogenous information revelation and the variety of channels through which such revelation may lead to multiplicity or enhance uniqueness. In addition, these and other dynamic applications of global games raise many other important methodological issues, such as the interaction between the global game uniqueness logic and 'herding' – informational externalities in dynamic settings without payoff complementarities – and 'signalling' – biasing choices from static best responses in order to influence opponents' beliefs in the future.

STEPHEN MORRIS

See also **purification.**

Bibliography

Angeletos, G.-M., Hellwig, C. and Pavan, A. 2006. Signalling in a global game: coordination and policy traps. *Journal of Political Economy* 114, 452–84.

Angeletos, G.-M., Hellwig, C. and Pavan, A. 2007. Dynamic global games of regime change: learning, multiplicity and timing of attacks. *Econometrica* 75, 711–56.

Angeletos, G.-M. and Werning, I. 2006. Crises and prices: information aggregation, multiplicity and volatility. *American Economic Review* 96, 1720–36.

Carlsson, H. 1989. Global games and the risk dominance criterion. Unpublished manuscript, Department of Economics, University of Lund.

Carlsson, H. and van Damme, E. 1993. Global games and equilibrium selection. *Econometrica* 61, 989–1018.

Corsetti, G., Dasgupta, A., Shin, H.S. and Morris, S. 2004. Does one Soros make a difference? The role of a large trader in currency crises. *Review of Economic Studies* 71, 87–114.

Dasgupta, A. 2007. Coordination and delay in global games. *Journal of Economic Theory* 134, 195–225.

Frankel, D., Morris, S. and Pauzner, A. 2003. Equilibrium selection in global games with strategic complementarities. *Journal of Economic Theory* 108, 1–44.

Fukao, K. 1994. Coordination failures under incomplete information and global games. Discussion Paper No. A299, Institute of Economic Research, Hitotsubashi University.

Goldstein, I. and Pauzner, A. 2005. Demand-deposit contracts and the probability of bank runs. *Journal of Finance* 60, 1293–327.

Guimaeres, B. and Morris, S. 2004. Risk and wealth in a model of self-fulfilling currency attacks. Discussion Paper No. 1433R, Cowles Foundation.

Harsanyi, J. and Selten, R. 1988. *A General Theory of Equilibrium Selection in Games*. Cambridge, MA: MIT Press.

Heinemann, F. 2000. Unique equilibrium in a model of self-fulfilling currency attacks: comment. *American Economic Review* 90, 316–8.

Hellwig, C., Mukherji, A. and Tsyvinski, A. 2006. Self-fulfilling currency crises: the role of interest rates. *American Economic Review* 96, 1769–87.

Kajii, A. and Morris, S. 1997. The robustness of equilibria to incomplete information. *Econometrica* 65, 1283–309.

Morris, S. and Shin, H.S. 1995. Informational events that trigger currency attacks. Working Paper No. 95-24, Federal Reserve Bank of Philadelphia.

Morris, S. and Shin, H.S. 1998. Unique equilibrium in a model of self-fulfilling currency attacks. *American Economic Review* 88, 587–97.

Morris, S. and Shin, H.S. 2000. Rethinking multiple equilibria in macroeconomics. In *NBER Macroeconomics Annual 2000*, ed. B.S. Bernanke and K. Rogoff. Cambridge, MA: MIT Press.

Morris, S. and Shin, H.S. 2003. Global games: theory and applications. In *Advances in Economics and Econometrics (Proceedings of the Eighth World Congress of the Econometric Society)*, ed. M. Dewatripont, L. Hansen and S. Turnovsky. Cambridge: Cambridge University Press.

Morris, S. and Shin, H.S. 2004. Coordination risk and the price of debt. *European Economic Review* 48, 133–53.

Morris, S. and Ui, T. 2005. Generalized potentials and robust sets of equilibria. *Journal of Economic Theory* 124, 45–78.

Rubinstein, A. 1989. The electronic mail game: strategic behavior under almost common knowledge. *American Economic Review* 79, 385–91.

Tarashev, N. 2003. Speculative attacks and informational role of the interest rate. *Journal of the European Economic Association.*

van Zandt, T. and Vives, X. 2007. Monotone equilibria in Bayesian games of strategic complementarities. *Journal of Economic Theory* 134, 339–60.

Weinstein, J. and Yildiz, M. 2007. A structure theorem for rationalizability with application to robust predictions of refinements. *Econometrica* 75, 365–400.

graphical games

Graphical games are a general parametric model for multi-player games that is most appropriate for settings in which not all players directly influence the payoffs of all others, but rather there is some notion of 'locality' to the direct strategic interactions. These interactions are represented as an undirected graph or network, where we assume that each player is identified with a vertex, and that the payoff of a given player is a function of only his or her own action and those of his or her immediate neighbours in the network. Specification of a graphical game thus consists of the graph or network, along with the local payoff function for each player. Graphical games offer a number of representational and algorithmic advantages over the normal form, have permitted the development of a theory relating network topology to equilibrium properties, and have played a central role in recent results on the computational complexity of computing Nash equilibria. They have also been generalized to exchange economies, evolutionary game theory, and other strategic settings.

Definitions

A graphical game begins with an undirected graph or network $G=(V, E)$, where V is the set of players or vertices, and E is a set of edges or unordered pairs of vertices/players. The assumed semantics of this graph are that the payoffs of players are determined only by their local neighbourhoods. More precisely, if we define the neighbour set of a player u as $N(u)=\{v\text{: } (u, v) \text{ is in } E\}$, the payoff of u is assumed to be a function not of the joint action of the entire population of players, but only the actions of u itself and the players in $N(u)$. Complete specification of a graphical game thus consists of the graph G, and the local payoff functions for each player. Note that at equilibrium, it remains the case that the strategy of a player may be indirectly influenced by players arbitrarily distant in the network; it is simply that such influences are effected by the propagation of the local, direct payoff influences.

In the case that G is the complete network, in which all pairs of vertices have an edge between them, the graphical game simply reverts to the multi-player normal form. However, in the interesting cases the graph may exhibit considerable asymmetry and structure, and also be much more succinct than the normal form. For instance, if $|N(u)| \leq = d$ for all players u, then the total number of parameters of the graphical game grows exponentially only in the degree bound d, as opposed to exponentially in n for the normal form. Thus when d is much smaller than n (a reasonable expectation in a large-population game with only local interactions), the graphical game representation is exponentially more parsimonious than the normal form. Qualitatively, one can think of graphical games as a good model for games in which there may be many players, but each player may be directly and strongly influenced by only a small number of others. Graphical games should be contrasted with other parametric models such as congestion and potential games, in which each player has global influence, but often of a highly specific and weak form. They can also be viewed as a natural generalization of more specific network-based games studied in the game theory and economics literature (Jackson, 2007).

Computational properties

In addition to the aforementioned potential for representational parsimony, graphical games permit a family of natural and sometimes provably efficient algorithms for the computation of Nash and other equilibria. It should be emphasized here that by 'efficient' we mean an algorithm whose running time is a 'slowly' growing function of *the number of parameters of the graphical game representation* (which may be considerably more challenging than a slowly growing function of the number of parameters in the normal form representation, which may be much larger). As is standard in computer science, 'slowly' growing typically means a polynomial function (ideally of low degree).

For instance, in the special case that the graph structure is a tree (or can be modified to a tree via a small number of standard topological operations involving, for instance, the merging of vertices), there is an algorithm running in time polynomial in the number of parameters that computes approximations to (all) Nash equilibria of the given graphical game (Kearns, Littman and Singh, 2001.) This algorithm is based on dynamic programming, and is decentralized in the sense that communication need take place only between neighbouring vertices in the network. In even more restrictive topologies, efficient algorithms for computing exact Nash equilibria are known (Elkind, Goldberg and Goldberg, 2006). For non-tree topologies, a generalization of this algorithm known as NashProp (Ortiz and Kearns, 2002) has been developed that is provably convergent, but has weaker guarantees of computational efficiency. Provably efficient algorithms have also been developed for computing correlated equilibria (again in the sense of the computation time being polynomial in the number of parameters) for general graphical games (Papadimitriou, 2005; see also Kakade et al., 2003). These algorithmic results are in sharp contrast to the status of computing equilibria for games represented in the normal form, where the results are either negative or remain unresolved. Graphical games have also proven valuable in establishing computational barriers to computing Nash equilibria efficiently, and certain classes of graphical games have been shown to be just as hard as the normal form in this regard (Daskalakis and Papadimitriou, 2005; 2006; Daskalakis, Goldberg and Papadimitriou, 2006; Schoenebeck and Vadhan, 2006).

Extensions of the model

Since the introduction of graphical games, a number of related models have been introduced and studied. In each case, the model again begins with an undirected network in which the edges represent the pairs of participants that are permitted to interact directly in some strategic or economic setting.

For instance, the model known as *graphical economics* (Kakade, Kearns and Ortiz, 2004) provides a network generalization of the classical exchange economies studied by Arrow and Debreu and others. As in the classical models, each consumer has an initial endowment over k commodities, and a subjective utility function describing his or her preferences over bundles of commodities. However, unlike the classical model, not all pairs of consumers may engage in trade. Instead, each consumer or vertex u may only trade with his or her neighbours $N(u)$, and there is no resale permitted. Equilibria in prices and consumption plans can still be shown to always exist, but now the equilibrium prices may need to be *local*, in the sense that two consumers may charge different prices per unit for the same commodity, and these prices may depend strongly on the network topology. This introduces variation in equilibrium wealth dependent on a consumer's position in the overall network (see below). As with graphical games, the graphical economics model permits efficient computation of equilibria under certain topological restrictions on the network.

More recently, a network version of evolutionary game theory (EGT) has also been examined (Kearns and Suri, 2006). In classical EGT, there are random encounters between all pairs of organisms; in the network generalization, such encounters are restricted to the edges of an undirected network. Thus the evolutionary fitness of an organism represented by vertex u is once again determined only by the strategies of its neighbours $N(u)$. More than one reasonable generalization of the evolutionary stable strategy (ESS) of classical EGT is possible in the network setting.

As with graphical games, both graphical economics and the network EGT model revert to their classical counterparts in the special case of the complete graph over all participants, and thus represent strict generalizations, but which are most interesting in cases where the underlying graph has some non-trivial structure.

Network structure and equilibrium properties

Aside from the algorithmic properties discussed above, one of the most interesting aspects of the various models under discussion is their ability to permit the study of how equilibrium properties are influenced by the structure or topology of the underlying network, and indeed there is a growing body of results in this direction.

In the case of graphical games, a tight connection can be drawn between the topology of the underlying graph G of the game and the structure of any minimal (in a certain natural technical sense) correlated equilibrium (CE), regardless of the details of the local payoff functions. Among other consequences, this result implies that CE in graphical games can be implemented using only local, distributed sources of randomization throughout the network in order effect the needed coordination, rather than the centralized randomization of classical CE. The result also provides broad conditions under which the play of two 'distant' players in the network may be conditionally independent – for instance, at CE, the play of vertices u and v is always conditionally independent given the pure strategies of any vertex cutset between them (Kakade et al., 2003).

In the graphical economics model, for certain simple cases one can give precise relationships between equilibrium wealth, price variation and network structure. For instance, in the case of an arbitrary bipartite network for a simple two-commodity buyer-seller economy with symmetric endowments and utilities (thus deliberately rendering network position the only source of asymmetry between consumers), there is no price or wealth variation at equilibrium if and only if the underlying network has a perfect matching sub-graph between buyers and sellers. More generally, a purely structural property of the network characterizes the ratio between the greatest and least consumer wealth at equilibrium. These structural results have been applied to analyse price and wealth variation in certain probabilistic models of buyer–seller network formation. For instance, it has been shown that whereas truly random networks with a certain minimum number of edges generally exhibit no variation in prices or wealth, those generated by recent models of social network formation such as preferential attachment lead to a power-law distribution of wealth (Kakade et al., 2004).

In the networked EGT setting, it has been proven that even networks with rather sparse connectivity (in which each organism directly interacts with only a small fraction of the total population), but in which the connections are formed randomly, classical ESS are always preserved, even if the initial locations of the invading population are arbitrary. Alternatively, if the network is arbitrary but the initial locations of the invading population are selected randomly, classical ESS are again preserved (Kearns and Suri, 2006). Related network models include those of Blume (1995) and Ellison (1993).

MICHAEL KEARNS

See also **learning and evolution in games: ESS; stochastic adaptive dynamics.**

Bibliography

Blume, L.E. 1995. The statistical mechanics of best-response strategy revision. *Games and Economic Behavior* 11, 111–45.

Daskalakis, C. and Papadimitriou, C. 2005. Computing pure Nash equilibria via Markov random fields. In *Proceedings of the 6th ACM Conference on Electronic Commerce.*

Daskalakis, C. and Papadimitriou, C. 2006. The complexity of games on highly regular graphs. In the 13th Annual European Symposium on Algorithms.

Daskalakis, C., Goldberg, P.W. and Papadimitriou, C.H. 2006. The complexity of computing a Nash equilibrium. In *Proceedings of the 38th ACM Symposium on Theory of Computing.*

Elkind, E., Goldberg, L.A. and Goldberg, P.W. 2006. Nash equilibria in graphical games on trees revisited. In *Proceedings of the 7th ACM Conference on Electronic Commerce.*

Ellison, G. 1993. Learning, local interaction, and coordination. *Econometrica* 61, 1047–71.

Jackson, M. 2007. The study of social networks ineconomics. In *The Missing Links: Formation and Decay of Economic Networks*, eds. J. Podolny, and J.E. Rauch. Russell Sage Foundation.

Kakade, S., Kearns, M. and Ortiz, L. 2004. Graphical economics. In *Proceedings of the 17th Conference on Computational Learning Theory.*

Kakade, S., Kearns, M., Langford, J. and Ortiz, L. 2003. Correlated equilibria in graphical games. In *Proceedings of the 4th ACM Conference on Electronic Commerce.*

Kakade, S., Kearns, M., Ortiz, L., Pemantle, R. and Suri, S. 2004. Economic properties of social networks. In *Neural Information Processing Systems 18.*

Kearns, M. and Suri, S. 2006. Networks preserving evolutionary stability and the power of randomization. In *Proceedings of the 7th ACM Conference on Electronic Commerce.*

Kearns, M., Littman, M. and Singh, S. 2001. Graphical models for game theory. In *Proceedings of the 17th Conference Uncertainty in Artificial Intelligence.*

Ortiz, L. and Kearns, M. 2002. Nash propagation for loopy graphical games. In *Neural Information Systems Processing 16.*

Papadimitriou, C. 2005. Computing correlated equilibria in multi-player games. In *Proceedings of the 37th ACM Symposium on the Theory of Computing.*

Schoenebeck, G. and Vadhan, S. 2006. The computational complexity of Nash equilibria in concisely represented games. In *Proceedings of the 7th ACM Conference on Electronic Commerce.*

incentive compatibility

Allocation mechanisms, organizations, voting procedures, regulatory bodies, and many other institutions are designed to accomplish certain ends such as the Pareto-efficient allocation of resources or the equitable resolution of disputes. In many situations it is relatively easy to conceive of feasible processes; processes which will accomplish the goals if all participants follow the rules and are capable of handling the informational requirements. Examples of such mechanisms include marginal cost pricing, designed to attain efficiency, and equal division, designed to attain equity. Of course once a feasible mechanism is found, the important question then becomes whether such a mechanism is also informationally feasible and compatible with 'natural' incentives of the participants. Incentive compatibility is the concept introduced by Hurwicz (1972, p. 320) to characterize those mechanisms for which participants in the process would not find it advantageous to violate the rules of the process.

The historical roots of the idea of incentive compatibility are many and deep. As was pointed out in one of a number of recent surveys,

> the concept of incentive compatibility may be traced to the 'invisible hand' of Adam Smith who claimed that in following individual self-interest the interests of society might be served. Related issues were a central concern in the 'Socialist Controversy' which arose over the viability of a decentralized socialist society. It was argued by some that such societies would have to rely on individuals to follow the rules of the system. Some believed this reliance was naive; others did not. (Groves and Ledyard, 1986, p. 1).

Further, the same issues have arisen in the design of voting procedures. Concepts and problems related to incentives were already identified and documented in the 18th century in discussions of proposals by Borda to provide alternatives to majority rule committee decisions. (See STRATEGY-PROOF ALLOCATION MECHANISMS for further information on voting procedures.)

Incentive compatibility is both desirable and elusive. The desirability of incentive compatibility can be easily illustrated by considering public goods, goods such that one consumer's consumption of them does not detract from another consumer's simultaneous consumption of that good. The existence of these collective consumption commodities creates a classic situation of *market failure*; the inability of markets to arrive at a Pareto-optimal allocation. It was commonly believed, prior to Groves and Ledyard (1977), that in economies with public goods it would be impossible to devise a decentralized process that would allocate resources efficiently since agents would have an incentive to 'free ride' on others' provision of those goods in order to reduce their own share of providing them. Of course Lindahl (1919) had proposed a feasible process which mimicked markets by creating a separate price for each individual's consumption of the public good. This designed process was, however, rejected as unrealistic by those who recognized that these 'synthetic markets' would be shallow (essentially monopsonistic) and therefore buyers would have no incentive to treat prices as fixed and invariant to their demands. The classic quotation is '... it is in the selfish interest of each person to give *false* signals, to pretend to have less interest in a given collective consumption activity than he really has...' (Samuelson, 1954, pp. 388–9). Allocating public goods efficiently through Lindahl pricing would be feasible and successful if consumers followed the rules; but, it would not be successful since the mechanism is not incentive compatible. If buyers do not follow the rules, efficient resource allocation will not be achieved and the goals of the design will be subverted because of the motivations of the participants. Any institution or rule, designed to accomplish group goals, must be incentive compatible if it is to perform as desired.

The elusiveness of incentive compatibility can be most easily illustrated by considering a situation with only private goods. Economists generally model behaviour in private goods markets by assuming that buyers and sellers 'follow the rules' and take prices as given. It is now known, however, that as long as the number of agents is finite then any one of them can still gain by misbehaving and, furthermore, can do so in a way which can not be detected by anyone else. The explanation is provided in two steps. First, if there are a finite number of traders, and none have a perfectly elastic offer curve (which will be true if preferences are non-linear) then one trader can gain by being able to control prices. For example, a buyer would want to set price where his marginal benefit equalled his marginal outlay and thereby gain monopsonistic benefits. Of course, if the others know that buyer's demand curve (either directly or through inferences based on revealed preference) then they would know that the buyer was not 'taking prices as given' and could respond with a suitable punishment against him. This brings us to our second step. Even though others can monitor and prohibit price setting behaviour, our benefit-seeking monopsonist has another strategy which can circumvent this supervision. He calculates a (false) demand curve which, when added to the others' offer curves, produces an equilibrium price equal to that which he would have set if he had direct control. He then calculates a set of preferences which yields that demand curve and participates in the process *as if he had these (false) preferences.* Usually this involves simply acting as if one has a slightly lower demand curve than one really does. Since preferences are not able to be observed by others, he can follow this behaviour which looks like it is price-taking, and therefore 'legal', and can do individually better. The unfortunate implication of such concealed misbehaviour is that the mechanism performs other than as intended. In this case, resources are artificially limited and too little is traded to attain efficiency.

In 1972 Hurwicz established the validity of the above intuition. His theorem can be precisely stated after the introduction of some notation and a framework for further discussion.

The impossibility theorem

The key concepts include economic environments, allocation mechanisms, incentive compatibility, the no-trade option, and Pareto-efficiency. We take up each in turn.

An *economic environment*, those features of an economy which are to be taken as given throughout the analysis, includes a description of the agents, the feasible allocations they have available and their preferences for those allocations. While many variations are possible, I concentrate here on a simple model. Agents (consumers, producers, politicians, etc.) are indexed by $i = 1, \ldots, n$. X is the set of feasible allocations where $x = (x^i, \ldots, x^n)$ is a typical element of X. (An exchange environment is one in which X is the set of all $x = (x^1, \ldots, x^n)$ such that $x^i \geq 0$ and $\sum x^i = \sum w^i$, where w^i is i's initial endowment of commodities.) Each agent has a selfish utility function $u^i(x^i)$. The environment is $e = [I, X, u^1, \ldots, u^n]$. A crucial fact is that initially *information is dispersed* since i, and only i, knows u^i. We identify the specific knowledge i initially has as i's *characteristic*, e^i. In our model, $e^i = u^i$.

Although there are many variations in models of allocation mechanisms, I begin with the one introduced by Hurwicz (1960). An *allocation mechanism* requests information from the agents and then computes a feasible allocation. It requests information in the form of messages m^i from agent i through a *response* function $f^i(m^i, \ldots, m^n)$. Agent i is told to report $f^i(m, e^i)$ if others have reported m and i's characteristic is e^i. An equilibrium of these response rules, for the environment e, is a joint message m such that $m^i = f^i(m, e^i)$ for all i. Let $\mu(e, f)$ be the set of equilibrium messages for the response functions f in the environment e. The allocation mechanism computes a feasible allocation x by using an *outcome* function $g(m)$ on equilibrium messages. The net result of all of this in the environment e is the allocation $g[\mu(e, f)] = x$ *if all i follow the rules, f.* Thus, for example, the *competitive mechanism* requests agents to send their demands as a function of prices which are in turn computed on the basis of the aggregate demands reported by the consumers. In equilibrium, each agent is simply allocated their stated demand. (An alternative mechanism, yielding exactly the same allocation in one iteration, would request the demand function and then compute the equilibrium price and allocation for the reported demand functions.) It is well known, for exchange economies with only private goods, that if agents report their true demands then the allocations computed by the competitive mechanism will be Pareto-optimal.

It is obviously important to be able to identify those mechanisms, those rules of communication, that have the property that they are self-enforcing. We do that by focusing on a class of mechanisms in which each agent gains nothing, and perhaps even loses, by misbehaving. While a multitude of misbehaviours could be considered it is sufficient for our purposes to consider a slightly restricted range. In particular we can concentrate on undetectable behaviour, behaviour which no outside agent can distinguish from that prescribed by the mechanism. We model this limitation on behaviour by requiring the agent to restrict his misrepresentations to those which are consistent with some characteristic he might have. An allocation mechanism is said to be *incentive compatible* for all environments in the class E if there is no agent i and no environment e in E and no characteristic e^{*i} such that (e/e^{*i}) is in E (where (e/e^{*i}) is the environment derived from e by replacing e^i with e^{*i}) and such that

$$u^i\{g[\mu(e,f)],e^i\} < u^i\{g[\mu(e/e^{*i},f)],e^i\}$$

where $u^i(x^*, e^i)$ is i's utility function in the environment e. That is, no agent can manipulate the mechanism by pretending to have a characteristic different from the true one and do better than acting according to the truth. The agent has an incentive to follow the rules and the rules are compatible with his motivations.

Incentive compatibility is at the foundation of the modern *theory of implementation.* In that theory, one tries to identify conditions under which a particular social choice rule or performance standard, $P : E \rightarrow X$, can be recreated by an allocation mechanism under the hypothesis that individuals will follow their self-interest when they participate in the implementation process. In our language, the rule P is implementable if and only if there is an incentive compatible mechanism (f, g) such that $g[\mu(e,f)] = P(e)$ for all e in E. The theory of implementation seeks to answer the question 'which P are implementable?' We will see some of the answers below for P which select from the set of Pareto-efficient allocations. Those interested in more general goals and performance standards should consult Dasgupta, Hammond and Maskin (1979) or Postlewaite and Schmeidler (1986).

An allocation mechanism is said to have the *no trade-option* if there is an allocation θ at which each participant may remain. In exchange environments the initial endowment is usually such an allocation. Mechanisms with a no-trade option are non-coercive in a limited sense. If an allocation mechanism possesses the no-trade option then the allocation it computes for an environment e, if agents follow the rules, must leave everyone at least as well off, using the utility functions for e, as they are at θ. That is, for all i and all e in E

$$u^i\{g[\mu(e,f)],e^i\} > u^i(\theta,e^i).$$

An allocation mechanism is said to be *Pareto-efficient in E* if the allocations selected by the mechanism, when agents follow the rules, are Pareto-optimal in e. That is, for each e in E, there is no allocation x^* in X such that, for all i,

$$u^i(x^*,e^i) \geq u^i\{g[\mu(e,f)],e^i\}$$

with strict inequality for some i.

With this language and notation, Hurwicz's theorem on the elusive nature of incentive compatibility in private markets, subsequently expanded by Ledyard and Roberts (1974) to include public goods environments, can now be easily stated. *Theorem:* In classical (public or private) economic environments with a finite number of agents, there is no incentive compatible allocation mechanism which possesses the no-trade option and is Pareto-efficient. (Classical environments include pure exchange environments with Cobb–Douglas utility functions.)

A more general version of this theorem, in the context of social choice theory, has been proven by Gibbard (1973) and Satterthwaite (1975) with the concept of a 'non-dictatorial social choice function' replacing that of a 'mechanism with the no-trade option'. (See STRATEGY-PROOF ALLOCATION MECHANISMS.)

There are a variety of possible reactions to this theorem. One is simply to give up the search for solutions to market failure since the theorem seems to imply that one should not waste any effort trying to create institutions to allocate resources efficiently. A second is to notice that, at least in private markets, if there are a very large number of individuals in each market then efficiency is 'almost' attainable (see Roberts and Postlewaite, 1976). A third is to recognize that the behaviour of individuals will generally be different from that implicitly assumed in the definition of incentive compatibility. A fourth is to accept the inevitable, lower one's sights, and look for the 'most efficient' mechanism among those which are incentive compatible and satisfy a voluntary participation constraint. We consider the last two options in more detail.

Other behaviour: Nash equilibrium

If a mechanism is incentive compatible, then each agent knows that his best strategy is to follow the rules according to his true characteristic, *no matter what the other agents will do.* Such a strategic structure is referred to as a dominant strategy game and has the property that no agent need know or predict anything about the others' behaviour. In mechanisms which are not incentive compatible, each agent must predict what others are going to do in order to decide what is best. In this situation agents' behaviour will not be as assumed in the definition of incentive compatibility. What it will be continues to be an active research topic and many models have been proposed. Since most of these are covered in Groves and Ledyard (1986), I will concentrate on the two which seem most sensible. Both rely on game-theoretic analyses of the strategic possibilities. The first concentrates on the outcome rule, g, and postulates that agents will not choose messages to follow the specifications of the response functions but to do the best they can against the messages sent by others. Implicitly this assumes that there is some type of iterative process (embodied in the response rules) which allows revision of one's message in light of the responses of others. We can formalize this presumed strategic behaviour in a new concept of incentive compatibility. An allocation mechanism (f, g) is called *Nash incentive compatible* for all environments in E if there is no environment e, no agent i, and no message m^{*i} which i can send such that

$$u^i(g[\mu(e,f)/m^{*i}, e^i]) > u^i(g[\mu(e,f), e^i])$$

where $\mu(e, f)$ is the 'equilibrium' message of the response rules f in the environment e, $g(m)$ is the outcome rule, and $[m/m^{*i}]$ is the vector m where m^{*i} replaces m^i. In effect this requires the equilibrium messages of the response rules to be Nash equilibria in the game in which messages are strategies and payoffs are given by $u[g(m)]$. It was shown in a sequence of papers written in the late 1970s, including those by Groves and Ledyard (1977), Hurwicz (1979), Schmeidler (1980), and Walker (1981), that Nash incentive compatibility is not elusive. The effective output of that work was to establish the following. *Theorem*: In classical (public or private) economic environments with a finite number of agents, there are many Nash incentive compatible mechanisms which possess the no-trade option and are Pareto-efficient.

With a change in the predicted behaviour of the participants in the mechanism, in recognition of the fact that in the absence of dominant strategies agents must follow some other self-interested strategies, the pessimism of the Hurwicz theorem is replaced by the optimistic prediction of a plethora of possibilities. (See Dasgupta, Hammond and Maskin, 1979, Postlewaite and Schmeidler (1986) and Groves and Ledyard (1986) for comprehensive surveys of these results including many for more general social choice environments.) Although it remains an unsettled empirical question whether participants will indeed behave this way, there is a growing body of experimental evidence that seems to me to support the behavioural hypotheses underpinning Nash incentive compatibility, especially in iterative tâtonnement processes.

Other behaviour: Bayes' equilibrium

The second approach to modelling strategic behaviour of agents in mechanisms, when dominant strategies are not available, is based on Bayesian decision theory. These models, called *games of incomplete information* (see Myerson, 1985), concentrate on the beliefs of the players about the situation in which they find themselves. In the simplest form, it is postulated that there is a common knowledge (everyone knows that everyone knows that...) probability function, $\pi(e)$, which describes everyone's prior beliefs. Each agent is then assumed to choose that message which is best against the expected behaviour of the other agents. The expected behaviour of the other agents is also constrained to be 'rational' in the sense that it should be best against the behaviour of others. This presumed strategic behaviour is embodied in a third type of incentive compatibility. (It could be argued that the concept of incentive compatibility remains the same, based on non-cooperative behaviour in the game induced by the mechanism, while only the presumed information structure and sequence of moves required to implement the allocation mechanism are changed. Such a view is not inconsistent with that which follows.) An allocation mechanism (f, g) is called *Bayes incentive compatible* for all environments in E given π on E if there is no environment e^*, no agent i, and no message m^{*i} which i can send such that

$$\int u^i\{g[\mu(e,f)/m^{*i}], e^{*i}\}\mathrm{d}\pi(e|e^{*i}) > \int u^i\{g[\mu(e,f), e^{*i}]\mathrm{d}\pi(e,|e^{*i})\}$$

where, as before, μ is the equilibrium message vector and g is the outcome rule. Further, $\pi(e/e^{*i})$ *is the conditional probability measure on* e given e^{*i}, and u^i is a von Neumann–Morgenstern utility function. In effect, this requires the equilibrium messages of the response rules to be Bayes equilibrium outcomes of the incomplete information game with messages as strategies, payoffs $u[g(m)]$ and common knowledge prior π.

There are two types of results which deal with the possibilities for Bayes incentive compatible design of allocation mechanisms, neither of which is particularly encouraging. The first type deals with the possibilities for incentive compatible design which is independent of the beliefs. The typical theorem is illustrated by the following result proven by Ledyard (1978). *Theorem*: In classical economic environments with a finite number of agents, there is no Bayes incentive compatible mechanism which possesses the no-trade option and is Pareto-efficient *for all* π *on E*. Understanding this result is easy when one realizes that any mechanism (f, g) is Bayes incentive compatible for all π for all e in E if and only if it is (Hurwicz) incentive compatible for all e in E. Thus the Hurwicz impossibility theorem again applies.

The second type of result is directed towards the possibilities for a specific prior π; that is, towards what can be done if the mechanism can depend on the common knowledge beliefs. The most general characterizations of the possibilities for Bayes incentive compatible design can be found in Palfrey and Srivastava (1987) and Postlewaite and Schmeidler (1986). They have shown that two conditions, called monotonicity and self-selection, are necessary and sufficient for a social choice correspondence to be implementable in the sense that there is a Bayes incentive compatible mechanism that reproduces that correspondence. The details of these conditions are not important. What is important is that many correspondences do not satisfy them. In particular, there appear to be many priors π and many sets of environments E for which there is no mechanism which is Bayes incentive compatible, provides a no-trade option and is Pareto-efficient. Thus, impossibility still usually occurs even if one allows the mechanism to depend on the prior.

One recent avenue of research which promises some optimistic counterweight to these negative results can be found in Palfrey and Srivastava (1987). In much the same way that the natural move from Hurwicz incentive compatibility to Nash incentive compatibility created opportunities for incentive compatible design, these authors have shown that a move back towards dominant strategies may also open up possibilities. Refinements arise by varying the equilibrium concept in a way that reduces the number of (Bayes or Nash) equilibria for a given e or π. Moore and Repullo use subgame perfect Nash equilibria. Palfrey and Srivastava eliminate weakly dominated strategies from the set of Nash equilibria. They have discovered that, in pure exchange environments, virtually all performance correspondences are implementable if behaviour satisfies these refinements. In particular, any selection from the Pareto-correspondence is implementable for these refinements, and so there are many refined-Nash incentive compatible mechanisms which are Pareto-efficient and allow a no-trade option. It is believed that these results will transfer naturally to refinements of Bayes equilibria, but the research remains to be done.

Incentive compatibility as a constraint

Another of the reactions to the Hurwicz impossibility result is to accept the inevitable, to view incentive compatibility as a constraint, and to design mechanisms to attain the best level of efficiency one can. If full efficiency is possible, it will occur as the solution. If not, then one will at least find the second-best allocation mechanism. Examples of this rapidly expanding research literature include work on optimal auctions (Harris and Raviv, 1981; Matthews, 1983; Myerson, 1981), the design of optimal contracts for the principle-agent problem, and the theory of optimal regulation (Baron and Myerson, 1982). As originally posed by Hurwicz (1972, pp. 299–301), the idea is to adopt a social welfare function $W(x, e)$, a measure of the social welfare attained from the allocation x if the environment is e and then to choose the mechanism (f, g) to maximize the (expected) value of W subject to the 'incentive compatibility constraints', the constraint that the rules (f, g) be consistent with the motivations of the participants. One chooses (f, g) to

$$\text{maximize} \int W\{g[\mu(e, f)], e\} \mathrm{d}\pi(e)$$

subject to, for every i, every e, and every e^{*i},

$$\int u^i\{g[\mu(e/e^{*i}, f)], e^i\} \mathrm{d}\pi(e|e^i) \leq \int u^i(g[\mu(e, f), e^i]) \mathrm{d}\pi(e|e^i).$$

As formalized here the incentive compatibility constraints embody the concept of Bayes incentive compatibility. Of course, other behavioural models could be substitued as appropriate.

Sometimes a voluntary participation constraint, related to the no-trade option of Hurwicz, is added to the optimal design problem. One form of this constraint requires that (f, g) also satisfy, for every i and every e,

$$\int u^i\{g[\mu(e)], e^i\} \mathrm{d}\pi(e|e^i) \geq \int u^i(\theta[e], e^i) \mathrm{d}\pi(e|e^i).$$

In practice this optimization can be a difficult problem since there are a large number of possible mechanisms (f, g). However, an insight due to Gibbard (1973) can be employed to reduce the range of alternatives and simplify the analysis. Now called the *revelation principle*, the observation he made was that, to find the maximum, it is sufficient to consider only mechanisms, called direct revelation mechanisms, in which agents are asked to report their own characteristics. The reason is easy to see. Suppose that (f^*, g^*) solves the maximum problem. Let (F^*, G^*) be a new (direct revelation) mechanism defined by $F^{*i}(m, e^i) = e^i$ and $G^*(m) = g[\mu(m, f)]$. Each i is told to report his characteristic and then G^* computes the allocation by computing that which would have been chosen if the original mechanism (f, G^*) had been used honestly in the reported environment. (F^*, G^*) yields the same allocation as (f^*, g^*), *if each agent reports the truth.* But the incentive compatibility constraints, which (f^*, g^*) satisfied, ensure that each agent will want to report truthfully. Thus, whatever can be done, by any arbitrary mechanism subject to the Bayes incentive compatibility constraints, can be done with direct revelation mechanisms subject to the constraint that each agent wants to report their true characteristic. One need only choose a function $G : E \to X$ to

$$\text{maximize} \int W\{G(e), e\} \mathrm{d}\pi(e)$$

subject to, for every i, e and e^i,

$$\int u^i\{G(e/e^{*i}), e^i\}\mathrm{d}\pi(e|e^i) \leq \int u^i[G(e), e^i]\mathrm{d}\pi(e|e^i),$$

and

$$\int u^i[G(e), e^i]\mathrm{d}\pi(e|e) \geq \int u^i(\theta[e], e^i)\mathrm{d}\pi(e|e^i).$$

There are at least two problems with this approach to organizational design. The first is that the choice of mechanism depends crucially on the prior beliefs, π. This is a direct result of the use of Bayes incentive compatibility in the constraints. Since the debate is still open let me simply summarize some of the arguments. One is that if the mechanism chosen for a given situation does not depend on common knowledge beliefs then we would not be using all the information at our disposal to pursue the desired goals and would do less than is possible. Further, since the beliefs are common knowledge we can all agree as to their validity (misrepresentation is not an issue) and therefore to their legitimate inclusion in the calculations. An argument is made against this on the practical grounds that one need only consider actual situations, such as the introduction of new technology by a regulated utility or the acquisition of a major new weapons system by the government, to understand the difficulties involved in arriving at agreements about the particulars of common knowledge. Another argument against is based on the feeling that mechanisms should be robust. A 'good' mechanism should be able to be described in terms of its mechanics and, while it probably should have the capacity to incorporate the common knowledge relevant to the current situation, it should be capable of being used in many situations. How to capture these criteria in the constraints or the objective function of the designer remains an open research question.

The second problem with the optimal auction approach to organizational design is the reliance on the revelation principle. Restricting attention to direct revelation mechanisms, in which an agent reports his entire characteristic, is an efficient way to prove theorems, but it provides little guidance for those interested in actual organization design. For example it completely ignores the informational requirements of the process and any limitations, if any, in the information processing capabilities of the agents or the mechanism. Writing down one's preferences for all possible consumption patterns is probably harder than writing down one's entire demand surface which is certainly harder than simply reacting to a single price vector and reporting only the quantities demanded at that price. A failure to recognize the information processing constraints in the optimization problem is undoubtedly one of the reasons there has been limited success in using the theory of optimal auctions to explain the existence of pervasive institutions, such as the first-price sealed-bid auction used in competitive contracting or the posted price institution used in retailing.

Summary

Incentive compatibility captures the fundamental positivist notion of self-interested behaviour that underlies almost all economic theory and application. It has proven to be an organizing principle of great scope and power. Combined with the modern theory of mechanism design, it provides a framework in which to analyse such diverse topics as auctions, central planning, regulation of monopoly, transfer pricing, capital budgeting, and public enterprise management. Incentive compatibility provides a basic constraint on the possibilities for normative analysis. As such it serves as the fundamental interface between what is desirable and what is possible in a theory of organizations.

JOHN O. LEDYARD

Bibliography

Baron, D. and Myerson, R. 1982. Regulating a monopolist with unknown costs. *Econometrica* 50, 911–30.

Dasgupta, P., Hammond, P. and Maskin, E. 1979. The implementation of social choice rules: some general results on incentive compatibility. *Review of Economic Studies* 46, 185–216.

Gibbard, A. 1973. Manipulation of voting schemes: a general result. *Econometrica* 41, 587–602.

Groves, T. and Ledyard, J. 1977. Optimal allocation of public goods: a solution to the 'free rider' problem. *Econometrica* 45, 783–809.

Groves, T. and Ledyard, J. 1986. Incentive compatibility ten years later. In *Information, Incentives, and Economic Mechanisms*, ed. T. Groves, R. Radner and S. Reiter. Minneapolis: University of Minnesota Press.

Harris, M. and Raviv, A. 1981. Allocation mechanisms and the design of auctions. *Econometrica* 49, 1477–99.

Hurwicz, L. 1960. Optimality and informational efficiency in resource allocation processes. In *Mathematical Methods in the Social Sciences*, ed. K. Arrow, S. Karlin and P. Suppes. Stanford: Stanford University Press, 27–46.

Hurwicz, L. 1972. On informationally decentralized systems. In *Decision and Organization: A Volume in Honor of Jacob Marschak*, ed. R. Radner and C. B. McGuire. Amsterdam: North-Holland, 297–336.

Hurwicz, L. 1979. Outcome functions yielding Walrasian and Lindahl allocations at Nash equilibrium points. *Review of Economic Studies* 46, 217–25.

Ledyard, J. 1978. Incomplete information and incentive compatibility. *Journal of Economic Theory* 18, 171–89.

Ledyard, J. and Roberts, J. 1974. On the incentive problem with public goods. Discussion Paper No. 116, Center for Mathematical Studies in Economics and Management Science, Northwestern University.

Lindahl, E. 1919. *Die Gerechtigkeit der Besteuerung*. Lund. Partial translation in*Classics in the Theory of Public finance*, ed. R. A. Musgrave and A. T. Peacock. London: Macmillan, 1958.

Matthews, S. 1983. Selling to risk averse buyers with unobservable tastes. *Journal of Economic Theory* 30, 370–400.

Moore, J. and Repullo, R. 1986. Subgame perfect implementation. Working paper, London School of Economics.

Myerson, R.B. 1981. Optimal auction design. *Mathematics of Operations Research* 6, 58–73.

Myerson, R.B. 1985. Bayesian equilibrium and incentive compatibility: an introduction. In *Social Goals and Social Organization: Essays in Memory of Elisha Pazner*, ed. L. Hurwicz, D. Schmeidler and H. Sonnenschein. Cambridge: Cambridge University Press.

Palfrey, T. and Srivastava, S. 1986. Implementation in exchange economies using refinements of Nash equilibrium. Graduate School of Industrial Administration, Carnegie-Mellon University, July 1986.

Palfrey, T. and Srivastava, S. 1987. On Bayesian implementable allocations. *Review of Economic Studies* 54, 193–208.

Postlewaite, A. and Schmeidler, D. 1986. Implementation in differential information economics. *Journal of Economic Theory* 39, 14–33.

Roberts, J. and Postlewaite, A. 1976. The incentives for price-taking behavior in large economies. *Econometrica* 44, 115–28.

Samuelson, P. 1954. The pure theory of public expenditure. *Review of Economics and Statistics* 36, 387–9.

Satterthwaite, M. 1975. Strategy-proofness and Arrow's conditions: existence and correspondence theorems for voting procedures and social welfare functions. *Journal of Economic Theory* 10, 187–217.

Schmeidler, D. 1980. Walrasian analysis via strategic outcome functions. *Econometrica* 48, 1585–93.

Walker, M. 1981. A simple incentive compatible scheme for attaining Lindahl allocations. *Econometrica* 49, 65–71.

large games (structural robustness)

Earlier literature on large (many players) cooperative games is surveyed in Aumann and Shapley (1974). For large strategic games, see Schmeidler (1973) and the follow-up literature on the purification of Nash equilibria. There is also substantial literature on large games with special structures, for example large auctions as reported in Rustichini, Satterthwaite, and Williams (1994).

Unlike the above, this survey concentrates on the structural robustness of (general) Bayesian games with many semi-anonymous players, as developed in Kalai (2004; 2005). (For additional notions of robustness in game theory, see Bergemann and Morris, 2005.)

Main message and examples

In simultaneous-move Bayesian games with many semi-anonymous players, all Nash equilibria are structurally robust. The equilibria survive under structural alterations that relax the simultaneous-play assumptions, and permit information transmission, revisions of choices, communication, commitments, delegation, and more.

Large economic and political systems and distributive systems such as the Web are examples of environments that give rise to such games. Immunity to alterations means that Nash equilibrium predictions are valid even in games whose structure is largely unknown to modellers or to players.

The next example illustrates immunity of equilibrium to revisions, or being *ex post* Nash, see Cremer and McLean (1985), Green and Laffont (1987) and Wilson (1987) for early examples.

Example 1 *Ex post stability illustrated in match pennies*

Simultaneously, each of k males and k females chooses one of two options, H or T. The payoff of every male is the proportion of females his choice matches and the payoff of every female is the proportion of males her choice *mis*matches. (When $k=1$ this is the familiar match-pennies game.) Consider the mixed-strategy equilibrium where every player chooses H or T with equal probabilities.

Structural robustness implies that the equilibrium must be *ex post* Nash: it should survive in alterations that allow players to revise their choices after observing their opponents' choices. Clearly this is not the case when k is small. But as k becomes large, the equilibrium becomes arbitrarily close to being *ex post* Nash. More precisely, the Prob[some player can improve his payoff by more than ε *ex post*] decreases to zero at an exponential rate as k becomes large.

Example 2 *Invariance to sequential play illustrated in a computer choice game*

Simultaneously, each of n players chooses one of two computers, I or M. But before choosing, with 0.50 – 0.50 i.i.d. probabilities, every player is privately informed that she is an I-type or an M-type. The payoff of every player is 0.1 if she chooses the computer of her type (zero otherwise) plus 0.9 times the proportion of opponents whose choices she matches. (Identical payoffs and prior probabilities are assumed only to ease the presentation. The robustness property holds without these assumptions.) Consider the favourite-computer equilibrium (FC) where every player chooses the computer of her type.

Structural robustness implies that the equilibrium must *be invariant to sequential play*: it should survive in alterations in which the (publicly observed) computer choices are made sequentially. Clearly this is not the case for small n, where any equilibrium must involve herding. But as n becomes large, the structural robustness theorem below implies that FC becomes an equilibrium in all sequential alterations. More precisely, the Prob[some player, by deviating to her non favorite computer, can achieve an ε-improvement at her turn] decreases to zero at an exponential rate.

The general definition of structural robustness, presented next, accommodates the above examples and much more.

Structural robustness

A mixed-strategy (Nash) equilibrium $\sigma = (\sigma_1, \ldots, \sigma_n)$ of a one-simultaneous-move n-person strategic game G is structurally robust if it *remains an equilibrium* in *every* structural *alteration* of G. Such an alteration is described by an extensive game, $\mathscr{A}$, and for σ to remain an equilibrium in $\mathscr{A}$ means that *every adaptation* of σ to $\mathscr{A}$, $\sigma^{\mathscr{A}}$, must be an equilibrium in $\mathscr{A}$.

Consider any n-person one-simultaneous-move Bayesian game G, like the Computer Choice game above.

Definition 1 *A (structural) alteration of* G is any finite extensive game $\mathscr{A}$ with the following properties:

1. *$\mathscr{A}$ includes the (original) G-players*: The players of $\mathscr{A}$ constitute a superset of the G-players (the players of G).
2. *Unaltered type structure*: At the first stage of $\mathscr{A}$, the G-players are assigned a profile of types by the same prior probability distribution as in G. Every player is informed of his own type.
3. *Playing $\mathscr{A}$ means playing G*: with every final node of $\mathscr{A}$, z, there is an associated unique profile of G pure-strategies, $a(z) = (a_1(z), \ldots, a_n(z))$.
4. *Unaltered payoffs*: the payoffs of the G-players at every final node z are the same as their payoffs in G (at the profile of realized types and final pure-strategies $a(z)$).
5. *Preservation of original strategies*: every pure-strategy a_i of a G-player i has at least one $\mathscr{A}$ adaptation. That is, an $\mathscr{A}$-strategy $a_i^{\mathscr{A}}$ that guarantees (w.p. 1) ending at a final node z with $a_i(z) = a_i$ (no matter what strategies are used by the opponents).

In the computer choice example, every play of an alteration $\mathscr{A}$ must produce a profile of computer allocations for the G-players. Their preferences in $\mathscr{A}$ are determined by their preferences over profiles of computer allocations in G. Moreover, every G-player i has at least one $\mathscr{A}$-strategy $I_i^{\mathscr{A}}$ (which guarantees ending at a final node where she is allocated I), and at least one $\mathscr{A}$-strategy $M_i^{\mathscr{A}}$ (which guarantees ending at a final node where she is allocated M).

Definition 2 *An $\mathscr{A}$ (mixed) strategy-profile, $\sigma^{\mathscr{A}}$, is an adaptation of* a G (mixed) strategy-profile σ, if for every G-player i, every $\sigma_i^{\mathscr{A}}$ is an $\mathscr{A}$-adaptation of σ_i. That is, for every G pure-strategy a_i, $\sigma_i(a_i) = \sigma_i^{\mathscr{A}}(a_i^{\mathscr{A}})$ for some $\mathscr{A}$-adaptation $a_i^{\mathscr{A}}$ of a_i.

In the computer choice example, for a G-strategy where player i randomizes 0.20 to 0.80 between I and M, an $\mathscr{A}$ adaptation must randomize 0.20–0.80 between a strategy of the type $I_i^{\mathscr{A}}$ and a strategy of the type $M_i^{\mathscr{A}}$.

Definition 3 An equilibrium σ of G *is structurally robust* if in every alteration of G, $\mathscr{A}$, and in every adaptation of σ, $\sigma^{\mathscr{A}}$, the strategy of every G-player i, $\sigma_i^{\mathscr{A}}$, is best response to $\sigma_{-i}^{\mathscr{A}}$.

Remark 1 The structural robustness theorem, discussed later, presents an asymptotic result: the equilibria are structurally robust up to two positive numbers (ε, ρ), which can be made arbitrarily small as n becomes large. The notion of approximate robustness is the following.

An equilibrium is (ε, ρ)-structurally robust if in every alteration and every adaptation as above, Prob[visiting an information set where a G-player can improve his payoff by more than ε] $\leq \rho$. (ε-improvement is computed conditional on being at the information set. To gain such improvement the player may coordinate his deviation: he may make changes at the information set under consideration together with changes at forthcoming ones.)

For the sake of brevity, the next section discusses full structural robustness. But all the observations presented there also hold for the properly defined approximate counterparts. For example, the fact that structural robustness implies *ex post* Nash also implies that approximate structural robustness implies approximate *ex post* Nash. The implications of approximate (as opposed to full) structural robustness are important, due to the asymptotic nature of the structural robustness theorem.

Implications of structural robustness

Structural robustness of an equilibrium σ in a game G is a strong property, because the set of G-alterations that σ must survive is rich. The simple examples below are meant to suggest the richness of its implications, with the first two examples showing how it implies the notions already discussed (see Dubey and Kaneko, 1984 for related issues).

Remark 2 *Ex post Nash and being information-proof*

G with revisions, $\mathscr{GR}$, is the following n-person extensive game. The n players are assigned types as in G (using the prior type distribution of G and informing every player of his own type). In a first round of simultaneous play, every player chooses one of his G pure strategies; the types realized and pure strategies chosen are all made public knowledge. Then, in a second round of simultaneous play, the players again choose pure strategies of G (to revise their first round choices). The payoffs are as in G, computed at the profile of realized types with the profile of pure strategies chosen in the *second* round.

Clearly $\mathscr{GR}$ satisfies the definition of an alteration (with no additional players), and every equilibrium σ of G has the following $\mathscr{GR}$ adaptation, σ^{NoRev}: in the first round the players choose their pure strategies according to σ, just as they do in G; in the second round nobody revises his first round choice.

Structural robustness of σ implies that σ^{NoRev} must be an equilibrium of $\mathscr{GR}$, that is, σ is *ex post* Nash.

Moreover, the above reasoning continues to hold even if the information revealed between the two rounds is partial and different for different players. The fact that σ^{NoRev} is an equilibrium in all such alterations shows that σ is *information-proof*: no revelation of information (even if strategically coordinated by G-players and outsiders) could give any player an incentive to revise. Thus, structural robustness is substantially stronger than all the variants of the *ex post* Nash condition. (In the non-approximate notions, being *ex post* Nash is equivalent to being information proof. But in the approximate notions information proofness is substantially stronger.)

Remark 3 *Invariance to order of play*

G played sequentially, $\mathscr{GS}$, is the following n-person extensive game. The n players are assigned types as in G. The play progresses sequentially, according to a fixed publicly known order. Every player, at his turn, knows all earlier choices.

Clearly, $\mathscr{GS}$ is an alteration of G, and every equilibrium σ of G has the following $\mathscr{GS}$ adaptation: At his turn, every player i chooses a pure-strategy with the same probability distribution σ_i as he does in the simultaneous-move game G. Structural robustness of σ implies that this adaptation of σ must be an equilibrium in every such $\mathscr{GS}$.

Moreover, the above reasoning continues to hold even if the order of play is determined dynamically, and even if it is strategically controlled by G-players and outsiders. Thus, a structurally robust equilibrium is invariant to the order of play in a strong sense.

Remark 4 *Invariance to revelation and delegation*

G with delegation, $\mathscr{GD}$, is the following $(n+1)$-players game. The original n G-players are assigned types as in G. In a first round of simultaneous play, every G-player chooses between (1) self-play and (2) delegate-the-play and report a type to an outsider, player $n+1$. In a second round of simultaneous play all the self-players choose their own G pure strategies, and the outsider chooses a profile of G pure strategies for all the delegators. The payoffs of the G players are as in G; the outsider may be assigned any payoffs.

Clearly, $\mathcal{GD}$ is an alteration of G, and every equilibrium σ of G has adaptations that involve no delegation.

In the computer choice game, for example, consider an outsider with incentives to coordinate: his payoff equals one when he chooses the same computer for all delegators, zero otherwise. This alteration has a new (more efficient) equilibrium, not available in G: everybody delegates and the outsider chooses the most-reported type.

Nevertheless, as structural robustness implies, FC remains an equilibrium in $\mathcal{GD}$ (nobody delegates in the first round and they choose their favorite computers in the second). Moreover, FC remains an equilibrium under any scheme that involves reporting and voluntary delegation of choices.

Remark 5 *Partially specified games*

Structurally robust equilibria survive under significantly more complex alterations than the ones above. For example, one could have multiple opportunities to revise, to delegate, to affect the order of play, to communicate, and more. Because of these strong invariance properties, such equilibria may be used in games which are only partially specified as illustrated by the following example.

Example 3 *A game played on the Web*

Suppose that instead of being played in one simultaneous move, the Computer Choice game has the following instruction: 'Go to Web site xyz before the end of the week, and click in your computer choice.' This instruction involves substantial structural uncertainty: In what order would the players choose? Who can observe whom? Who can talk to whom? Can players sign binding agreements? Can players revise their choices? Can players delegate their choices? And so forth.

Because it is unaffected by the answers to such questions, a structurally robust equilibrium σ of the one-simultaneous-move game can be played on the Web in a variety of ways without losing the equilibrium property. For example, players may make their choices according to their σ_i probabilities prior to the beginning of the click-in period, then go to the Web and click in their realized choices at individually selected times.

Remark 6 *Competitive prices in Shapley–Shubik market games*

For a simple illustration, consider the following n-trader market game (see Shapley and Shubik, 1977, and later references in Dubey and Geanakoplos, 2003, and McLean, Peck and Postlewaite, 2005). There are two fruits, apples and bananas, and a finite number of trader types. A type describes the fruit a player owns and the fruit he likes to consume. The players' types are determined according to individual independent prior probability distributions. Each trader knows his own type, and his payoff depends on his own type and the fruit he ends up with, as well as on the distribution of types and fruit ownership of his opponents (externalities are allowed, for example, a player may wish to own the fruit that most opponents like). In one simultaneous move, every player has to choose between (1) keeping his fruit and (2) trading it for the other kind.

The banana/apple price is determined proportionately (with one apple and banana added in to avoid division by zero). For example, if 199 bananas and 99 apples are traded, the price of bananas to apples would be $(199+1)/(99+1)=2$, that is, every traded apple brings back two bananas and every traded banana brings back 0.5 apples.

With a small number of traders, the price is unlikely to be competitive. If players are allowed to re-trade after the realized price becomes known, they would, and a new price would emerge.

However, when n is large, approximate structural robustness implies being approximately information-proof. So even when the realized price becomes known, no player has significant incentive to re-trade, that is, the price is approximately competitive (Prob[some player can ε-improve his expected payoff by re-trading at the observed price] $\leq \rho$).

This is stronger than classical results relating Nash equilibrium to Walras equilibrium (for example, Dubey, Mas-Colell and Shubik, 1980). First, being conducted under incomplete information, the above relates Bayesian equilibria to rational expectations equilibria (rather than Walras). Also the competitive property described here is substantially stronger, due to the immunity of the equilibria to alterations represented by extensive games. If allowance is made for spot markets, coordinating institutions, trade on the Web, and so on, the Nash-equilibrium prices of the simple simultaneous-move game are sustained through the intermediary steps that may come up under such possibilities.

Remark 7 *Embedding a game in bigger worlds*

Alterations allow the inclusion of outside players who are not from G. Moreover, the restrictions imposed on the strategies and payoffs of the outsiders are quite limited. This means that alterations may describe bigger worlds in which G is embedded. Structural robustness of an equilibrium means that the small-world (G) equilibrium remains an equilibrium even when the game is embedded in such bigger worlds.

Remark 8 *Self-purification*

Schmeidler (1973) shows that in a normal-form game with a continuum of anonymous players, every strategy can be purified, that is, for every mixed-strategy equilibrium one can construct a pure-strategy equilibrium (Ali Khan and Sun, 2002 survey some of the large follow-up literature).

The *ex post* Nash property above constitutes a stronger (but asymptotic) result. Since the resulting play of a mixed strategy equilibrium yields pure-strategy profiles that are Nash equilibria (of the perfect information game), one does not need to construct pure-strategy equilibria: simply playing a mixed-strategy equilibrium yields pure-strategy profiles that are equilibria.

The approximate statement is: for every (ε, ρ) for sufficiently large n, Prob[ending at a pure strategy profile that is not an ε Nash equilibrium of the realized perfect information game] $\leq \rho$. Since both ε and ρ can be made arbitrarily small, this is asymptotic purification. Note that the model of Schmeidler, with a continuum of players, requires non-standard techniques to describe a continuum of independent random variables (the mixed strategies of the players). The asymptotic result stated here, dealing always with finitely many players, does not require any non-standard techniques.

Remark 9 *'As if' learning*

Kalai and Lehrer (1993) show that in playing an equilibrium of a Bayesian repeated game, after a sufficiently long time the players best-respond as if they know their opponents' realized types and, hence, their mixed strategies.

But being information-proof, at a structurally robust equilibrium (even of a one shot game) players' best respond (immediately) as if they know their opponents' realized types, their mixed strategies and even the pure-strategies they end up with.

Sufficient conditions for structural robustness

Theorem 1 *Structural Robustness* (rough statement): the equilibria of large one-simultaneous-move Bayesian games are (approximately) structurally robust if

(a) the players' types are drawn independently, and
(b) payoff functions are anonymous and continuous.

Payoff anonymity means that in addition to his own type and pure-strategy, every player's payoff may depend only on aggregate data of the opponents' types and pure-strategies. For example, in the computer choice game a player's payoff may depend on her own type and choice, and on the *proportions* of opponents in the four groups: *I*-types who chose *I*, *I*-types who chose *M*, *M*-types who chose *I*, and *M*-types who chose *M*.

The players in the games above are only semi-anonymous, because there are no additional symmetry or anonymity restrictions other than the restriction above. In particular, players may have different individual payoff functions and different prior probabilities (publicly known).

The continuity condition relates games of different sizes and rules out games of the type below.

Example 4 *Match the expert*

Each of n players has to choose one of two computers, *I* or *M*. Player 1 is equally likely to be one of two types: 'an expert who is informed that *I* is better' (*I*-better) or 'an expert who is informed that *M* is better' (*M*-better). Players 2, … , n are of one possible 'non-expert' type. Every player's payoff is one if he chooses the better computer, zero otherwise. (Stated anonymously: choosing computer X pays one, if the proportion of the *X-better* type is positive, zero otherwise.)

Consider the equilibrium where player 1 chooses the computer he was told was better and every other player chooses I or M with equal probabilities. This equilibrium fails to be *ex post* Nash (and hence, fails structural robustness), especially as n becomes large, because after the play approximately one-half of the players would want to revise their choices to match the observed choice of player 1. (With a small n there may be 'accidental *ex post* Nash', but it becomes extremely unlikely as n becomes large.)

This failure is due to discontinuity of the payoff functions. The proportions of I-better types and M-better types in this game must be either $(1/n, 0)$ or $(0, 1/n)$, because only one of the n players is to be one of these types. Yet, whatever n is, every player's payoff is drastically affected (from 0 to 1 or from 1 to 0) when we switch from $(1/n, 0)$ to $(0, 1/n)$ (keeping everything else the same).

As n becomes large, this change in the type proportions becomes arbitrarily small, yet it continues to have a drastic effect on players' payoffs. This violates a condition of uniform equicontinuity imposed simultaneously on all the payoff functions in the games with $n = 1, 2, \ldots$ players.

EHUD KALAI

See also **purification.**

Bibliography

Aumann, R.J. and Shapley, L.S. 1974. *Values of Nonatomic Games.* Princeton, NJ: Princeton University Press.

Bergemann, D. and Morris, S. 2005. Robust mechanism design. *Econometrica* 73, 1771–813.

Cremer, J. and McLean, R.P. 1985. Optimal selling strategies under uncertainty for a discriminating monopolist when demands are interdependent. *Econometrica* 53, 345–61.

Dubey, P., Mas-Colell, A. and Shubik, M. 1980. Efficiency properties of strategic market games: an axiomatic approach. *Journal of Economic Theory* 22, 339–62.

Dubey, P. and Kaneko, M. 1984. Information patterns and Nash equilibria in extensive games: 1. *Mathematical Social Sciences* 8, 111–39.

Dubey, P. and Geanakoplos, J. 2003. From Nash to Walras via Shapley–Shubik. *Journal of Mathematical Economics* 39, 391–400.

Green, J.R. and Laffont, J.J. 1987. Posterior implementability in a two-person decision problem. *Econometrica* 55, 69–94.

Kalai, E. 2004. Large robust games. *Econometrica* 72, 1631–66.

Kalai, E. 2005. Partially-specified large games. *Lecture Notes in Computer Science* 3828, 3–13.

Kalai, E. and Lehrer, E. 1993. Rational learning leads to Nash equilibrium. *Econometrica* 61, 1019–45.

Khan, A.M. and Sun, Y. 2002. Non-cooperative games with many players. In *Handbook of Game Theory with Economic Applications*, vol. 3, ed. R.J. Aumann and S. Hart. Amsterdam: North-Holland.

McLean, R., Peck, J. and Postlewaite, A. 2005. On price-taking behavior in asymmetric information economies. In *Essays in Dynamic General Equilibrium: Festschrift for David Cass*, ed. A. Citanna, J. Donaldson, H. Polemarchakis, P. Siconolfi and S. Spear. Berlin: Springer. Repr. in *Studies in Economic Theory* 20, 129–42.

Rustichini, A., Satterthwaite, M.A. and Williams, S.R. 1994. Convergence to efficiency in a simple market with incomplete information. *Econometrica* 62, 1041–64.
Schmeidler, D. 1973. Equilibrium points of nonatomic games. *Journal of Statistical Physics* 17, 295–300.
Shapley, L.S. and Shubik, M. 1977. Trade using one commodity as a means of payment. *Journal of Political Economy* 85, 937–68.
Wilson, R. 1987. Game-theoretic analyses of trading processes. In *Advances in Economic Theory: Fifth World Congress*, ed. T. Bewley. Cambridge: Cambridge University Press.

learning and evolution in games: adaptive heuristics

'Adaptive heuristics' are simple behavioural rules that are directed towards payoff improvement but may be less than fully rational. The number and variety of such rules are virtually unlimited; here we survey several prominent examples drawn from psychology, computer science, statistics and game theory. Of particular interest are the informational inputs required by different learning rules and the forms of equilibrium to which they lead. We shall begin by considering very primitive heuristics, such as reinforcement learning, and work our way up to more complex forms, such as hypothesis testing, which still, however, fall well short of perfectly rational learning.

One of the simplest examples of a learning heuristic is *cumulative payoff matching*, in which the subject plays actions next period with probabilities proportional to their cumulative payoffs to date. Specifically, consider a finite stage game G that is played infinitely often, where all payoffs are assumed to be strictly positive. Let $a_{ij}(t)$ denote the cumulative payoff to player i over all those periods $0 \leq t' \leq t$ when he played action j, including some *initial propensity* $a_{ij}(0) > 0$. The cumulative payoff matching rule stipulates that in period $t+1$, player i chooses action j with probability

$$p_{ij}(t+1) = a_{ij}(t) / \sum_k a_{ik}(t). \qquad (1)$$

Notice that the distribution has full support given the assumption that the initial propensities are positive. This idea was first proposed by the psychologist Nathan Herrnstein (1970) to explain certain types of animal behaviour, and falls under the more general rubric of *reinforcement learning* (Bush and Mosteller, 1951; Suppes and Atkinson, 1960; Cross, 1983). The key feature of a reinforcement model is that the probability of choosing an action increases monotonically with the total payoff it has generated in the past (on the assumption that the payoffs are positive). In other words, taking an action and receiving a positive payoff *reinforces* the tendency to take that same action again. This means, in particular, that play can become concentrated on certain actions simply because they were played early and often, that is, play can be *habit-forming* (Roth and Erev, 1995; Erev and Roth, 1998).

Reinforcement models differ in various details that materially affect their theoretical behaviour as well as their empirical plausibility. Under cumulative payoff matching, for example, the payoffs are not discounted, which means that current payoffs have an impact on current behaviour that diminishes as $1/t$. Laboratory experiments suggest, however, that recent payoffs matter more than those long past (Erev and Roth, 1998); furthermore, the rate of discounting has implications for the asymptotic properties of such models (Arthur, 1991).

Another variation in this class of models relies on the concept of an *aspiration level*. This is a level of payoffs, sometimes endogenously determined by past play, that triggers a change in a player's behaviour when current payoffs fall below the level and inertial behaviour when payoffs are above the level. The theoretical properties of these models have been studied for 2×2 games, but relatively little is known about their behaviour in general games (Börgers and Sarin, 2000; Cho and Matsui, 2005).

Next we turn to a class of adaptive heuristics based on the notion of minimizing *regret*, about which more is known in a theoretical sense. Fix a particular player and let $\alpha(t)$ denote the average per period payoff that she received over all periods $t' \leq t$. Let $\alpha_j(t)$ denote the average payoff she *would have* received by playing action j in every period through t, on the assumption that the opponents played as they actually did. The difference $r_j(t) = \alpha_j(t) - \alpha(t)$ is the subject's *unconditional regret* from not having played j in every period through t. (In the computer science literature this is known as *external regret*; see Greenwald and Gondek, 2002.)

The following simple heuristic was proposed by Hart and Mas-Colell (2000; 2001) and is known as *unconditional regret matching*: play each action with a probability that is proportional to the positive part of its unconditional regret, that is,

$$p_j(t+1) = [r_j(t)]_+ / \sum_k [r_k(t)]_+. \tag{2}$$

This learning rule has the following remarkable property: when used by any one player, his regrets become non-positive almost surely as t goes to infinity *irrespective of the behaviour of the other players.* When all players use the rule, their time average behaviour converges almost surely to a generalization of correlated equilibrium known as the *Hannan set* or the *coarse correlated equilibrium set* (Hannan, 1957; Moulin and Vial, 1978; Hart and Mas-Colell, 2000; Young, 2004). In general, a *coarse correlated equilibrium* (CCE) is a probability distribution over outcomes (joint actions) such that, given a choice between (*a*) committing *ex ante* to whatever joint action will be realized, and (*b*) committing *ex ante* to a fixed action, given that the others are committed to playing their part of whatever joint action will be realized, every player weakly prefers the former option. By contrast, a *correlated equilibrium* (CE) is a distribution such that, after a player's part of the realized joint action has been disclosed, he would just as soon play it as something else, given that the others are going to play their part of the realized joint action. It is straightforward to show that the coarse correlated equilibria form a convex set that contains the set of correlated equilibria (Young, 2004, ch. 3).

The heuristic specified in (2) belongs to a large family of rules whose time-average behaviour converges almost surely to the coarse correlated equilibrium set; equivalently, that assures no long-run regret for all players simultaneously. For example, this property holds if we let $p_j(t+1) = [r_j(t)]_+^\theta / \sum_k [r_k(t)]_+^\theta$ for some exponent $\theta > 0$; one may even take different exponents for different players. Notice that these heuristics put positive probability only on actions that would have done strictly better (on average) than the player's realized average payoff. These are sometimes called *better reply rules*. Fictitious play, by contrast, puts positive probability only on action(s) that would have done *best* against the opponents' frequency distribution of play.

Fictitious play does not necessarily converge to the coarse correlated equilibrium set (CCES); indeed, in some 2×2 coordination games fictitious play causes perpetual miscoordination, in which case both players have unconditional long-run regret (Fudenberg and Kreps, 1993; Young, 1993). By choosing θ to be very large, however, we see that there exist better reply rules that are arbitrarily close to fictitious play and that do converge almost surely to the CCES. Fudenberg and Levine (1995; 1998; 1999) and Hart and Mas-Colell (2001) give general conditions under which stochastic forms of fictitious play converge in time average to the CCES.

Without complicating the adjustment process too much, one can construct rules whose time average behaviour converges almost surely to the *correlated equilibrium set* (CES). To define this class of heuristics we need to introduce the notion of conditional regret. Given a history of play through time t and a player i, consider the change in per period payoff if i had played action k in all those periods $t' \leq t$ when he actually played action j (and the opponents played what they did). If the difference is positive, player i has conditional regret – he wishes he had played k instead of j. Formally, i's *conditional regret* at playing j instead of k up through time t, $r^i_{jk}(t)$, is $1/t$ times the increase in payoff that would have resulted from playing k instead of j in all periods $t' \leq t$. Notice that the average is taken over all t periods to date; hence, if j was not played very often, $r^i_{jk}(t)$ will be small.

Consider the following *conditional regret matching* heuristic proposed by Hart and Mas-Colell (2000): if a given agent played action j in period t, then in period $t+1$ he plays according to the distribution

$$q_k(t+1) = \varepsilon r_{jk}(t)_+ \text{ for all } k \neq j, \text{ and } q_j(t+1) = 1 - \varepsilon \sum\nolimits_{k \neq j} r_{jk}(t)_+. \tag{3}$$

In effect $1 - \varepsilon$ is the degree of inertia, which must be large enough that $q_k(t+1)$ is non-negative for all realizations of the conditional regrets $r_{jk}(t)$. If all players use conditional regret matching and ε is sufficiently small, then almost surely the joint frequency of play converges to the set of correlated equilibria (Hart and Mas-Colell, 2000). Notice that *pointwise* convergence is not guaranteed; the result says only that the empirical distribution converges to a convex *set*. In particular, the players' time-average behaviour may wander from one correlated equilibrium to another. It should also be remarked that, if a single player uses conditional regret matching, there is no assurance that his conditional regrets will become non-positive over time unless we assume that the other players use the same rule. This stands in contrast to unconditional regret matching, which assures non-positive unconditional regret for any player who uses it irrespective of the behaviour of the other players. One can, however, design more sophisticated updating procedures that unilaterally assure no conditional regret; see for example Foster and Vohra (1999), Fudenberg and Levine (1998, ch. 4), Hart and Mas-Colell (2000), and Young (2004, ch. 4).

A natural question now arises: do there exist simple heuristics that allow the players to learn *Nash* equilibrium instead of correlated or still coarser forms of equilibrium? The answer depends on how demanding we are about the long-run convergence properties of the learning dynamic. Notice that the preceding results on regret matching were concerned solely with time-average behaviour; no claim was made that period-by-period behaviour converges to any notion of equilibrium. Yet surely it is period-by-period behaviour that is most relevant if we want to assert that the players have 'learned' to play equilibrium. It turns out that it is very difficult to design adaptive learning rules under which period-by-period behaviour converges almost surely to Nash equilibrium in any finite game, unless one builds in some form of coordination among the players (Hart and Mas-Colell, 2003; 2006). The situation becomes even more problematic if one insists on fully rational, Bayesian learning. In this case it can be shown that there exist games of incomplete information in which no form of Bayesian rational learning causes period-by-period behaviours to come close to Nash equilibrium behaviour even in a probabilistic sense (Jordan, 1991, 1993; Foster and Young, 2001; Young, 2004; see also LEARNING AND EVOLUTION IN GAMES: BELIEF LEARNING).

If one does not insist on full rationality, however, one can design stochastic adaptive heuristics that cause period-by-period behaviours to come close to Nash equilibrium – indeed close to subgame perfect equilibrium – most of the time (without necessarily *converging* to an equilibrium). Here is one approach due to Foster and Young (2003); for related work see Foster and Young (2006) and Germano and Lugosi (2007). Let G be a finite n-person game that is played infinitely often. At each point in time, each player thinks that the others are playing i.i.d. strategies. Specifically, at time t player i thinks that j is playing the i.i.d strategy $p_j(t)$ on j's action space, and that the opponents are playing independently; that is, their joint strategies are given by the product distribution $p_{-i}(t) = \prod_{j \neq i} p_j(t)$. Suppose that i's best response is to play a smoothed best response to $p_{-i}(t)$. Specifically, assume that i plays each action j with a probability proportional to $e^{\beta u_i(j, p_{-i})}$, where $u_i(j, p_{-i})$ is i's expected utility from playing j in every period when the opponents play p_{-i}, and $\beta > 0$ is a *response parameter*. This is known as a *quantal* or *log linear* response function. For brevity, denote i's response in period t by $q_i^{\beta}(t)$; this depends, of course, on $p_{-i}(t)$. Player i views $p_{-i}(t)$ as a hypothesis that he wishes to test against data. After first adopting this hypothesis he waits for a number of periods (say s) while he observes the opponents' behaviour, all the while playing $q_i^{\beta}(t)$. After s periods have elapsed, he compares the empirical frequency distribution of the opponents' play during these periods with his hypothesis. Notice that both the empirical frequency distribution and the hypothesized distribution lie in the same compact subset of Euclidean space. If the two differ by more than some tolerance level τ (in the Euclidean metric), he rejects his current hypothesis and chooses a new one.

In choosing a new hypothesis, he may wish to take account of information revealed during the course of play, but we shall also assume he engages in some *experimentation*. Specifically, let us suppose that he chooses a new hypothesis according to a probability density that is uniformly bounded away from zero on the space of hypotheses. One can show the following: given any $\varepsilon > 0$, if the response parameter β is sufficiently large, the test tolerance τ is sufficiently small (given β), and the amount of data collected s is sufficiently large (given β and τ), then the players' *period-by-period* behaviours constitute an ε-equilibrium of the stage game G at least $1 - \varepsilon$ of the time (Foster and Young, 2003). In other words, classical statistical hypothesis testing is a heuristic for learning Nash equilibria of the stage game. Moreover, if the players adopt hypotheses that condition on history, they can learn complex equilibria of the repeated game, including forms of subgame perfect equilibrium.

The theoretical literature on strategic learning has advanced rapidly in recent years. A much richer class of learning models has been identified since the mid-1990s, and more is known about their long-run convergence properties. There is also a greater understanding of the various kinds of equilibrium that different forms of learning deliver. An important open question is how these theoretical proposals relate to the empirical behaviour of laboratory subjects. While there is no reason to think that any of these rules can fully explain subjects' behaviour, they can nevertheless play a useful role by identifying phenomena that experimentalists should look for. In particular, the preceding discussion suggests that weaker forms of equilibrium may turn out to be more robust predictors of long-run behaviour than is Nash equilibrium.

H. PEYTON YOUNG

See also **learning and evolution in games: belief learning.**

Bibliography

Arthur, W.B. 1991. Designing agents that act like human agents: a behavioral approach to bounded rationality. *American Economic Association, Papers and Proceedings* 81, 353–9.

Börgers, T. and Sarin, R. 2000. Naïve reinforcement learning with endogenous aspirations. *International Economic Review* 31, 921–50.

Bush, R.R. and Mosteller, F. 1951. A mathematical model for simple learning. *Psychological Review* 58, 313–23.

Cho, I.-K. and Matsui, A. 2005. Learning aspiration in repeated games. *Journal of Economic Theory* 124, 171–201.

Cross, J. 1983. *A Theory of Adaptive Economic Behavior.* Cambridge: Cambridge University Press.

Erev, I. and Roth, A.E. 1998. Predicting how people play games: reinforcement learning in experimental games with unique, mixed strategy equilibria. *American Economic Review* 88, 848–81.

Foster, D.P. and Vohra, R. 1999. Regret in the on-line decision problem. *Games and Economic Behavior* 29, 7–35.

Foster, D.P. and Young, H.P. 2001. On the impossibility of predicting the behavior of rational agents. *Proceedings of the National Academy of Sciences of the USA* 98(222), 12848–53.

Foster, D.P. and Young, H.P. 2003. Learning, hypothesis testing, and Nash equilibrium. *Games and Economic Behavior* 45, 73–96.

Foster, D.P. and Young, H.P. 2006. Regret testing: learning Nash equilibrium without knowing you have an opponent. *Theoretical Economics* 1, 341–67.

Fudenberg, D. and Kreps, D. 1993. Learning mixed equilibria. *Games and Economic Behavior* 5, 320–67.

Fudenberg, D. and Levine, D. 1995. Universal consistency and cautious fictitious play. *Journal of Economic Dynamics and Control* 19, 1065–90.

Fudenberg, D. and Levine, D. 1998. *The Theory of Learning in Games*. Cambridge MA: MIT Press.

Fudenberg, D. and Levine, D. 1999. Conditional universal consistency. *Games and Economic Behavior* 29, 104–30.

Germano, F. and Lugosi, G. 2007. Global Nash convergence of Foster and Young's regret testing. *Games and Economic Behavior* 60, 135–54.

Greenwald, A. and Gondek, D. 2002. On no-regret learning and game-theoretic equilibria. *Journal of Machine Learning* 1, 1–20.

Hannan, J. 1957. Approximation to Bayes risk in repeated plays. In *Contributions to the Theory of Games*, vol. 3, ed. M. Dresher, A.W. Tucker and P. Wolfe. Princeton, NJ: Princeton University Press.

Hart, S. and Mas-Colell, A. 2000. A simple adaptive procedure leading to correlated equilibrium. *Econometrica* 68, 1127–50.

Hart, S. and Mas-Colell, A. 2001. A general class of adaptive strategies. *Journal of Economic Theory* 98, 26–54.

Hart, S. and Mas-Colell, A. 2003. Uncoupled dynamics do not lead to Nash equilibrium. *American Economic Review* 93, 1830–6.

Hart, S. and Mas-Colell, A. 2006. Stochastic uncoupled dynamics and Nash equilibrium. *Games and Economic Behavior* 57, 286–303.

Herrnstein, R.J. 1970. On the law of effect. *Journal of the Experimental Analysis of Behavior* 13, 243–66.

Jordan, J.S. 1991. Bayesian learning in normal form games. *Games and Economic Behavior* 3, 60–81.

Jordan, J.S. 1993. Three problems in learning mixed-strategy equilibria. *Games and Economic Behavior* 5, 368–86.

Moulin, H. and Vial, J.P. 1978. Strategically zero-sum games: the class of games whose completely mixed equilibria cannot be improved upon. *International Journal of Game Theory* 7, 201–21.

Roth, A.E. and Erev, I. 1995. Learning in extensive-form games: experimental data and simple dynamic models in the intermediate term. *Games and Economic Behavior* 8, 164–212.

Suppes, P. and Atkinson, R. 1960. *Markov Learning Models for Multiperson Interaction*. Stanford CA: Stanford University Press.

Young, H.P. 1993. The evolution of conventions. *Econometrica* 61, 57–84.

Young, H.P. 2004. *Strategic Learning and Its Limits*. Oxford: Oxford University Press.

learning and evolution in games: an overview

The theory of learning and evolution in games provides models of disequilibrium behaviour in strategic settings. Much of the theory focuses on whether and when disequilibrium behaviour will resolve in equilibrium play, and, if it does, on predicting which equilibrium will be played. But the theory also offers techniques for characterizing perpetual disequilibrium play.

1. A taxonomy

Models from *evolutionary game theory* consider the behaviour of large populations in strategic environments. In the biological strand of the theory, agents are genetically programmed to play fixed actions, and changes in the population's composition are the result of natural selection and random mutations. In economic approaches to the theory, agents actively choose which actions to play using simple myopic rules, so that changes in aggregate behaviour are the end result of many individual decisions. *Deterministic evolutionary dynamics*, usually taking the form of ordinary differential equations, are used to describe behaviour over moderate time spans, while *stochastic evolutionary dynamics*, modelled using Markov processes, are more commonly employed to study behaviour over very long time spans.

Models of *learning in games* focus on the behaviour of small groups of players, one of whom fills each role in a repeated game. These models too can be partitioned into two categories. Models of *heuristic learning* (or *adaptive learning*) resemble evolutionary models, in that their players base their decisions on simple myopic rules. One sometimes can distinguish the two sorts of models by the inputs to the agents' decision rules. In both the stochastic evolutionary model of c, Kandori, Mailath and Rob (1993) and the heuristic learning model of Young (1993), agents' decisions take the form of noisy best responses. But in the former model agents evaluate each action by its performance against the population's current behaviour, while in the latter they consider performance against the time averages of opponents' past play.

In models of *coordinated Bayesian learning* (or *rational learning*), each player forms explicit beliefs about the repeated game strategies employed by other players, and plays a best response to those beliefs in each period. The latter models assume a degree of coordination of players' prior beliefs that is sufficient to ensure that play converges to Nash equilibrium. By dropping this coordination assumption, one obtains the more general class of *Bayesian learning* (or *belief learning*) models. Since such models can entail quite naive beliefs, belief learning models overlap with heuristic learning models – see Section 3 below.

2. Evolutionary game theory

The roots of evolutionary game theory lie in mathematical biology. Maynard Smith and Price (1973) introduced the equilibrium notion of an *evolutionarily stable strategy* (or ESS) to capture the possible stable outcomes of a dynamic evolutionary process by way of a static definition. Later, Taylor and Jonker (1978) offered the *replicator dynamic* as an explicitly dynamic model of the natural selection process. The decade that followed saw an explosion of research on the replicator dynamic and related models of animal behaviour, population ecology, and population genetics: see Hofbauer and Sigmund (1988).

In economics, evolutionary game theory studies the behaviour of populations of strategically interacting agents who actively choose among the actions available to them. Agents decide when to switch actions and which action to choose next using simple myopic rules known as *revision protocols* (see Sandholm, 2006). A population of agents, a game, and a revision protocol together define a stochastic process – in particular, a Markov process – on the set of population states.

2.1 Deterministic evolutionary dynamics

How the analysis proceeds depends on the time horizon of interest. Suppose that for the application in question, our interest is in moderate time spans. Then if the population size is large enough, the idiosyncratic noise in agent's choices is averaged away, so that the evolution of aggregate behaviour follows an almost deterministic path (Benaïm and Weibull, 2003). This path is described by a solution to an ordinary differential equation. For example, Björnerstedt and Weibull (1996) and Schlag (1998) show that if agents use certain revision protocols based on imitation of successful opponents, then the population's aggregate behaviour follows a solution to Taylor and Jonker's (1978) replicator dynamic. This argument provides an alternative, economic interpretation of this fundamental evolutionary model.

Much of the literature on deterministic evolutionary dynamics focuses on connections with traditional game theoretic solution concepts. For instance, under a wide range of deterministic dynamics, all Nash equilibria of the underlying game are rest points. While some dynamics (including the replicator dynamic) have additional non-Nash rest points, there are others under which rest points and Nash equilibria are identical (Brown and von Neumann, 1950; Smith, 1984; Sandholm, 2006).

A more important question, though, is whether Nash equilibrium will be approached from arbitrary disequilibrium states. For certain specific classes of games, general convergence results can be established (Hofbauer, 2000; Sandholm, 2007). But beyond these classes, convergence cannot be guaranteed. One can construct games under which no reasonable deterministic evolutionary dynamic will converge to equilibrium – instead, the population cycles through a range of disequilibrium states forever (Hofbauer and Swinkels, 1996; Hart and Mas-Colell, 2003). More surprisingly, one can construct games in which nearly all deterministic evolutionary dynamics not only cycle for ever, but also fail to eliminate strictly dominated strategies (Hofbauer and Sandholm, 2006). If we truly are interested in modelling the dynamics of behaviour, these results reveal that our predictions cannot always be confined to equilibria; rather, more complicated limit phenomena like cycles and chaotic attractors must also be permitted as predictions of play.

2.2 Stochastic evolutionary dynamics

If we are interested in behaviour over very long time horizons, deterministic approximations are no longer valid, and we must study our original Markov process directly. Under certain non-degeneracy assumptions, the long-run behaviour of this process is captured by its unique stationary distribution, which describes the proportion of time the process spends in each population state.

While stochastic evolutionary processes can be more difficult to analyse than their deterministic counterparts, they also permit us to make surprisingly tight predictions. By making the amount of noise in agents' choice rules vanishingly small, one can often ensure that all mass in the limiting stationary distribution is placed on a single population state. This *stochastically stable state* provides a unique prediction of play even in games with multiple strict equilibria (Foster and Young, 1990; Kandori, Mailath and Rob, 1993).

The most thoroughly studied model of stochastic evolution considers agents who usually play a best response to the current population state, but who occasionally choose a strategy at random. Kandori, Mailath and Rob (1993) show that if the agents are randomly matched to play a symmetric 2×2 coordination game, then taking the probability of 'mutations' to zero generates a unique stochastically stable state. In this state, called the *risk dominant equilibrium*, all agents play the action that is optimal against an opponent who is equally likely to choose each action.

Selection results of this sort have since been extended to cases in which the underlying game has an arbitrary number of strategies, as well as to settings in which agents are positioned on a fixed network, interacting only with neighbours (see Kandori and Rob, 1995; Blume, 2003; Ellison, 1993; 2000). Stochastic stability has also been employed in contexts where the underlying game has a nontrivial extensive form; these analyses have provided support for notions of backward induction (for example, subgame perfection) and forward induction (for example, signalling game equilibrium refinements): see Nöldeke and Samuelson (1993) and Hart (2002).

Still, these selection results must be interpreted with care. When the number of agents is large or the rate of 'mutation' is small, states that fail to be stochastically stable can be coordinated upon for great lengths of time (Binmore, Samuelson and Vaughan, 1995). Consequently, if the relevant time span for the application at hand is not long enough, the stochastically stable state may not be the only reasonable prediction of behaviour.

3. Learning in games

3.1 Heuristic learning

Learning models study disequilibrium adjustment processes in repeated games. Like evolutionary models, heuristic learning models assume that players employ simple myopic rules in deciding how to act. In the simplest of these models, each player decides how to act by considering the payoffs he has earned in the past. For instance, under reinforcement learning (Börgers and Sarin, 1997; Erev and Roth, 1998), agents choose each strategy with probability proportional to the total payoff that the strategy has earned in past periods.

By considering rules that look not only at payoffs earned, but also at payoffs foregone, one can obtain surprisingly strong convergence results. Define a player's *regret* for (not having played) action a to be the difference between the average payoff he would have earned had he always played a in the past, and the average payoff he actually received. Under *regret matchingt*, each action whose regret is positive is chosen with probability proportional to its regret. Hart and Mas-Colell (2000) show that regret matching is a *consistent* repeated game strategy: it forces a player's regret for each action to become nonpositive. If used by all players, regret matching ensures that their time-averaged behaviour converges to the set of *coarse correlated equilibria* of the underlying game. (*Coarse correlated equilibrium* is a generalization of correlated equilibrium under which players' incentive constraints must be satisfied at the *ex ante* stage rather than at the interim stage: see Young, 2004.)

Some of the most striking convergence results in the evolution and learning literature establish a stronger conclusion: namely, convergence of time-averaged behaviour to the set of *correlated equilibria*, regardless of the game at hand. The original result of this sort is due to Foster and Vohra (1997; 1998), who prove the result by constructing a calibrated procedure for forecasting opponents' play. A *forecasting procedure* produces probabilistic forecasts of how opponents will act. The procedure is *calibrated* if in those periods in which the forecast is given by the probability vector p, the empirical distribution of opponents' play is approximately p. It is not difficult to show that if players always choose myopic best responses to calibrated forecasts, then their time-averaged behaviour converges to the set of correlated equilibria.

Hart and Mas-Colell (2000) construct simpler procedures – in particular, procedures that define conditionally consistent repeated game strategies – also ensure convergence to correlated equilibrium. A repeated game strategy is *conditionally consistent* if for each frequently played action a, the agent would not have been better off had he always played an alternative action a' in place of a. As a matter of definition, the use of conditionally consistent strategies by all players leads time-averaged behavior to converge to the set of correlated equilibria.

Another variety of heuristic learning models, based on *random search and independent verification*, ensures a stochastic form of convergence to Nash equilibrium regardless of the game being played (Foster and Young, 2003). However, in these models the time required before equilibrium is first reached is quite long, making them most relevant to applications with especially long time horizons.

In some heuristic learning models, players use simple rules to predict how opponents will behave, and then respond optimally to those predictions. The leading examples of such models are *fictitious play* and its stochastic variants (Brown, 1951;

Fudenberg and Kreps, 1993): in these models, the prediction about an opponents' next period play is given by the empirical frequencies of his past plays. Beginning with Robinson (1951), many authors have proved convergence results for standard and stochastic fictitious play in specific classes of games (see Hofbauer and Sandholm (2002) for an overview). But as Shapley (1964) and others have shown, these models do not lead to equilibrium play in all games.

3.2 Coordinated Bayesian learning

The prediction rule underlying two-player fictitious play can be described by a belief about the opponent's repeated game strategy that is updated using Bayes's rule in the face of observed play. This belief specifies that the opponent choose his stage game actions in an i.i.d. fashion, conditional on the value of an unknown parameter. (In fact, the player's beliefs about this parameter must come from the family of Dirichlet distributions, the conjugate family of distributions for multinomial trials.) Evidently, each player's beliefs about his opponent are wrong: player 1 believes that player 2 chooses actions in an i.i.d. fashion, whereas player 2 actually plays optimally in response to his own (i.i.d.) predictions about player 1's behaviour. It is therefore not surprising that fictitious play processes do not converge in all games.

In models of *coordinated Bayesian learning* (or *rational learning*), it is not only supposed that players form and respond optimally to beliefs about the opponent's repeated game strategy; it is also assumed that the players' initial beliefs are coordinated in some way. The most studied case is one in which prior beliefs satisfy an absolute continuity condition: if the distribution over play paths generated by the players' actual strategies assigns positive probability to some set of play paths, then so must the distribution generated by each player's prior. A strong sufficient condition for absolute continuity is that each player's prior assigns a positive probability to his opponent's actual strategy.

The fundamental result in this literature, due to Kalai and Lehrer (1993), shows that under absolute continuity, each player's forecast along the path of play is asymptotically correct, and the path of play is asymptotically consistent with Nash equilibrium play in the repeated game. Related convergence results have been proved for more complicated environments in which each player's stage game payoffs are private information (Jordan, 1995; Nyarko, 1998). If the distributions of players types are continuous, then the sense in which play converges to equilibrium can involve a form of purification: while actual play is pure, it appears random to an outside observer.

How much coordination of prior beliefs is needed to prove convergence to equilibrium play? Nachbar (2005) proves that for a large class of repeated games, for any belief learning model, there are no prior beliefs that satisfy three criteria: learnability, consistency with optimal play, and diversity. Thus, if players can learn to predict one another's behaviour, and are capable of responding optimally to their updated beliefs, then each player's beliefs about his opponents must rule out some seemingly natural strategies a priori. In this sense, the assumption of coordinated prior beliefs that ensures convergence to equilibrium in rational learning models does not seem dramatically weaker than a direct assumption of equilibrium play.

For additional details about the theory of learning and evolution in games, we refer the reader to the entries on specific topics listed in the cross-references below.

WILLIAM H. SANDHOLM

See also **deterministic evolutionary dynamics; learning and evolution in games: adaptive heuristics; learning and evolution in games: belief learning; learning and evolution in games: ESS; stochastic adaptive dynamics.**

The author thanks John Nachbar for a number of helpful conversations and for sharing his expertise on coordinated Bayesian learning. Financial support under NSF Grants SES-0092145 and SES-0617753 is gratefully acknowledged.

Bibliography

Benaïm, M. and Weibull, J.W. 2003. Deterministic approximation of stochastic evolution in games. *Econometrica* 71, 873–903.

Binmore, K., Samuelson, L. and Vaughan, R. 1995. Musical chairs: modeling noisy evolution. *Games and Economic Behavior* 11, 1–35.

Björnerstedt, J. and Weibull, J.W. 1996. Nash equilibrium and evolution by imitation. In *The Rational Foundations of Economic Behavior*, ed. K.J. Arrow, E. Colombatto, M. Perlman, and C. Schmidt. New York: St. Martin's Press.

Blume, L.E. 2003. How noise matters. *Games and Economic Behavior* 44, 251–71.

Börgers, T. and Sarin, R. 1997. Learning through reinforcement and the replicator dynamics. *Journal of Economic Theory* 77, 1–14.

Brown, G.W. 1951. Iterative solutions of games by fictitious play. In *Activity Analysis of Production and Allocation*, ed. T.C. Koopmans. New York: Wiley.

Brown, G.W. and von Neumann, J. 1950. Solutions of games by differential equations. In *Contributions to the Theory of Games I*, ed. H.W. Kuhn and A.W. Tucker, vol. 24 of *Annals of Mathematics Studies*. Princeton: Princeton University Press.

Ellison, G. 1993. Learning, local interaction, and coordination. *Econometrica* 61, 1047–71.

Ellison, G. 2000. Basins of attraction, long run equilibria, and the speed of step-by-step evolution. *Review of Economic Studies* 67, 17–45.

Erev, I. and Roth, A.E. 1998. Predicting how people play games: reinforcement learning in experimental games with unique, mixed strategy equilibria. *American Economic Review* 88, 848–81.

Foster, D.P. and Vohra, R. 1997. Calibrated learning and correlated equilibrium. *Games and Economic Behavior* 21, 40–55.

Foster, D.P. and Vohra, R. 1998. Asymptotic calibration. *Biometrika* 85, 379–90.

Foster, D.P. and Young, H.P. 1990. Stochastic evolutionary game dynamics. *Theoretical Population Biology* 38, 219–32.

Foster, D.P. and Young, H.P. 2003. Learning, hypothesis testing, and Nash equilibrium. *Games and Economic Behavior* 45, 73–96.

Fudenberg, D. and Kreps, D.M. 1993. Learning mixed equilibria. *Games and Economic Behavior* 5, 320–67.

Fudenberg, D. and Levine, D.K. 1998. *Theory of Learning in Games*. Cambridge: MIT Press.

Hart, S. 2002. Evolutionary dynamics and backward induction. *Games and Economic Behavior* 41, 227–64.

Hart, S. and Mas-Colell, A. 2000. A simple adaptive procedure leading to correlated equilibrium. *Econometrica* 68, 1127–50.

Hart, S. and Mas-Colell, A. 2003. Uncoupled dynamics do not lead to Nash equilibrium. *American Economic Review* 93, 1830–6.

Hofbauer, J. 2000. From Nash and Brown to Maynard Smith: equilibria, dynamics, and ESS. *Selection* 1, 81–8.

Hofbauer, J. and Sandholm, W.H. 2002. On the global convergence of stochastic fictitious play. *Econometrica* 70, 2265–94.

Hofbauer, J. and Sandholm, W.H. 2006. Survival of dominated strategies under evolutionary dynamics. Working paper. University of Vienna and University of Wisconsin.

Hofbauer, J. and Sigmund, K. 1988. *Theory of Evolution and Dynamical Systems.* Cambridge: Cambridge University Press.

Hofbauer, J. and Swinkels, J.M. 1996. A universal Shapley example. Working paper. University of Vienna and Northwestern University.

Jordan, J.S. 1995. Bayesian learning in repeated games. *Games and Economic Behavior* 9, 8–20.

Kalai, E. and Lehrer, E. 1993. Rational learning leads to Nash equilibrium. *Econometrica* 61, 1019–45.

Kandori, M., Mailath, G.J. and Rob, R. 1993. Learning, mutation, and long run equilibria in games. *Econometrica* 61, 29–56.

Kandori, M. and Rob, R. 1995. Evolution of equilibria in the long run: a general theory and applications. *Journal of Economic Theory* 65, 383–414.

Maynard Smith, J. and Price, G.R. 1973. The logic of animal conflict. *Nature* 246, 15–8.

Nachbar, J.H. 2005. Beliefs in repeated games. *Econometrica* 73, 459–80.

Nöldeke, G. and Samuelson, L. 1993. An evolutionary analysis of backward and forward induction. *Games and Economic Behavior* 5, 425–54.

Nyarko, Y. 1998. Bayesian learning and convergence to Nash equilibria without common priors. *Economic Theory* 11, 643–55.

Robinson, J. 1951. An iterative method of solving a game. *Annals of Mathematics* 54, 296–301.

Sandholm, W.H. 2006. Pairwise comparison dynamics and evolutionary foundations for Nash equilibrium. Working paper. University of Wisconsin.

Sandholm, W.H. 2007. Population games and evolutionary dynamics. Cambridge, MA: MIT Press.

Schlag, K.H. 1998. Why imitate, and if so, how? A boundedly rational approach to multi-armed bandits. *Journal of Economic Theory* 78, 130–56.

Shapley, L.S. 1964. Some topics in two person games. In *Advances in Game Theory*, ed. M. Dresher, L.S. Shapley and A.W. Tucker, vol. 52 of *Annals of Mathematics Studies*. Princeton: Princeton University Press.

Smith, M.J. 1984. The stability of a dynamic model of traffic assignment: an application of a method of Lyapunov. *Transportation Science* 18, 245–52.

Taylor, P.D. and Jonker, L. 1978. Evolutionarily stable strategies and game dynamics. *Mathematical Biosciences* 40, 145–56.

Young, H.P. 1993. The evolution of conventions. *Econometrica* 61, 57–84.

Young, H.P. 2004. *Strategic Learning and Its Limits*. Oxford: Oxford University Press.

learning and evolution in games: belief learning

In the context of learning in games, *belief learning* refers to models in which players are engaged in a dynamic game and each player optimizes, or ε optimizes, with respect to a *prediction rule* that gives a forecast of next period opponent behaviour as a function of the current history. This article focuses on the most studied class of dynamic games, two-player discounted repeated games with finite stage game action sets and perfect monitoring. An important example of a dynamic game that violates perfect monitoring and therefore falls outside this framework is Fudenberg and Levine (1993). For a more comprehensive survey of belief learning, see Fudenberg and Levine (1998).

The earliest example of belief learning is the *best-response dynamics* of Cournot (1838). In Cournot's model, each player predicts that her opponent will repeat next period whatever action her opponent chose in the previous period.

The most studied belief learning model is *fictitious play* (Brown, 1951), and its variants. In fictitious play, each player predicts that the probability that her opponent will play an action, say L, next period is a weighted sum of an initial probability on L and the frequency with which L has been chosen to date. The weight on the frequency is $t/(t+k)$, where t is the number of periods thus far and $k>0$ is a parameter. The larger is k, the more periods for which the initial probability significantly affects forecasting.

The remainder of this article discusses four topics: (1) belief learning versus Bayesian learning, (2) convergence to equilibrium, (3) special issues in games with payoff uncertainty, and (4) sensible beliefs.

Belief learning versus Bayesian learning

Recall that, in a repeated game, a behaviour strategy gives, for every history, a probability over the player's stage game actions next period. In a Bayesian model, each player chooses a behaviour strategy that best responds to a *belief*, a probability distribution over the opponent's behaviour strategies.

Player 1's prediction rule about player 2 is mathematically identical to a behaviour strategy for player 2. Thus, any belief learning model is equivalent to a Bayesian model in which each player optimizes with respect to a belief that places probability 1 on her prediction rule, now reinterpreted as the opponent's behaviour strategy.

Conversely, any Bayesian model is equivalent to a belief learning model. Explicitly, for any belief over player 2's behaviour strategies there is a degenerate belief, assigning probability 1 to a particular behaviour strategy, that is equivalent in the sense that both beliefs induce the same distributions over play in the game, no matter what behaviour strategy player 1 herself adopts. This is a form of Kuhn's theorem (Kuhn, 1964). I refer to the behaviour strategy used in the degenerate belief as a *reduced form* of the original belief. Thus, any Bayesian model is equivalent to a Bayesian model in which each player's belief places probability 1 on the reduced form, and any such Bayesian model is equivalent to a belief learning model.

As an example, consider fictitious play. I focus on stage games with just two actions, L and R. By an i.i.d. strategy for player 2, I mean a behaviour strategy in which player 2 plays L with probability q, independent of history. Thus, if $q = 1/2$, then player 2 always randomizes 50:50 between L and R. Fictitious play is equivalent to a degenerate Bayesian model in which each player places probability 1 on the fictitious play prediction rule, and one can show that this is equivalent in turn to a non-degenerate Bayesian model in which the belief is represented as a beta distribution over q. The uniform distribution over q, for example, corresponds to taking the initial probability of L to be 1/2 and the parameter k to be 2.

There is a related but distinct literature in which players optimize with respect to *stochastic* prediction rules. In some cases (for example, Foster and Young, 2003), these models have a quasi-Bayesian interpretation: most of the time, players optimize with respect to fixed prediction rules, as in a Bayesian model, but occasionally players switch to new prediction rules, implicitly abandoning their priors.

Convergence to equilibrium

Within the belief learning literature, the investigation of convergence to equilibrium play splits into two branches. One branch investigates convergence within the context of specific classes of belief learning models. The best-response dynamics, for example, converge to equilibrium if the stage game is solvable by the iterated deletion of strictly dominated strategies. See Bernheim (1984) and, for a more general class of models, Milgrom and Roberts (1991). For an ε optimizing variant of fictitious play, convergence to approximate equilibrium play obtains for all zero-sum games, all games with an interior ESS, and all common interest games, in addition to all games that are strict dominance solvable, with the approximation closer the smaller is ε. Somewhat weaker convergence results are available for supermodular games. These claims follow from results in Hofbauer and Sandholm (2002).

In the results surveyed above, convergence is to repeated play of a single-stage game Nash equilibrium; in the case of ε fictitious play, this equilibrium may be mixed. There is a large body of work on convergence that is weaker than what I am considering here. In particular, there has been much work on convergence of the empirical marginal or joint distributions. For mixed strategy equilibrium, it is possible for empirical distributions to converge to equilibrium even though play does not resemble repeated equilibrium play; play may exhibit obvious cycles, for example. The study of convergence to equilibrium play is relatively recent and was catalysed by Fudenberg and Kreps (1993).

There are classes of games that cause convergence problems for many standard belief learning models, even when one considers only weak forms of convergence, such as convergence of the empirical marginal distributions (see Shapley, 1962; Jordan,

1993). Hart and Mas-Colell (2003; 2006) (hereafter HM) shed light on non-convergence by investigating learning models, including but not limited to belief learning models, that are *decoupled*, meaning that player 1's behaviour does not depend directly on player 2's stage game payoffs. A continuous time version of fictitious play fits into the framework of Hart and Mas-Colell (2003). The HM results imply that universal convergence is impossible for large classes of decoupled belief learning models: for any such model there exist stage games and initial conditions for which play fails to converge to equilibrium play.

The second branch of the literature, for which Kalai and Lehrer (1993a) (hereafter KL) is the central paper, takes a Bayesian perspective and asks what conditions on beliefs are sufficient to give convergence to equilibrium play. I find it helpful to characterize this literature in the following way. Say that a belief profile (giving a belief for each player) has the *learnable best-response property* (LBR) if there is a profile of best-response strategies (LBR strategies) such that, if the LBR strategies are played, then each player learns to predict the play path.

A player *learns to predict the play path* if her prediction of next period's play is asymptotically as good as if she knew her opponent's behaviour strategy. If the behaviour strategies call for randomization then players accurately predict the distribution over next period's play rather than the realization of next period's play. For example, consider a 2×2 game in which player 1 has stage game actions T and B and player 2 has stage game actions L and R. If player 2 is randomizing 50:50 every period and player 1 learns to predict the path of play, then for every ε there is a time, which depends on the realization of player 2's strategy, after which player 1's next period forecast puts the probability of L within ε of 1/2. (This statement applies to a set of play paths that arises with probability 1 with respect to the underlying probability model; I gloss over this sort of complication both here and below.) For a more complicated example, suppose that in period t player 2 plays L with probability $1 - \alpha$, where α is the frequency that the players have played the profile (B, R). If player 1 learns to predict the play path, then for any ε there is a time, which now depends on the realization of both players' strategies, after which player 1's next period forecast puts the probability of L within ε of $1 - \alpha$.

Naively, if LBR holds, and players are using their LBR strategies, then, in the continuation game, players are optimizing with respect to posterior beliefs that are asymptotically correct and so continuation behaviour strategies should asymptotically be in equilibrium. This intuition is broadly correct, but there are three qualifications.

First, in general, convergence is to Nash equilibrium play in the *repeated* game, not necessarily to repeated play of a single stage game equilibrium. If players are myopic (meaning that players optimize each period as though their discount factors were zero), then the set of equilibrium play paths comprises all possible sequences of stage game Nash equilibria, which is a very large set if the stage game has more than one equilibrium. If players are patient, then the folk theorem applies and the set of possible equilibrium paths is typically even larger.

Second, convergence is to an equilibrium play path, not necessarily to an equilibrium of the repeated game. The issue is that LBR implies accurate forecasting only along the play path. A player's predictions about how her opponent would respond to deviations may be grossly in error, for ever. Therefore, posterior beliefs need *not* be asymptotically correct and, unless players are myopic, continuation behaviour strategies need *not* be asymptotically in equilibrium. Kalai and Lehrer (1993b) shows that behaviour strategies can be doctored at information sets off the play path so that the modified behaviour strategies are asymptotically in equilibrium yet still generate the same play path. This implies that the play path of the original strategy profile was asymptotically an equilibrium play path.

Third, the exact sense in which play converges to equilibrium play depends on the strength of learning. See KL and also Sandroni (1998).

KL shows that a strong form of LBR holds if beliefs satisfy an absolute continuity condition: each player assigns positive probability to any (measurable) set of play paths that has positive probability given the players' actual strategies. A sufficient condition for this is that each player assigns positive, even if extremely low, probability to her opponent's actual strategy, a condition that KL call *grain of truth*. Nyarko (1998) provides the appropriate generalization of absolute continuity for games with type space structures, including the games with payoff uncertainty discussed below.

Games with payoff uncertainty

Suppose that, at the start of the repeated game, each player is privately informed of his or her stage game payoff function, which remains fixed throughout the course of the repeated game. Refer to player *i*'s stage game payoff function as her *payoff type*. Assume that the joint distribution over payoff functions is independent (to avoid correlation issues that are not central to my discussion) and commonly known.

Each player can condition her behaviour strategy in the repeated game on her realized payoff type. A mathematically correct way of representing this conditioning is via distributional strategies (see Milgrom and Weber, 1985).

For any belief about player 2, now a probability distribution over player 2's distributional strategies, and given the probability distribution over player 2's payoff types, there is a behaviour strategy for player 2 in the repeated game that is equivalent in the sense that it generates the same distribution over play paths. Again, this is essentially Kuhn's theorem. And again, I refer to this behaviour strategy as a *reduced form*.

Say that a player *learns to predict the play path* if her forecast of next period's play is asymptotically as good as if she knew the reduced form of her opponent's distributional strategy. This definition specializes to the previous one if the distribution over types is degenerate. If distributional strategies are in equilibrium then, in effect, each player is optimizing with respect to a degenerate belief that puts probability one on her opponent's actual distributional strategy and in this case players trivially learn to predict the path of play.

One can define LBR for distributional strategies and, as in the payoff certainty case, one can show that LBR implies convergence to equilibrium play in the repeated game with payoff types. More interestingly, there is a sense in which play converges to equilibrium play of the *realized* repeated game – the repeated game determined by the realized type profile. The central paper is Jordan (1991). Other important papers include KL (cited above), Jordan (1995), Nyarko (1998), and Jackson and Kalai (1999) (which studies recurring rather than repeated games).

Suppose first that the realized type profile has positive probability. In this case, if a player learns to predict the play path, then, as shown by KL, her forecast is asymptotically as good as if she knew both her opponent's distributional strategy *and* her opponent's realized type. LBR then implies that actual play, meaning the distribution over play paths generated by the realized behaviour strategies, converges to equilibrium play of the realized repeated game. For example, suppose that the type profile for matching pennies gets positive probability. In the unique equilibrium of repeated matching pennies, players randomize 50:50 in every period. Therefore, LBR implies that, if the matching pennies type profile is realized, then each player's behaviour strategy in the realized repeated game involves 50:50 randomization asymptotically.

If the distribution over types admits a continuous density, so that no type profile receives positive probability, then the form of convergence is more subtle. Suppose that players are myopic and that the realized stage game is like matching pennies, with a unique and fully mixed equilibrium. Given myopia, the unique equilibrium of the realized repeated game calls for repeated play of the stage game equilibrium. In particular, it calls for players to randomize. It is not hard to show, however, that in a type space game with a continuous density, optimization calls for each player to play a pure strategy for almost every realized type. Thus, for almost every realized type profile in a neighbourhood of a game like matching pennies, actual play (again meaning the distribution over play paths generated by the realized behaviour strategies) cannot converge to equilibrium play, *even if the distributional strategies are in equilibrium.* Foster and Young (2001) provides a generalization for non-myopic players.

There is, however, a weaker sense in which play nevertheless does converge to equilibrium play in the realized repeated game. For simplicity, assume that each player knows the other's distributional strategy and that these strategies are in equilibrium. One can show that to an outsider observed play looks asymptotically like equilibrium play in the realized repeated game. In particular, if the realized game is like repeated matching pennies then observed play looks random. Moreover, to a player in the game, opponent behaviour looks random because, even though she knows her opponent's distributional strategy, she does not know her opponent's type. As play proceeds, each player in effect learns more about her opponent's type, but never enough to zero in on her opponent's realized, pure, behaviour strategy. Thus, when the distribution over types admits a continuous density, convergence to equilibrium involves a form of purification in the sense of Harsanyi (1973), a point that has been emphasized by Nyarko (1998) and Jackson and Kalai (1999).

Sensible beliefs

A number of papers investigate classes of prediction rules that are sensible in that they exhibit desirable properties, such as the ability to detect certain kinds of patterns in opponent behaviour (see Aoyagi, 1996; Fudenberg and Levine, 1995; 1999; Sandroni, 2000).

Nachbar (2005) instead studies the issue of sensible beliefs from a Bayesian perspective. For simplicity, focus on learning models with known payoffs. Fix a belief profile, fix a subset of behaviour strategies for each player, and consider the following criteria for these subsets.

- *Learnability* – given beliefs, if players play a strategy profile drawn from these subsets then they learn to predict the play path.
- *Richness.* Informally (the formal statement is tedious), richness requires that if a behaviour strategy is included in one of the strategy subsets then certain variations on that strategy must be included as well. Richness, called CSP in Nachbar (2005), is satisfied automatically if the strategy subsets consist of all strategies satisfying a standard complexity bound, the same bound for both players. Thus richness holds if the subsets consist of all strategies with k-period memory, or all strategies that are automaton implementable, or all strategies that are Turing implementable, and so on.
- *Consistency* – each player's subset contains a best response to her belief.

The motivating idea is that, if beliefs are probability distributions over strategy subsets satisfying learnability, richness, and consistency, then beliefs are sensible, or at least are candidates for being considered sensible. Nachbar (2005) studies whether any such beliefs exist.

Consider, for example, the Bayesian interpretation of fictitious play in which beliefs are probability distributions over the i.i.d. strategies. The set of i.i.d. strategies satisfies learnability and richness. But for any stage game in which neither player has a weakly dominant action, the i.i.d. strategies violate consistency: any player who is optimizing will not be playing i.i.d.

Nachbar (2005) shows that this feature of Bayesian fictitious play extends to all Bayesian learning models. For large classes of repeated games, for *any* belief profile there are *no* strategy subsets that simultaneously satisfy learnability, richness, and consistency. Thus, for example, if each player believes the other is playing a strategy that has a k-period memory, then one can show that learnability and richness hold but consistency fails: best responding in this setting requires using a strategy with a memory of more than k periods. The impossibility result generalizes to ε optimization and ε consistency, for ε sufficiently small. The result also generalizes to games with payoff uncertainty (with learnability, richness, and consistency now defined in terms of distributional strategies) (see Nachbar, 2001).

I conclude with four remarks. First, since the set of all strategies satisfies richness and consistency, it follows that the set of all strategies is not learnable for *any* beliefs: for any belief profile there is a strategy profile that the players will not learn to predict.

This can also be shown directly by a diagonalization argument along the lines of Oakes (1985) and Dawid (1985). The impossibility result of Nachbar (2005) can be viewed as a game theoretic version of Dawid (1985). For a description of what subsets *are* learnable, see Noguchi (2005).

Second, if one constructs a Bayesian learning model satisfying learnability and consistency then LBR holds and, if players play their LBR strategies, play converges to equilibrium play. This identifies a potentially attractive class of Bayesian models in which convergence obtains. The impossibility result says, however, that if learnability and consistency hold, then player beliefs must be partially equilibrated in the sense of, in effect, excluding some of the strategies required by richness.

Third, consistency is not *necessary* for LBR or convergence. For example, for many stage games, variants of fictitious play satisfy LBR and converge even though these learning models are inconsistent. The impossibility result is a statement about the ability to construct Bayesian models with certain properties; it is not a statement about convergence per se.

Last, learnability, richness, and consistency may be too strong to be taken as necessary conditions for beliefs to be considered sensible. It is an open question whether one can construct Bayesian models satisfying conditions that are weaker but still strong enough to be interesting.

JOHN NACHBAR

See also **deterministic evolutionary dynamics; learning and evolution in games: adaptive heuristics; learning and evolution in games: an overview; learning and evolution in games: ESS; purification; repeated games; stochastic adaptive dynamics.**

Bibliography

Aoyagi, M. 1996. Evolution of beliefs and the Nash equilibrium of normal form games. *Journal of Economic Theory* 70, 444–69.

Bernheim, B.D. 1984. Rationalizable strategic behavior. *Econometrica* 52, 1007–28.

Brown, G.W. 1951. Iterative solutions of games by fictitious play. In *Activity Analysis of Production and Allocation*, ed. T.J. Koopmans. New York: Wiley.

Cournot, A. 1838. Researches into the Mathematical Principles of the Theory of Wealth. Trans. N.T. Bacon, New York: Kelley, 1960.

Dawid, A.P. 1985. The impossibility of inductive inference. *Journal of the American Statistical Association* 80, 340–1.

Foster, D. and Young, P. 2001. On the impossibility of predicting the behavior of rational agents. *Proceedings of the National Academy of Sciences* 98, 12848–53.

Foster, D. and Young, P. 2003. Learning, hypothesis testing, and Nash equilibrium. *Games and Economic Behavior* 45, 73–96.

Fudenberg, D. and Kreps, D. 1993. Learning mixed equilibria. *Games and Economic Behavior* 5, 320–67.

Fudenberg, D. and Levine, D. 1993. Steady state learning and Nash equilibrium. *Econometrica* 61, 547–74.

Fudenberg, D. and Levine, D. 1995. Universal consistency and cautious fictitious play. *Journal of Economic Dynamics and Control* 19, 1065–89.

Fudenberg, D. and Levine, D. 1998. *Theory of Learning in Games*. Cambridge, MA: MIT Press.

Fudenberg, D. and Levine, D. 1999. Conditional universal consistency. *Games and Economic Behavior* 29, 104–30.

Harsanyi, J. 1973. Games with randomly disturbed payoffs: a new rationale for mixed-strategy equilibrium points. *International Journal of Game Theory* 2, 1–23.

Hart, S. and Mas-Colell, A. 2003. Uncoupled dynamics do not lead to Nash equilibrium. *American Economic Review* 93, 1830–6.

Hart, S. and Mas-Colell, A. 2006. Stochastic uncoupled dynamics and Nash equilibrium. *Games and Economic Behavior* 57, 286–303.

Hofbauer, J. and Sandholm, W. 2002. On the global convergence of stochastic fictitious play. *Econometrica* 70, 2265–94.

Jackson, M. and Kalai, E. 1999. False reputation in a society of players. *Journal of Economic Theory* 88, 40–59.

Jordan, J.S. 1991. Bayesian learning in normal form games. *Games and Economic Behavior* 3, 60–81.

Jordan, J.S. 1993. Three problems in learning mixed-strategy Nash equilibria. *Games and Economic Behavior* 5, 368–86.

Jordan, J.S. 1995. Bayesian learning in repeated games. *Games and Economic Behavior* 9, 8–20.

Kalai, E. and Lehrer, E. 1993a. Rational learning leads to Nash equilibrium. *Econometrica* 61, 1019–45.

Kalai, E. and Lehrer, E. 1993b. Subjective equilibrium in repeated games. *Econometrica* 61, 1231–40.

Kuhn, H.W. 1964. Extensive games and the problem of information. In *Contributions to the Theory of Games*, vol. 2, ed. M. Dresher, L.S. Shapley and A.W. Tucker. Princeton: Princeton University Press.

Milgrom, P. and Roberts, J. 1991. Adaptive and sophisticated learning in repeated normal form games. *Games and Economic Behavior* 3, 82–100.

Milgrom, P. and Weber, R. 1985. Distributional strategies for games with incomplete information. *Mathematics of Operations Research* 10, 619–32.

Nachbar, J.H. 2001. Bayesian learning in repeated games of incomplete information. *Social Choice and Welfare* 18, 303–26.

Nachbar, J.H. 2005. Beliefs in repeated games. *Econometrica* 73, 459–80.

Noguchi, Y. 2005. Merging with a set of probability measures: a characterization. Working paper, Kanto Gakuin University.

Nyarko, Y. 1998. Bayesian learning and convergence to Nash equilibria without common priors. *Economic Theory* 11, 643–55.

Oakes, D. 1985. Self-calibrating priors do not exist. *Journal of the American Statistical Association* 80, 339–42.

Sandroni, A. 1998. Necessary and sufficient conditions for convergence to Nash equilibrium: the almost absolute continuity hypothesis. *Games and Economic Behavior* 22, 121–47.

Sandroni, A. 2000. Reciprocity and cooperation in repeated coordination games: the principled-player approach. *Games and Economic Behavior* 32, 157–82.

Shapley, L. 1962. On the nonconvergence of fictitious play. Discussion Paper RM-3026, RAND.

learning and evolution in games: ESS

1. Introduction

According to John Maynard Smith in his influential book *Evolution and the Theory of Games* (1982, p.10), an ESS (that is, an *evolutionarily stable strategy*) is 'a strategy such that, if all members of the population adopt it, then no mutant strategy could invade the population under the influence of natural selection'. The ESS concept, based on static fitness comparisons, was originally introduced and developed in the biological literature (Maynard Smith and Price, 1973) as a means to predict the eventual outcome of evolution for individual behaviours in a single species. It avoids the complicated dynamics of the evolving population that may ultimately depend on spatial, genetic and population size effects.

To illustrate the Maynard Smith (1982) approach, suppose individual fitness is the expected payoff in a random pairwise contest. The ESS strategy p^* must then do at least as well as a mutant strategy p in their most common contests against p^* and, if these contests yield the same payoff, then p^* must do better than p in their rare contests against a mutant. That is, Maynard Smith's definition applied to a symmetric two-player game says p^* is an ESS if and only if, for all $p \neq p^*$,

$$\begin{array}{lll} \text{(i)} & \pi(p,p^*) \leq \pi(p^*,p^*) & \text{(equilibrium condition)} \\ \text{(ii)} & \text{if } \pi(p,p^*) = \pi(p^*,p^*),\ \ \pi(p,p) < \pi(p^*,p) & \text{(stability condition)} \end{array} \quad (1)$$

where $\pi(p,\hat{p})$ is the payoff of p against $\hat{p}$. One reason the ESS concept has proven so durable is that it has equivalent formulations that are equally intuitive (see especially the concepts of invasion barrier and local superiority in Section 2.1).

By (1) (i), an ESS is a Nash equilibrium (NE) with the extra refinement condition (ii) that seems heuristically related to dynamic stability. In fact, there is a complex relationship between the static ESS conditions and dynamic stability, as illustrated throughout this article with specific reference to the replicator equation. It is this relationship that formed the initial basis of what has come to be known as 'evolutionary game theory'.

ESS theory (and evolutionary game theory in general) has been extended to many classes of games besides those based on a symmetric two-player game. This article begins with ESS theory for symmetric normal form games before briefly describing the additional features that arise in each of several types of more general games. The unifying principle of local (or neighborhood) superiority will emerge in the process.

2. ESS for symmetric games

In a symmetric evolutionary game, there is a single set S of pure strategies available to the players, and the payoff to pure strategy e_i is a function π_i of the system's strategy distribution. In the following subsections we consider two-player symmetric games with S finite in normal and extensive forms (Sections 2.1 and 2.2 respectively) and with S a continuous set (Section 2.3).

2.1 Normal form games

Let $S \equiv \{e_1, \ldots, e_n\}$ be the set of pure strategies. A player may also use a mixed strategy $p \in \Delta^n \equiv \{p = (p_1, \ldots, p_n) | \sum p_i = 1, p_i \geq 0\}$ where p_i is the proportion of the time this individual uses pure strategy e_i. Pure strategy e_i is identified with the ith unit vector in Δ^n. The population state is $\hat{p} \in \Delta^n$ whose components are the current frequencies of strategy use in the population (that is, the strategy distribution). We assume the expected payoff to p is the bilinear function $\pi(p, \hat{p}) = \sum_{i,j=1}^{n} p_i \pi(e_i, e_j) \hat{p}_j$ resulting from random two-player contests.

Suppose the resident population is monomorphic at p^* (that is, all members adopt strategy p^*) and a monomorphic sub-population of mutants using p appears in the system. These mutants will not invade if there is a positive *invasion barrier* $\varepsilon_0(p)$ (Bomze and Pötscher, 1989). That is, if the proportion ε of mutants in the system is less than $\varepsilon_0(p)$, then the mutants will eventually die out due to their lower replication rate. In mathematical terms, $\varepsilon = 0$ is a (locally) asymptotically stable rest point of the corresponding resident-mutant invasion dynamics. For invasion dynamics based on replication, Bomze and Pötscher show p^* is an ESS (that is, satisfies (1)) if and only if every $p \neq p^*$ has a positive invasion barrier.

Important and somewhat surprising consequences of an ESS p^* are its asymptotic stability for many evolutionary dynamics beyond these monomorphic resident systems invaded by a single type of mutant. For instance, p^* is asymptotically stable when simultaneously invaded by several types of mutants and when a polymorphic resident system consisting of several (mixed) strategy types whose average strategy is p^* is invaded (see the 'strong stability' concept developed in Cressman, 1992). In particular, p^* is asymptotically stable for the replicator equation (Taylor and Jonker, 1978; Hofbauer, Schuster and Sigmund, 1979; Zeeman, 1980)

$$\dot{p}_i = p_i(\pi(e_i, p) - \pi(p, p)) \tag{2}$$

when each individual player is a pure strategist.

Games that have a completely mixed ESS (that is, p^* is in the interior of Δ^n) enjoy further dynamic stability properties since these games are *strictly stable* (that is, $\pi(p - \hat{p}, p - \hat{p}) < 0$ for all $p \neq \hat{p}$) (Sandholm, 2006). The ESS of a strictly stable game is also globally asymptotically stable for the best response dynamics (the continuous-time version of fictitious play) (Hofbauer and Sigmund, 1998) and for the Brown–von Neumann–Nash dynamics (related to Nash's, 1951, proof of existence of NE) (Hofbauer and Sigmund, 2003).

The preceding two paragraphs provide a strong argument that an ESS will be the ultimate outcome of the evolutionary adjustment process. The proofs of these results use two other equivalent characterizations of an ESS p^* of a symmetric normal form game; namely,

(a) p^* has a *uniform* invasion barrier (i.e. $\varepsilon_0(p) > 0$ is independent of p)
(b) for all p sufficiently close (but not equal) to p^*

$$\pi(p, p) < \pi(p^*, p). \tag{3}$$

It is this last characterization, called 'local superiority' (Weibull, 1995), that proves so useful for other classes of games (see below). Heuristically, (3) suggests p^* will be asymptotically stable since there is an incentive to shift towards p^* whenever the system is slightly perturbed from p^*.

Unfortunately, there are many normal form games that have no ESS. These include most three-strategy games classified by Zeeman (1980) and Bomze (1995). No mixed strategy p^* can be an ESS of a symmetric zero-sum game (that is, $\pi(\hat{p}, p) = -\pi(p, \hat{p})$ for all $p, \hat{p} \in \Delta^n$) since $\pi(p^*, p) = \pi(p^* - p, p) \leq 0 = \pi(p, p)$ for all $p \in \Delta^n$ in some direction from p^*. Thus, the classic zero-sum Rock–Scissors–Paper Game in Table 1 has no ESS since its only NE $(\frac{1}{3}, \frac{1}{3}, \frac{1}{3})$ is interior. An early attempt to relax the ESS conditions to rectify this replaces the strict inequality in (1) (ii) by $\pi(p, p) \leq \pi(p^*, p)$. The NE p^* is then called a *neutrally stable strategy* (NSS) (Maynard Smith, 1982; Weibull, 1995). The only NE of the Rock–Scissors–Paper Game is a NSS.

Also, the normal forms of most interesting extensive form games have no ESS, especially when NE outcomes do not specify choices off the equilibrium path and so correspond to NE components. In general, when NE are not isolated, the *ESSet* introduced by Thomas (1985) is more important. This is a set E of NSS so that (1) (ii) holds for all $p^* \in E$ and $p \notin E$. An ESSet is a finite union of disjoint NE components, each of which must be an ESSet in its own right. Each ESSet has setwise dynamic stability consequences analogous to an ESS (Cressman, 2003). The ES structure of a game refers to its collection of ESSs and ESSets.

There are then several classes of symmetric games that always have an ESSet. Every two-strategy game has an ESSet (Cressman, 2003) which generically (that is, unless $\pi(\hat{p}, \hat{p}) - \pi(p, \hat{p})$ for all $p, \hat{p} \in \Delta^2$) is a finite set of ESSs. All games with symmetric payoff function (that is, $\pi(\hat{p}, p) = \pi(p, \hat{p})$ for all $p, \hat{p} \in \Delta^n$) have an ESSet corresponding to the set of local maxima of $\pi(p, p)$ which generically is a set of isolated ESSs). These are called partnership games (Hofbauer and Sigmund, 1998) or common interest games (Sandholm, 2006).

Symmetric games with payoff, $\pi_i(\hat{p})$, of pure strategy e_i nonlinear in the population state $\hat{p}$ are quite common in biology and in economics (Maynard Smith, 1982; Sandholm, 2006), where they are called playing-the-field models or population games. With $\pi(p, \hat{p}) = \sum_i p_i \pi_i(\hat{p})$, nonlinearity implies (1) is a weaker condition than (3), as

Table 1 *The payoff matrix for the Rock–Scissors–Paper Game*

Rock	0	1	−1
Scissors	−1	0	1
Paper	1	−1	0

Each entry is the payoff to the row player when column players are listed in the same order.

examples in Bomze and Pötscher (1989) show. Local superiority (3) is then taken as the operative definition of an ESS p^* (Hofbauer and Sigmund, 1998) and it is equivalent to the existence of a uniform invasion barrier for p^*.

2.2 Extensive form games

The application of ESS theory to finite extensive form games has been less successful (see Figure 1). Every ESS can have no other realization equivalent strategies in its normal form (van Damme, 1991) and so, in particular, must be *pervasive strategy* (that is, it must reach every information set when played against itself). To ease these problems, Selten (1983) defined a *direct ESS* in terms of behaviour strategies (that is, strategies that specify the local behaviour at each player information set) as a b^* that satisfies (1) for any other behaviour strategy b. He showed each such b^* is subgame perfect and arises from the backward induction technique applied to the ES structure of the subgames and their corresponding truncations.

Consider backward induction applied to Figure 1. Its second-stage subgame $\begin{matrix} \ell \\ r \end{matrix}\begin{bmatrix} -5 & 5 \\ -4 & 4 \end{bmatrix}$ has mixed ESS $b_2^* = (\frac{1}{2}, \frac{1}{2})$ and, when the second decision point of player 1 is replaced by the payoff 0 from b_2^*, the truncated single-stage game $\begin{matrix} L \\ R \end{matrix}\begin{bmatrix} 0 & 1 \\ 1 & 0 \end{bmatrix}$ also has a mixed ESS $b_1^* = (\frac{1}{2}, \frac{1}{2})$. Since both stage games have a mixed ESS (and so a unique NE since they are strictly stable), (b_1^*, b_2^*) is the only NE of Figure 1 and it is pervasive.

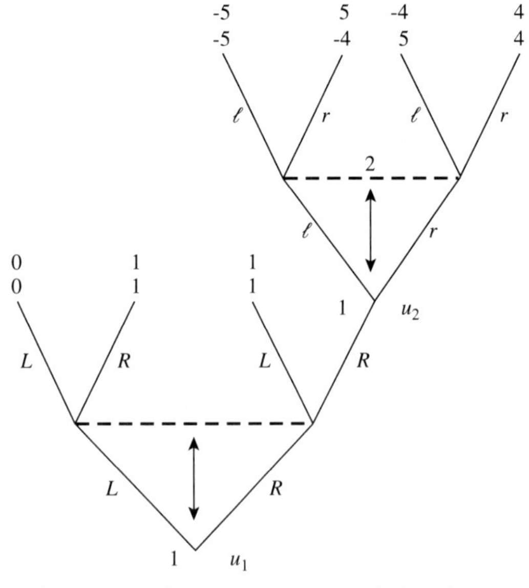

Figure 1 The extensive form tree of the van Damme example. For the construction of the tree of a symmetric extensive form game, see Selten (1983) or van Damme (1991)

Surprisingly, this example has no direct ESS as Selten originally hoped since (b_1^*, b_2^*) can be invaded by the pure strategy that plays Rr (van Damme, 1991).

The same technique applied to Figure 1 with second-stage subgame replaced by $\begin{matrix} \ell \\ r \end{matrix}\begin{bmatrix} -1 & 0 \\ 0 & -1 \end{bmatrix}$ yields $b_2^* = (\frac{1}{2}, \frac{1}{2})$ and truncated single-stage game $\begin{matrix} L \\ R \end{matrix}\begin{bmatrix} 0 & 1 \\ 1 & -1/2 \end{bmatrix}$ with $b_1^* = (\frac{3}{5}, \frac{2}{5})$. This is an example of a two-stage War of Attrition with base game $\begin{bmatrix} 0 & 1 \\ 1 & 0 \end{bmatrix}$ where a player remains (R) at the first stage in the hope the opponent will leave (L) but incurs a waiting cost of one payoff unit if both players remain. This (b_1^*, b_2^*) is a direct ESS since all N-stage War of Attrition games are strictly stable (Cressman, 2003).

The examples in the preceding two paragraphs show that, although backward induction determines candidates for the ES structure, it is not useful for determining which candidates are actually direct ESSs. The situation is more discouraging for non-pervasive NE. For example, the only NE outcome of the two-stage repeated Prisoner's Dilemma game (Nachbar, 1992) with cumulative payoffs is mutual defection at each stage. This NE outcome cannot be an isolated behaviour strategy (that is, there is a corresponding NE component) and so there is no direct ESS. Worse, for typical single-stage payoffs such as $\begin{matrix} \text{Defect} \\ \text{Cooperate} \end{matrix}\begin{bmatrix} -1 & 10 \\ -2 & 5 \end{bmatrix}$, this component does not satisfy setwise extensions of the ESS (for example, it is not an ESSet).

Characterization of NE found by backward induction with respect to dynamically stable rest points of the subgames and their truncations shows more promise. Each direct ESS b^* yields an ESSet in the game's normal form (Cressman, 2003) and so is dynamically stable. Furthermore, for the class of simultaneity games where both players know all player actions at earlier stages, Cressman shows that, if b^* is a pervasive NE, then it is asymptotically stable with respect to the replicator equation if and only if it comes from this backward induction process. In particular, the NE for Figure 1 and for the N-stage War of Attrition are (globally) asymptotically stable. Although the subgame perfect NE for the N-stage Prisoner's Dilemma game that defects at each decision point is not asymptotically stable, the eventual outcome of evolution is in the NE component (Nachbar, 1992; Cressman, 2003).

2.3 Continuous strategy space

Evolutionary game theory for symmetric games with a continuous set of pure strategies S has been slower to develop. Most recent work examines static payoff comparisons that predict an $x^* \in S$ is the evolutionary outcome. There are now fundamental differences between the ESS notion (1) and that of local superiority (3) as well as between invasion by monomorphic mutant sub-populations and the polymorphic model of the replicator equation. Here, we illustrate these differences when S is a subinterval of real numbers and $\pi(x, y)$ is a continuous payoff function of $x, y \in S$.

First, consider an $x^* \in S$ that satisfies (3). In particular,

$$\pi(x,x) < \pi(x^*,x) \tag{4}$$

for all $x \in S$ sufficiently close (but not equal) to x^*. This is the *neighbourhood invader strategy* (NIS) condition of Apaloo (1997) that states x^* can invade any nearby monomorphism x. On the other hand, from (1), x^* cannot be invaded by these x if it is a *neighbourhood strict NE*, that is

$$\pi(x,x^*) < \pi(x^*,x^*) \tag{5}$$

for any other x sufficiently close to x^*. Inequalities (4) (5) are independent of each other and combine to assert that x^* strictly dominates x in all these two-strategy games $\{x^*, x\}$.

In the polymorphic model, populations are described by a P in the infinite dimensional set $\Delta(S)$ of probability distributions with support in S. When the expected payoff $\pi(x, P)$ is given through random pairwise contests, Cressman (2005) shows that strict domination implies x^* is *neighbourhood superior* (that is,

$$\pi(x^*,P) > \pi(P,P) \tag{6}$$

for all other $P \in \Delta(S)$ with support sufficiently close to x^*) and conversely, neighbourhood superiority implies weak domination. Furthermore, a neighborhood superior monomorphic population x^* (that is, the Dirac delta probability distribution δ_{x^*}) is asymptotically stable for all initial P with support sufficiently close to x^* (and containing x^*) under the replicator equation. This is now a dynamic on $\Delta(S)$ (Oechssler and Riedel, 2002) that models the evolution of the population distribution.

In the monomorphic model, the population is a monomorphism $x(t) \in S$ at all times. If a nearby mutant strategy $y \in S$ can invade x, the whole population is shifted in this direction. This intuition led Eshel (1983) to define a *continuously stable strategy* (CSS) as a neighbourhood strict NE x^* that satisfies, for all x sufficiently close to x^*,

$$\pi(y,x) > \pi(x,x) \tag{7}$$

for all y between x^* and x that are sufficiently close to x. Later, Dieckmann and Law (1996) developed the canonical equation of adaptive dynamics to model the evolution of this monomorphism and showed a neighbourhood strict NE x^* is a CSS if and only if it is an asymptotically stable rest point. Cressman (2005) shows x^* is a CSS if and only if it is *neighbourhood half-superior* (that is, there is a uniform invasion barrier of at least $\frac{1}{2}$ in the two-strategy games $\{x^*, x\}$) (see also the half-dominant concept of Morris, Rob and Shin, 1995).

For example, take $S = \mathbf{R}$ and payoff function

$$\pi(x,y) = -x^2 + bxy \tag{8}$$

that has strict NE $x^* = 0$ for all values of the fixed parameter b. x^* is a NIS (CSS) if and only if $b < 1$ $(b < 2)$ (Cressman and Hofbauer, 2005). Thus, there are strict NE when $b > 2$ that are not 'evolutionarily stable'.

3. Asymmetric games

Following Selten (1980) and van Damme (1991), in a two-player asymmetric game with two roles (or species), pairwise contests may involve players in the same or in opposite roles. First, consider ESS theory when there is a finite set of pure strategies $S = \{e_1, \ldots, e_n\}$ and $T = \{f_1, \ldots, f_m\}$ for players in role 1 and 2 respectively. Assume payoff to a mixed strategist is given by a bilinear payoff function and let $\pi_1(p; \hat{p}, \hat{q})$ be the payoff to a player in role one using $p \in \Delta^n$ when the current state of the population in roles 1 and 2 are $\hat{p}$ and $\hat{q}$ respectively. Similarly, $\pi_2(q; \hat{p}, \hat{q})$ is the payoff to a player in role 2 using $q \in \Delta^m$. For a discussion of resident-mutant invasion dynamics, see Cressman (1992), who shows the monomorphism (p^*, q^*) is uninvadable by any other mutant pair (p, q) if and only if it is a *two-species ESS*, that is, for all (p, q) sufficiently close (but not equal) to (p^*, q^*),

$$\text{either } \pi_1(p; p, q) < \pi_1(p^*; p, q) \text{ or} \pi_2(q; p, q) < \pi_2(q^*; p, q). \tag{9}$$

The ESS condition (9) is the two-role version of local superiority (3) and has an equivalent formulation analogous to (1) (Cressman, 1992). This ESS also enjoys similar stability properties to the ESS of Subsection 2.1 such as its asymptotic stability under the (two-species) replicator equation (Cressman, 1992; 2003).

A particularly important class of asymmetric games consists of truly asymmetric games that have no contests between players in the same role (that is, there are no intraspecific contests). These are bimatrix games (that is, given by an $n \times m$ matrix whose *ij*th entry is the pair of payoffs $(\pi_1(e_i, f_j), \pi_2(e_i, f_j))$ for the interspecific contest between e_i and f_j). The ESS concept is now quite restrictive since Selten (1980) showed that (p^*, q^*) satisfies (9) if and only if it is a strict NE. This is also equivalent to asymptotic stability under the (two-species) replicator equation (Cressman, 2003). Standard examples (Cressman, 2003), with two strategies for each player include the Buyer–Seller Game that has no ESS since its only NE is in the interior. Another is the Owner–Intruder Game that has two strict NE Maynard Smith (1982) called the bourgeois ESS where the owners defend their territory and the paradoxical ESS where owners retreat.

Asymmetric games with continuous sets of strategies have recently received a great deal of attention (Leimar, 2006). For a discussion of neighbourhood (half) superiority conditions that generalize (6) and (7) to two-role truly asymmetric games with continuous payoff functions, see Cressman (2005). He also shows how these conditions are related to NIS and CSS concepts based on (9) and to equilibrium selection results for games with discontinuous payoff functions such as the Nash Demand Game (Binmore, Samuelson and Young, 2003).

ROSS CRESSMAN

See also **deterministic evolutionary dynamics; learning and evolution in games: an overview.**

Bibliography

Apaloo, J. 1997. Revisiting strategic models of evolution: the concept of neighborhood invader strategies. *Theoretical Population Biology* 52, 71–7.

Binmore, K., Samuelson, L. and Young, P. 2003. Equilibrium selection in bargaining models. *Games and Economic Behavior* 45, 296–328.

Bomze, I. 1995. Lotka–Volterra equation and replicator dynamics: new issues in classification. *Biological Cybernetics* 72, 447–53.

Bomze, I. and Pötscher, B. 1989. *Game Theoretical Foundations of Evolutionary Stability.* Lecture notes in economics and mathematical systems 324. Berlin: Springer-Verlag.

Cressman, R. 1992. *The Stability Concept of Evolutionary Games (A Dynamic Approach).* Lecture notes in biomathematics 94. Berlin: Springer-Verlag.

Cressman, R. 2003. *Evolutionary Dynamics and Extensive Form Games.* Cambridge, MA: MIT Press.

Cressman, R. 2005. Continuously stable strategies, neighborhood superiority and two-player games with continuous strategy space. Mimeo.

Cressman, R. and Hofbauer, J. 2005. Measure dynamics on a one-dimensional continuous trait space: theoretical foundations for adaptive dynamics. *Theoretical Population Biology* 67, 47–59.

Dieckmann, U. and Law, R. 1996. The dynamical theory of coevolution: a derivation from stochastic ecological processes. *Journal of Mathematical Biology* 34, 579–612.

Eshel, I. 1983. Evolutionary and continuous stability. *Journal of Theoretical Biology* 103, 99–111.

Hofbauer, J., Schuster, P. and Sigmund, K. 1979. A note on evolutionarily stable strategies and game dynamics. *Journal of Theoretical Biology* 81, 609–12.

Hofbauer, J. and Sigmund, K. 1998. *Evolutionary Games and Population Dynamics.* Cambridge: Cambridge University Press.

Hofbauer, J. and Sigmund, K. 2003. Evolutionary game dynamics. *Bulletin of the American Mathematical Society* 40, 479–519.

Leimar, O. 2006. Multidimensional convergence stability and the canonical adaptive dynamics. In *Elements of Adaptive Dynamics*, ed. U. Dieckmann and J. Metz. Cambridge University Press.

Maynard Smith, J. 1982. *Evolution and the Theory of Games.* Cambridge: Cambridge University Press.

Maynard Smith, J. and Price, G. 1973. The logic of animal conflicts. *Nature* 246, 15–18.

Morris, S., Rob, R. and Shin, H. 1995. Dominance and belief potential. *Econometrica* 63, 145–57.

Nachbar, J. 1992. Evolution in the finitely repeated Prisoner's Dilemma. *Journal of Economic Behavior and Organization* 19, 307–26.

Nash, J. 1951. Non-cooperative games. *Annals of Mathematics* 54, 286–95.

Oechssler, J. and Riedel, F. 2002. On the dynamic foundation of evolutionary stability in continuous models. *Journal of Economic Theory* 107, 223–52.

Sandholm, W. 2006. *Population Games and Evolutionary Dynamics.* Cambridge, MA: MIT Press.

Selten, R. 1980. A note on evolutionarily stable strategies in asymmetrical animal contests. *Journal of Theoretical Biology* 84, 93–101.

Selten, R. 1983. Evolutionary stability in extensive two-person games. *Mathematical Social Sciences* 5, 269–363.

Taylor, P. and Jonker, L. 1978. Evolutionarily stable strategies and game dynamics. *Mathematical Biosciences* 40, 145–156.

Thomas, B. 1985. On evolutionarily stable sets. *Journal of Mathematical Biology* 22, 105–15.

van Damme, E. 1991. *Stability and Perfection of Nash Equilibria*, 2nd edn. Berlin: Springer-Verlag.

Weibull, J. 1995. *Evolutionary Game Theory.* Cambridge, MA: MIT Press.

Zeeman, E. 1980. Population dynamics from game theory. In *Global Theory of Dynamical Systems*, eds. Z. Nitecki and C. Robinson. Lecture notes in mathematics 819. Berlin: Springer.

mechanism design

Overview

A mechanism is a specification of how economic decisions are determined as a function of the information that is known by the individuals in the economy. In this sense, almost any kind of market institution or economic organization can be viewed, in principle, as a mechanism. Thus mechanism theory can offer a unifying conceptual structure in which a wide range of institutions can be compared, and optimal institutions can be identified.

The basic insight of mechanism theory is that *incentive constraints* should be considered coequally with *resource constraints* in the formulation of the economic problem. In situations where individuals' private information and actions are difficult to monitor, the need to give people an incentive to share information and exert efforts may impose constraints on economic systems just as much as the limited availability of raw materials. The theory of mechanism design is the fundamental mathematical methodology for analysing these constraints.

The study of mechanisms begins with a special class of mechanisms called *direct-revelation* mechanisms, which operate as follows. There is assumed to be a mediator who can communicate separately and confidentially with every individual in the economy. This mediator may be thought of as a trustworthy person, or as a computer tied into a telephone network. At each stage of the economic process, each individual is asked to report all of his private information (that is, everything that he knows that other individuals in the economy might not know) to the mediator. After receiving these reports confidentially from every individual, the mediator may then confidentially recommend some action or move to each individual. A direct-revelation mechanism is any rule for specifying how the mediator's recommendations are determined, as a function of the reports received.

A direct-revelation mechanism is said to be *incentive compatible* if, when each individual expects that the others will be honest and obedient to the mediator, then no individual could ever expect to do better (given the information available to him) by reporting dishonestly to the mediator or by disobeying the mediator's recommendations. That is, if honesty and obedience is an equilibrium (in the game-theoretic sense), then the mechanism is incentive compatible.

The analysis of such incentive-compatible direct-revelation mechanisms might at first seem to be of rather narrow interest, because such fully centralized mediation of economic systems is rare, and incentives for dishonesty and disobedience are commonly observed in real economic institutions. The importance of studying such mechanisms is derived from two key insights: (i) for any equilibrium of any general mechanism, there is an incentive-compatible direct-revelation mechanism that is essentially equivalent; and (ii) the set of incentive-compatible direct-revelation mechanisms has simple mathematical properties that often make it easy to characterize, because it can be defined by a set of linear inequalities. Thus, by analysing incentive-compatible direct-revelation mechanisms, we can characterize what can be accomplished in all possible equilibria of all possible mechanisms, for a given economic situation.

Insight (i) above is known as the *revelation principle.* It was first recognized by Gibbard (1973), but for a somewhat narrower solution concept (dominant strategies, instead of Bayesian equilibrium) and for the case where only informational honesty is problematic (no moral hazard). The formulation of the revelation principle for the broader solution concept of Bayesian equilibrium, but still in the case of purely informational problems, was recognized independently by many authors around 1978 (see Dasgupta, Hammond and Maskin, 1979; Harris and Townsend, 1981; Holmstrom, 1977; Myerson, 1979; Rosenthal, 1978). Aumann's (1974; 1987) concept of *correlated equilibrium* gave the first expression to the revelation principle in the case where only obedient choice of actions is problematic (pure moral hazard, no adverse selection). The synthesis of the revelation principle for general Bayesian games with incomplete information, where both honesty and obedience are problematic, was given by Myerson (1982). A generalization of the revelation principle to multistage games was stated by Myerson (1986).

The intuition behind the revelation principle is as follows. First, a central mediator who has collected all relevant information known by all individuals in the economy could issue recommendations to the individuals so as to simulate the outcome of any organizational or market system, centralized or decentralized. After the individuals have revealed all of their information to the mediator, he can simply tell them to do whatever they would have done in the other system. Second, the more information that an individual has, the harder it may be to prevent him from finding ways to gain by disobeying the mediator. So the incentive constraints will be least binding when the mediator reveals to each individual only the minimal information needed to identify his own recommended action, and nothing else about the reports or recommendations of other individuals. So, if we assume that the mediator is a discrete and trustworthy information-processing device, with no costs of processing information, then there is no loss of generality in assuming that each individual will confidentially reveal all of his information to the mediator (maximal revelation to the trustworthy mediator), and the mediator in return will reveal to each individual only his own recommended action (minimal revelation to the individuals whose behaviour is subject to incentive constraints).

The formal proof of the revelation principle is difficult only because it is cumbersome to develop the notation for defining, in full generality, the set of all general mechanisms, and for defining equilibrium behaviour by the individuals in any given mechanism. Once all of this notation is in place, the construction of the equivalent incentive-compatible direct-revelation mechanism is straightforward. Given any mechanism and any equilibrium of the mechanism, we simply specify that the mediator's recommended actions are those that would result in the given mechanism if everyone behaved as specified in the given equilibrium when his actual private information was as reported to the mediator. To check that this constructed direct-revelation mechanism is incentive compatible, notice that any player who could gain by disobeying the mediator could also gain by similarly disobeying his own strategy in the given equilibrium of the given mechanism, which is impossible (by definition of equilibrium).

Mathematical formulations

Let us offer a precise general formulation of the proof of the revelation principle in the case where individuals have private information about which they could lie, but there is no question of disobedience of recommended actions or choices. For a general model, suppose that there are n individuals, numbered 1 to n. Let C denote the set of all possible combinations of actions or resource allocations that the individuals may choose in the economy. Each individual in the economy may have some private information about his preferences and endowments, and about his beliefs about other individuals' private information. Following Harsanyi (1967), we may refer to the state of an individual's private information as his *type*. Let T_i denote the set of possible types for any individual i, and let $T = T_i \times \cdots \times T_n$ denote the set of all possible combinations of types for all individuals.

The preferences of each individual i may be generally described by some *payoff function* $u_i : C \times T \to \mathbb{R}$, where $u_i(c,(t_i, \ldots, t_n))$ denotes the payoff, measured in some von Neumann–Morgenstern utility scale, that individual i would get if c was the realized resource allocation in C when $(t_i, \ldots, t_n)$ denotes the actual types of the individuals $1, \ldots, n$ respectively. For short, we may write $t = (t_i, \ldots, t_n)$ to describe a combination of types for all individuals.

The beliefs of each individual i, as a function of his type, may be generally described by some function $p_i(\cdot\,|\,\cdot)$, where $p_i,(t_1, \ldots, t_{i-1}, t_{i+1}, \ldots, t_n|t_i)$ denotes the probability that individual i would assign to the event that the other individuals have types as in $(t_1, \ldots, t_{i-1}, t_{i+1}, \ldots, t_n)$, when i knows that his own type is t_i. For short, we may write $t_{-i} = (t_1, \ldots, t_{i-1}, t_{i+1}, \ldots, t_n)$, to describe a combination of types for all individuals other than i. We may let $T_{-i} = T_1 \times \cdots \times T_{i-1} \times T_{i+1} \times \cdots \times T_n$ denote the set of all possible combinations of types for the individuals other than i.

The general model of an economy defined by these structures $(C, T_1, \ldots, T_n, u_1, \ldots, u_n, p_1, \ldots, p_n)$ is called a Bayesian collective-choice problem.

Given a Bayesian collective-choice problem, a general mechanism would be any function of the form γ: $S_1 \times \cdots \times S_n, \to C$, where, for each i, S_i is a nonempty set that denotes the set of strategies that are available for individual i in this mechanism. That is, a general mechanism specifies the strategic options that each individual may choose among, and the social choice or allocation of resources that would result from any combination of strategies that the individuals might choose. Given a mechanism, an equilibrium is any specification of how each individual may choose his strategy in the mechanism as a function of his type, so that no individual, given only his own information, could expect to do better by unilaterally deviating from the equilibrium. That is, $\sigma = (\sigma_1, \ldots, \sigma_n)$ is an equilibrium of the mechanism γ if, for each individual i, σ_i is a function from T_i to S_i, and, for every t_i in T_i and every s_i in S_i,

$$\Sigma_{t_{-i} \in T_{-i}} \; p_i(t_{-i}|t_i)u_i(\gamma(\sigma(t)), t) \geq \Sigma_{t_{-i} \in T_{-i}} \; p_i(t_{-i}|t_i)u_i(\gamma(\sigma_{-i}(t_{-i}), s_i), t).$$

(Here $\sigma(t) = (\sigma_1(t_1), \ldots, \sigma_n(t_n))$ and $(\sigma_{-i}(t_{-i}), s_i) = (\sigma_1(t_1), \ldots, \sigma_{i-1}(t_{i-1}), s_i, \sigma_{i+1}(t_{i+1}), \ldots, \sigma_n(t_n))$.) Thus, in an equilibrium σ, no individual i, knowing only his own type t_i, could increase his expected payoff by changing his strategy from $\sigma_i(t_i)$ to some other strategy s_i, when he expects all other individuals to behave as specified by the equilibrium σ. (This concept of equilibrium is sometimes often called *Bayesian equilibrium* because it respects the assumption that each player knows only his own type when he chooses his strategy in S_i. For a comparison with other concepts of equilibrium, see Dasgupta, Hammond and Maskin, 1979, and Palfrey and Srivastava, 1987).

In this context, a direct-revelation mechanism is any mechanism such that the set S_i of possible strategies for each player i is the same as his set of possible types T_i. A direct-revelation mechanism is (Bayesian) incentive-compatible iff it is an equilibrium (in the Bayesian sense defined above) for every individual always to report his true type. Thus, μ: $T_1 \times \ldots \times T_n \to C$ is an incentive-compatible direct-revelation mechanism if, for each individual i and every pair of types t_i and r_i in T_i,

$$\Sigma_{t_{-i} \in T_{-i}} \; p_i(t_{-i}|t_i)u_i(\mu(t), t) \geq \Sigma_{t_{-i} \in T_{-i}} \; p_i(t_{-i}|t_i)u_i(\mu(t_{-i}, r_i), t).$$

(Here $(t_{-i}, r_i) = (t_1, \ldots, t_{i-1}, r_i, t_{i+1}, \ldots, t_n)$.) We may refer to these constraints as the *informational incentive constraints* on the direct-revelation mechanism μ. These informational incentive constraints are the formal representation of the economic problem of *adverse selection*, so they may also be called adverse-selection constraints (or self-selection constraints).

Now, to prove the revelation principle, given any general mechanism γ and any Bayesian equilibrium σ of the mechanism γ, let μ be the direct-revelation mechanism μ defined so that, for every t in T,

$$\mu(t) = \gamma(\sigma(t)).$$

Then this mechanism μ always leads to the same social choice as γ does, when the individuals behave as in the equilibrium σ. Furthermore, μ is incentive compatible because, for any individual i and any two types t_i and r_i in T_i,

$$\begin{aligned}\Sigma_{t_{-i} \in T_{-i}} \; p_i(t_{-i}|t_i)u_i(\mu(t), t) &= \Sigma_{t_{-i} \in T_{-i}} \; p_i(t_{-i}|t_i)u_i(\gamma(\sigma(t)), t) \\ &\geq \Sigma_{t_{-i} \in T_{-i}} \; p_i(t_{-i}|t_i)u_i(\gamma(\sigma_{-i}(t_{-i}), \sigma_i(r_i)), t) \\ &= \Sigma_{t_{-i} \in T_{-i}} \; p_i(t_{-i}|t_i)u_i(\mu(t_{-i}, r_i), t).\end{aligned}$$

Thus, μ is an incentive-compatible direct-revelation mechanism that is equivalent to the given mechanism γ with its equilibrium σ.

Notice that the revelation principle asserts that any pair consisting of a mechanism *and* an equilibrium is equivalent to an incentive-compatible direct-revelation mechanism. Thus, a general mechanism that has several equilibria may correspond to several different incentive-compatible mechanisms, depending on which equilibrium is considered.

Furthermore, the same general mechanism will generally have different equilibria in the context of different Bayesian collective-choice problems, where the structure of individuals' beliefs and payoffs are different. For example, consider a first-price sealed-bid auction where there are five potential bidders who are risk-neutral with independent private values drawn from the same distribution over \$0 to \$10. If the bidders' values are drawn from a uniform distribution over this interval, then there is an equilibrium in which each bidder bids 4/5 of his value. On the other hand, if the bidders' values are drawn instead from a distribution with a probability density that is proportional to the square of the value, then there is an equilibrium in which each bidder bids 8/9 of his value. So in one situation the first-price sealed-bid auction (a general mechanism) corresponds to an incentive-compatible mechanism in which the bidder who reports the highest value gets the object for 4/5 of his reported value; but in the other situation it corresponds to an incentive-compatible mechanism in which the bidder who reports the highest value gets the object for 8/9 of his reported value. There is no incentive-compatible direct-revelation mechanism that is equivalent to the first-price sealed-bid auction in all situations, independently of the bidders' beliefs about each others' values. Thus, if we want to design a mechanism that has good properties in the context of many different Bayesian collective-choice problems, we cannot necessarily restrict our attention to incentive-compatible direct-revelation mechanisms, and so our task is correspondingly more difficult. (See Wilson, 1985, for a remarkable effort at this kind of difficult question.)

Even an incentive-compatible mechanism itself may have other dishonest equilibria that correspond to different incentive-compatible mechanisms. Thus, when we talk about selecting an incentive-compatible mechanism and assume that it will then be played according to its honest equilibrium, we are implicitly making an assumption about the selection of an equilibrium as well as of a mechanism or communication structure. Thus, for example, when we say that a particular incentive-compatible mechanism maximizes a given individual's expected utility, we mean that, if you could choose any general mechanism for coordinating the individuals in the economy and if you could also (by some public statement, as a focal arbitrator, using Schelling's, 1960, *focal-point effect*) designate the equilibrium that the individuals would play in your mechanism, then you could not give this given individual a higher expected utility than by choosing this incentive-compatible mechanism and its honest equilibrium.

In many situations, an individual may have a right to refuse to participate in an economic system or organization. For example, a consumer generally has the right to refuse to participate in any trading scheme and instead just consume his initial endowment. If we let $w_i(t_i)$ denote the utility payoff that individual i would get if he refused to participate when his type is t_i, and if we assume that an individual can make the choice not to participate after learning his type, then an incentive-compatible mechanism μ must also satisfy the following constraint, for every individual i and every possible type t_i:

$$\Sigma_{t_{-i}\in T_{-i}}\ p_i(t_{-i}|t_i)u_i(\mu(t),t) \geq w_i(t_i).$$

These constraints are called *participational incentive constraints*, or *individual-rationality* constraints.

In the analysis of Bayesian collective-choice problems, we have supposed that the only incentive problem was to get people to share their information, and to agree to participate in the mechanism in the first place. More generally, a social choice may be privately controlled by one or more individuals who cannot be trusted to follow some pre-specified plan when it is not in their best interests. For example, suppose now that the choice in C is privately controlled by some individual (call him 'individual 0') whose choice of an action in C cannot be regulated. To simplify matters here, let us suppose that this individual 0 has no private information. Let $p_0(t)$ denote the probability that this individual would assign to the event that $t=(t_1,\ldots, t_n)$ is the profile of types for the other n individuals, and let $u_0(c, t)$ denote the utility payoff that this individual receives if he chooses action c when t is the actual profile of types. Then, to give this active individual an incentive to obey the recommendations of a mediator who is implementing the direct-revelation mechanism μ, μ must satisfy

$$\Sigma_{t\in T}\ p_0(t)u_0(\mu(t),t) \geq \Sigma_{t\in T}\ p_0(t)u_0(\delta(\mu(t)),t)$$

for every function δ: $C\rightarrow C$. These constraints assert that obeying the actions recommended by the mediator is better for this individual than any disobedient strategy δ under which he would choose $\delta(c)$ if the mediator recommended c. Such constraints are called *strategic incentive constraints* or *moral-hazard constraints*, because they are the formal representation of the economic problem of moral hazard.

For a formulation of general incentive constraints that apply when individuals both have private information and control private actions, see Myerson (1982) or (1985).

Applications

In general, the mechanism-theoretic approach to economic problems is to list the constraints that an incentive-compatible mechanism must satisfy, and to try to characterize the incentive-compatible mechanisms that have properties of interest.

For example, one early contribution of mechanism theory was the derivation of general *revenue equivalence* theorems in auction theory. Ortega-Reichert (1968) found that, when bidders are risk-neutral and have private values for the object being sold that are independent and drawn from the same distribution, then a remarkably diverse collection of different auction mechanisms all generate the same expected revenue to the seller, when bidders use equilibrium strategies. In all of these different mechanisms and equilibria, it turned out that the bidder whose value for the object was highest would always end up getting the object, while a bidder whose value for the object was zero would never pay anything. By analysing the incentive constraints, Harris and Raviv (1981), Myerson (1981) and Riley and Samuelson (1981) showed that all incentive-compatible mechanisms with these properties would necessarily generate the same expected revenue, in such economic situations.

Using methods of constrained optimization, the problem of finding the incentive-compatible mechanism that maximizes some given objective (one individual's expected utility, or some social welfare function) can be solved for many examples. The resulting optimal mechanisms often have remarkable qualitative properties.

For example, suppose a seller, with a single indivisible object to sell, faces five potential buyers or bidders, whose private values for the object are independently drawn from a uniform distribution over the interval from \$0 to \$10. If the objective is to maximize the sellers' expected revenue, optimal auction mechanisms exist and all have the property that the object is sold to the bidder with the highest value for it, except that the seller keeps the object in the event that the bidders' values are all less than \$5. Such a result may seem surprising, because this event could occur with positive probability (1/32) and in this event the seller is getting no revenue in an 'optimal' auction, even though any bidder would almost surely be willing to pay him a positive price for the object. Nevertheless, no incentive-compatible mechanism (satisfying the participational and informational incentive constraints) can offer the seller higher expected utility than these optimal auctions, and thus no equilibrium of any general auction mechanism can offer higher expected revenue either. Maximizing expected revenue requires a positive probability of seemingly wasteful allocation.

The threat of keeping the object, when all bidders report values below \$5, increases the seller's expected revenue because it gives the bidders an incentive to bid higher and pay more when their values are above \$5. In many other economic environments, we can similarly prove the optimality of mechanisms in which seemingly wasteful threats are carried out with positive probability. People have intuitively understood that costly threats are often made to give some individual an incentive to reveal some information or choose some action, and the analysis of incentive constraints allows us to formalize this understanding rigorously.

In some situations, incentive constraints imply that such seemingly wasteful allocations may have to occur with positive probability in all incentive-compatible mechanisms, and so also in all equilibria of all general mechanisms. For example, Myerson and Satterthwaite (1983) considered bilateral bargaining problems between a seller of some object and a potential buyer, both of whom are risk-neutral and have independent private values for the object that are drawn out of distributions that have continuous positive probability densities over some pair of intervals that have an intersection of positive length. Under these technical (but apparently quite weak) assumptions, it is impossible to satisfy the participational and informational incentive constraints with any mechanism in which the buyer gets the object whenever it is worth more to him than to the seller. Thus, we cannot hope to guarantee the attainment of full *ex post* efficiency of resource allocations in bilateral bargaining problems where the buyer and seller are uncertain about each other's reservation prices. If we are concerned with welfare and efficiency questions, it may be more productive to try to characterize the incentive-compatible mechanisms that maximize the expected total gains from trade, or that maximize the probability that a mutually beneficial trade will occur. For example, in the bilateral bargaining problem where the seller's and buyer's private values for the object are independent random variables drawn from a uniform distribution over the interval from \$0 to \$10, both of these objectives are maximized subject to incentive constraints by mechanisms in which the buyer gets the object if and only if his value is greater than the seller's value by \$2.50 or more. Under such a mechanism, the event that the seller will keep the object when it is actually worth more to the buyer has probability 7/32, but no equilibrium of any general mechanism can generate a lower probability of this event.

The theory of mechanism design has fundamental implications about the domain of applicability of Coase's (1960) theorem, which asserts the irrelevance of initial property rights to efficiency of final allocations. The unavoidable possibility of failure to realize mutually beneficial trades, in such bilateral trading problems with two-sided uncertainty, can be interpreted as one of the 'transaction costs' that limits the validity of Coase's theorem. Indeed, as Samuelson (1985) has emphasized, reassignment of property rights generally changes the payoffs that individuals can guarantee themselves without selling anything, which changes the right-hand sides of the participational incentive constraints, which in turn can change the maximal social welfare achievable by an optimal incentive-compatible mechanism.

For example, consider again the case where there is one object and two individuals who have private values for the object that are independent random variables drawn from a uniform distribution over the interval from \$0 to \$10. When we assumed above that one was the 'seller', we meant that he had the right to keep the object and pay nothing to anyone, until he agreed to some other arrangement. Now, let us suppose instead that the rights to the object are distributed equally between the two individuals. Suppose that the object is a divisible good and each individual has a right to take half of the good and pay nothing, unless he agrees to some other arrangement. (Assume that, if an individual's value for the whole good is t_i, then his value for half would be $t_i/2$.) With this symmetric assignment of property rights, we can design incentive-compatible mechanisms in which the object always ends up being owned entirely by the individual who has the higher value for it, as Cramton, Gibbons and Klemperer (1987) have shown.

For example, consider the game in which each individual independently puts money in an envelope, and then the individual who put more money in his envelope gets the object, while the other individual takes the money in both envelopes. This game has an equilibrium in which each individual puts into his envelope an amount equal to one-third of his value for the whole good. This equilibrium of this game is equivalent to an incentive-compatible direct-revelation mechanism in which the individual who reports the higher value pays one-third of his value to buy out the other individual's half-share. This mechanism would violate the participational incentive constraints if one individual had a right to the whole good (in which case, for example, if his value were $10 then he would be paying $3.33 under this mechanism for a good that he already owned). But with rights to only half of the good, no type of either individual could expect to do better (at the beginning of the game, when he knows his own value but not the other's) by keeping his half and refusing to participate in this mechanism.

More generally, redistribution of property rights tends to reduce the welfare losses caused by incentive constraints when it creates what Lewis and Sappington (1989) have called *countervailing incentives*. In games where one individual is the seller and the other is the buyer, if either individual has an incentive to lie, it is usually because the seller wants to overstate his value or the buyer wants to understate his value. In the case where either individual may buy the other's half-share, neither individual can be sure at first whether he will be the buyer or the seller (unless he has the highest or lowest possible value). Thus, a buyer-like incentive to understate values, in the event where the other's value is lower, may help to cancel out a seller-like incentive to overstate values, in the event where the other's value is higher.

The theory of mechanism design can also help us to appreciate the importance of mediation in economic relationships and transactions. There are situations in which, if the individuals were required to communicate with each other only through perfect noiseless communication channels (for example, in face-to-face dialogue), then the set of all possible equilibria would be much smaller than the set of incentive-compatible mechanisms that are achievable with a mediator. (Of course, the revelation principle asserts that the former set cannot be larger than the latter.)

For example, consider the following 'sender–receiver game' due to J. Farrell. Player 1 has a privately known type that may be α or β, but he has no payoff-relevant action to choose. Player 2 has no private information, but he must choose an action from the set $\{x, y, z\}$. The payoffs to players 1 and 2 respectively depend on 1's type and 2's action as follows.

	x	y	z
α	$2, 3$	$1, 2$	$0, 0$
β	$4, -3$	$8, -1$	$0, 0$

At the beginning of the game, player 2 believes that each of 1's two possible types has probability 1/2.

Suppose that, knowing his type, player 1 is allowed to choose a message in some arbitrarily rich language, and player 2 will hear player 1's message (with no noise or distortion) before choosing his action. In every equilibrium of this game, including the randomized equilibria, player 2 must choose y with probability 1, after every message that player 1 may choose in equilibrium (see Farrell, 1993; Myerson, 1988). If there were some message that player 1 could use to increase the probability of player 2 choosing x (for example, 'I am α, so choosing x would be best for us both!'), then he would always send such a message when his type was α. (It can be shown that no message could ever induce player 2 to randomize between x and z.) So not receiving such a message would lead 2 to infer that 1's type was β, which implies that 2 would rationally choose z whenever such a message was not sent, so that both types of 1 should always send the message (any randomization between x and y is better than z for both types of 1). But a message that is always sent by player 1, no matter what his type is, would convey no information to player 2, so that 2 would rationally choose his *ex ante* optimal action y.

If we now allow the players to communicate through a mediator who uses a randomized mechanism, then we can apply the revelation principle to characterize the surprisingly large set of possible incentive-compatible mechanisms. Among all direct-revelation mechanisms that satisfy the relevant informational incentive constraints for player 1 and strategic incentive constraints for player 2, the best for player 2 is as follows: if player 1 reports to the mediator that his type is α then with probability 2/3 the mediator recommends x to player 2, and with probability 1/3 the mediator recommends y to player 2; if player 1 reports to the mediator that his type is β then with probability 2/3 the mediator recommends y to player 2, and with probability 1/3 the mediator recommends z to player 2. Notice that this mechanism is also better for player 1 than the unmediated equilibria when 1's type is α, although it is worse for 1 when his type is β.

Other mechanisms that player 2 might prefer would violate the strategic incentive constraint that player 2 should not expect to gain by choosing z instead of y when y is recommended. If player 2 could pre-commit himself always to obey the mediator's recommendations, then better mechanisms could be designed.

Efficiency

The concept of efficiency becomes more difficult to define in economic situations where individuals have different private information at the time when the basic decisions about production and allocation are made. A welfare economist or social planner who analyses the Pareto efficiency of an economic system must use the perspective of an outsider, so he cannot base his analysis on the individuals' private information. Otherwise, public testimony as to whether an economic mechanism or its outcome would be 'efficient' could implicitly reveal some individuals' private information to other individuals, which could in turn alter their rational behaviour and change the outcome of the mechanism! Thus, Holmstrom and Myerson (1983) argued that efficiency should be considered as a property of mechanisms, rather than of the outcome or allocation ultimately realized by the mechanism (which will depend on the individuals' private information).

Thus, a definition of Pareto efficiency in a Bayesian collective-choice problem must look something like this: 'a mechanism is efficient if there is no other feasible mechanism that may make some other individuals better off and will certainly not make other individuals worse off.' However, this definition is ambiguous in at least two ways.

First, we must specify whether the concept of feasibility takes incentive constraints into account or not. The concept of feasibility that ignores incentive constraints may be called *classical feasibility.* In these terms, the fundamental insight of mechanism theory is that incentive constraints are just as real as resource constraints, so that incentive compatibility may be a more fruitful concept than classical feasibility for welfare economics.

Second, we must specify what information is to be considered in determining whether an individual is 'better off' or 'worse off'. One possibility is to say that an individual is made worse off by a change that decreases his expected utility payoff as would be computed before his own type or any other individuals' types are specified. This is called the *ex ante* welfare criterion. A second possibility is to say that an individual is made worse off by a change that decreases his conditionally expected utility, given his own type (but not given the types of any other individuals). An outside observer, who does not know any individual's type, would then say that an individual may be made worse off, in this sense, if this conditionally expected utility were decreased for at least one possible type of the individual. This is called the *interim* welfare criterion. A third possibility is to say that an individual is made worse off by a change that decreases his conditionally expected utility given the types of all individuals. An outside observer would then say that an individual may be worse off in this sense if his conditionally expected utility were decreased for at least one possible combination of types for all the individuals. This is called the *ex post* welfare criterion.

If each individual knows his own type at the time when economic plans and decisions are made, then the interim welfare criterion should be most relevant to a social planner. Thus, Holmstrom and Myerson (1983) argue that, for welfare analysis in a Bayesian collective-choice problem, the most appropriate concept of efficiency is that which combines the interim welfare criterion and the incentive-compatible definition of feasibility. This concept is called *incentive efficiency*, or *interim incentive efficiency.* That is, a mechanism $\mu: T \rightarrow C$ is incentive efficient if it is an incentive-compatible mechanism and there does not exist any other incentive-compatible mechanism $\gamma: T \rightarrow C$ such that for every individual i and every type t_i in T_i,

$$\Sigma_{t_{-i} \in T_{-i}} \; p_i(t_{-i}|t_i) u_i(\gamma(t), t) \geq \Sigma_{t_{-i} \in T_{-i}} \; p_i(t_{-i}|t_i) u_i(\mu(t), t),$$

and there is at least one type of at least one individual for which this inequality is strict. If a mechanism is incentive efficient, then it cannot be common knowledge among the individuals, at the stage when each knows only his own type, that there is some other incentive-compatible mechanism that no one would consider worse (given his own information) and some might consider strictly better.

For comparison, another important concept is classical *ex post* efficiency, defined using the *ex post* welfare criterion and the classical feasibility concept. That is, a mechanism $\mu: T \rightarrow C$ is (*classically*) *ex post efficient* iff there does not exist any other mechanism $\gamma: T \rightarrow C$ (not necessarily incentive compatible) such that, for every individual i and every combination of individuals' types t in $T = T_1 \times \cdots \times T_n$,

$$u_i(\gamma(t), t) \geq u_i(\mu(t), t),$$

with strict inequality for at least one individual and at least one combination of individuals' types.

The appeal of *ex post* efficiency is that there may seem to be something unstable about a mechanism that sometimes leads to outcomes such that, if everyone could share their information, they could identify another outcome that would make them all better off. However, we have seen that bargaining situations exist where no incentive-compatible mechanisms are *ex post* efficient. In such situations, the incentive constraints imply that rational individuals would be unable to share their information to achieve these gains, because if everyone were expected to do so then at least one type of one individual would have an incentive to lie.

Thus, a benevolent outside social planner who is persuaded by the usual Paretian arguments should choose some incentive-efficient mechanism. To determine more specifically an 'optimal' mechanism within this set, a social welfare function is needed that defines tradeoffs, not only between the expected payoffs of different individuals but also between the expected payoffs of different types of each individual. That is, given any positive utility-weights $\lambda_i(t_i)$ for each type t_i of each individual i, one can generate an incentive-efficient mechanism by maximizing

$$\Sigma_{i=1}^{n} \ \Sigma_{t_i \in T_i} \lambda_i(t_i) \ \Sigma_{t_{-i} \in T_{-i}} \ p_i(t_{-i}|t_i) u_i(\mu(t), t)$$

over all $\mu: T \to C$ that satisfy the incentive constraints; but different vectors of utility weights may generate different incentive-efficient mechanisms.

Bargaining over mechanisms

A positive economic theory must go beyond welfare economics and try to predict the economic institutions that may actually be chosen by the individuals in an economy. Having established that a social planner can restrict his attention to incentive-compatible direct-revelation mechanisms, which is a mathematically simple set, it is natural to assume that rational economic agents who are themselves negotiating the structure of their economic institutions should be able to bargain over the set of incentive-compatible direct-revelation mechanisms. But if we assume that individuals know their types already at the time when fundamental economic plans and decisions are made, then we need a theory of mechanism selection by individuals who have private information.

When we consider bargaining games in which individuals can bargain over mechanisms, there should be no loss of generality in restricting our attention to equilibria in which there is one incentive-compatible mechanism that is selected with probability 1 independently of anyone's type. This proposition, called the *inscrutability principle*, can be justified by viewing the mechanism-selection process as itself part of a more broadly defined general mechanism and applying the revelation principle. For example, suppose that there is an equilibrium of the mechanism-selection game in which some mechanism μ would be chosen if individual 1's type were α and some other mechanism ν would be chosen if 1's type were β. Then there should exist an equivalent equilibrium of the mechanism-selection game in which the individuals always select a direct-revelation mechanism that coincides with mechanism μ when individual 1 confidentially reports type α to the mediator (in the implementation of the mechanism, after it has been selected), and that coincides with mechanism ν when 1 reports type β to the mediator.

However, the inscrutability principle does not imply that the possibility of revealing information during a mechanism-selection process is irrelevant. There may be some mechanisms that we should expect not to be selected by the individuals in such a process, precisely because some individuals would choose to reveal information about their types rather than let these mechanisms be selected. For example, consider the following Bayesian collective-choice problem, due to Holmstrom and Myerson (1983). There are two individuals, 1 and 2, each of whom has two possible types, α and β, which are independent and equally likely. There are three social choice options, called x, y and z. Each individual's utility for these options depends on his type according to the following table.

Option	$1,\alpha$	$1,\beta$	$2,\alpha$	$2,\beta$
x	2	0	2	2
y	1	4	1	1
z	0	9	0	−8

The incentive-efficient mechanism that maximizes the *ex ante* expected sum of the two individuals' utilities is as follows: if 1 reports type α and 2 reports α then choose x, if 1 reports type β and 2 reports α then choose z, and if 2 reports β then choose y (regardless of 1's report). However, Holmstrom and Myerson argue that such a mechanism would not be chosen in a mechanism-selection game that is played when 1 already knows his type, because, when 1 knows that his type is α, he could do better by proposing to select the mechanism that always chooses x, and 2 would always want to accept this proposal. That is, because 1 would have no incentive to conceal his type from 2 in a mechanism-selection game if his type were α (when his interests would then have no conflict with 2's), we should not expect the individuals in a mechanism-selection game to agree inscrutably to an incentive-efficient mechanism that implicitly puts as much weight on 1's type-β payoff as the mechanism described above.

For another example, consider again the sender–receiver game due to Farrell. Recall that y would be the only possible equilibrium outcome if the individuals could communicate only face-to-face, with no mediation or other noise in their communication channel. Suppose that the mechanism-selection process is as follows: first 2 proposes a mediator who is committed to implement some incentive-compatible mechanism; then 1 can either accept this mediator and communicate with 2 thereafter only through him, or 1 can reject this mediator and thereafter communicate with 2 only face-to-face. Suppose now that 2 proposes that they should use a mediator who will implement the incentive-compatible mediation plan that is best for 2 (recommending x with probability 2/3 and y with probability 1/3 if 1 reports α, recommending y with probability 2/3 and z with probability 1/3 if 1 reports β). We have seen that this mechanism is worse than y for 1 if his type is β. Furthermore, this mechanism would be worse than y for player 1 under the *ex ante* welfare criterion, when his expected payoffs for type α and type β are averaged, each with weight 1/2. However, it is an equilibrium of this mechanism-selection game for player 1 always to accept this proposal, no matter what his type is. If 1 rejected 2's proposed mediator, then 2 might reasonably infer that l's type was β, in which case 2's rational choice would be z instead of y, and z is the worse possible outcome for both of 1's types.

Now consider a different mechanism-selection process for this example, in which the informed player 1 can select any incentive-compatible mechanism himself, with only the restriction that 2 must know what mechanism has been selected by 1. For any incentive-compatible mechanism μ, there is an equilibrium in which 1 chooses μ for sure, no matter what his type is, and they thereafter play the honest and obedient equilibrium of this mechanism. To support such an equilibrium, it suffices to suppose that, if any mechanism other than μ were selected, then 2 would infer that 1's type was β and therefore choose z. Thus, concepts like sequential equilibrium from non-cooperative game theory cannot determine the outcome of this mechanism-selection game, beyond what we already know from the revelation principle; we cannot even say that 1's selected mechanism will be incentive-efficient. To get incentive efficiency as a result of mechanism-selection games, we need some further assumptions, like those of cooperative game theory.

An attempt to extend traditional solution concepts from cooperative game theory to the problem of bargaining over mechanisms has been proposed by Myerson (1983; 1984a; 1984b). In making such an extension, one must consider not only the traditional problem of how to define reasonable compromises between the conflicting interests of different individuals, but also the problem of how to define reasonable compromises between the conflicting interests of different types of the same individual. That is, to conceal his type in the mechanism-selection process, an individual should bargain for some inscrutable compromise between what he really wants and what he would have wanted if his type had been different; and we need some formal theory to predict what a reasonable inscrutable compromise might be. In the above sender–receiver game, where only type β of player 1 should feel any incentive to conceal his type, we might expect an inscrutable compromise to be resolved in favor of type α. That is, in the mechanism-selection game where 1 selects the mechanism, we might expect both types of 1 to select the incentive-compatible mechanism that is best for type α. (In this mechanism, the mediator recommends x with probability 0.8 and y with probability 0.2 if 1 reports α; and the mediator recommends x with probability 0.4, y with probability 0.4, and z with probability 0.2 if 1 reports β.) This mechanism is the *neutral optimum* for player 1, in the sense of Myerson (1983).

ROGER B. MYERSON

See also **incentive compatibility; mechanism design: new developments; revelation principle.**

Bibliography

Aumann, R.J. 1974. Subjectivity and correlation in randomized strategies. *Journal of Mathematical Economics* 1, 67–96.

Aumann, R.J. 1987. Correlated equilibrium as an expression of Bayesian rationality. *Econometrica* 55, 1–18.

Coase, R. 1960. The problem of social cost. *Journal of Law and Economics* 3, 1–44.

Cramton, P., Gibbons, R. and Klemperer, P. 1987. Dissolving a partnership efficiently. *Econometrica* 55, 615–32.

Dasgupta, P., Hammond, P. and Maskin, E. 1979. The implementation of social choice rules: some general results on incentive compatibility. *Review of Economic Studies* 46, 185–216.

Farrell, J. 1993. Meaning and credibility in cheap-talk games. *Games and Economic Behavior* 5, 514–31. Repr. in *Mathematical Models in Economics*, ed. M. Bacharach and M. Dempster. Oxford: Oxford University Press, 1997.

Gibbard, A. 1973. Manipulation of voting schemes: a general result. *Econometrica* 41, 587–602.

Harris, M. and Raviv, A. 1981. Allocation mechanisms and the design of auctions. *Econometrica* 49, 1477–99.

Harris, M. and Townsend, R.M. 1981. Resource allocation under asymmetric information. *Econometrica* 49, 33–64.

Harsanyi, J.C. 1967. Games with incomplete information played by Bayesian players. *Management Science* 14, 159–82, 320–34, 481–502.

Holmstrom, B. 1977. On incentives and control in organizations. Ph.D. thesis, Graduate School of Business, Stanford University.

Holmstrom, B. and Myerson, R.B. 1983. Efficient and durable decision rules with incomplete information. *Econometrica* 51, 1799–19.

Lewis, T.R. and Sappington, D.E.M. 1989. Countervailing incentives in agency problems. *Journal of Economic Theory* 49, 294–313.

Myerson, R.B. 1979. Incentive compatibility and the bargaining problem. *Econometrica* 47, 61–74.

Myerson, R.B. 1981. Optimal auction design. *Mathematics of Operation Research* 6, 58–73.

Myerson, R.B. 1982. Optimal coordination mechanisms in generalized principal–agent problems. *Journal of Mathematical Economics* 10, 67–81.

Myerson, R.B. 1983. Mechanism design by an informed principal. *Econometrica* 51, 1767–97.

Myerson, R.B. 1984a. Two-person bargaining problems with incomplete information. *Econometrica* 52, 461–87.

Myerson, R.B. 1984b. Cooperative games with incomplete information. *International Journal of Game Theory* 13, 69–86.

Myerson, R.B. 1985. Bayesian equilibrium and incentive compatibility. In *Social Goals and Social Organization*, ed. L. Hurwicz, D. Schmeidler and H. Sonnenschein. Cambridge: Cambridge University Press.

Myerson, R.B. 1986. Multistage games with communication. *Econometrica* 54, 323–58.

Myerson, R.B. 1988. Incentive constraints and optimal communication systems. In *Proceedings of the Second Conference on Theoretical Aspects of Reasoning about Knowledge*, ed. M.Y. Vardi. Los Altos: Morgan Kaufmann.

Myerson, R.B. and Satterthwaite, M. 1983. Efficient mechanisms for bilateral trading. *Journal of Economic Theory* 29, 265–81.

Ortega-Reichert, A. 1968. Models for competitive bidding under uncertainty. Ph.D. thesis, Department of Operations Research, Stanford University.

Palfrey, T. and Srivastava, S. 1987. On Bayesian implementable allocations. *Review of Economic Studies* 54, 193–208.

Riley, J.G. and Samuelson, W.F. 1981. Optimal auctions. *American Economic Review* 71, 381–92.

Rosenthal, R.W. 1978. Arbitration of two-party disputes under uncertainty. *Review of Economic Studies* 45, 595–604.
Samuelson, W. 1985. A comment on the Coase Theorem. In *Game-Theoretic Models of Bargaining*, ed. A.E. Roth. Cambridge: Cambridge University Press.
Schelling, T.C. 1960. *The Strategy of Conflict*. Cambridge, MA: Harvard University Press.
Wilson, R. 1985. Incentive efficiency of double auctions. *Econometrica* 53, 1101–15.

mechanism design (new developments)

1. Possibility results and robustness

Game theory provides methods to predict the outcome of a given game. Mechanism design concerns the *reverse* question: given some desirable outcome, can we design a game which produces it? Formally, the *environment* is $\langle$A, N, $\Theta\rangle$, where A is a set of feasible and verifiable alternatives or outcomes, $N = \{1, \ldots, n\}$ is a set of agents, and Θ is a set of possible *states of the world.* Except where indicated, we consider *private values* environments, where a state is $\theta = (\theta_1, \ldots, \theta_n) \in \times_i \Theta_i = \Theta$, each agent i knows his own 'type' $\theta_i \in \Theta_i$, and his payoff $u_i(a, \theta_i)$ depends only on the chosen alternative and his own type. (This does not rule out the possibility that the agents know something about each others' types.) If values are not private, then they are said to be *interdependent.* A *mechanism* or *contract* $\Gamma = (S, h)$ specifies a set of feasible actions S_i for each agent i, and an outcome function $h : S \equiv \times_{i=1}^{n} S_i \rightarrow A$. An outside party (a principal or social planner), or the agents themselves, want to design a mechanism which produces optimal outcomes. These are often represented by a *social choice rule* (SCR) $F : \Theta \rightarrow A$. A *social choice function* (SCF) is a single-valued SCR. Implicitly, it is assumed that the mechanism designer does not know the true θ, and this lack of information makes it impossible for her to directly choose an outcome in $F(\theta)$. Instead, she uses the more roundabout method of designing a mechanism which produces an outcome in $F(\theta)$, *whatever the true* θ *may be.*

In a *revelation mechanism*, each agent simply reports what he knows (so if agent i only knows θ_i then $S_i = \Theta_i$). By definition, an *incentive compatible* revelation mechanism has a *truthful* Bayesian–Nash equilibrium, that is, it achieves *truthful implementation.* Truthful implementation plays an important role in the theory because of the revelation principle (see the dictionary entry on MECHANISM DESIGN, which surveys the early literature on truthful implementation). The early literature produced powerful results on optimal mechanisms for auction design, bargaining problems, and other applications. However, it generally made quite strong assumptions, for example, that the agents and the principal share a common prior over Θ, that the principal can *commit* to a mechanism, that the agents cannot side-contract and always use equilibrium strategies, and so on. We survey the recent literature which deals with these issues. In addition, we note that the notion of truthful implementation has a drawback: it does not rule out the possibility that non-truthful equilibria also exist, and these may produce suboptimal outcomes. (A non-truthful equilibrium may even Pareto dominate the truthful equilibrium for the agents, and hence provide a natural focal point for coordinating their actions.) To rule out the possibility of suboptimal equilibria, we may require *full implementation*: for all $\theta \in \Theta$, the set of equilibrium outcomes should precisely equal $F(\theta)$.

Maskin (1999) assumed *complete information*: each agent knows the true θ. If $n \geq 3$ agents know θ, then any SCF can be truthfully implemented: let the agents report θ, and if at least $n - 1$ agents announce the same θ then implement the outcome $F(\theta)$. Unilateral deviations from a consensus are disregarded, so truth-telling is a Nash equilibrium. Of course, this revelation mechanism will also have non-truthful equilibria. For full implementation, more complex mechanisms are required. (Even if $n = 2$, any SCF can be truthfully implemented if the principal can credibly threaten to 'punish' both agents if they report different states; in an economic environment, this might be achieved by making each agent pay a fine.)

A necessary condition for full Nash implementation is (*Maskin*) *monotonicity* (Maskin, 1999). Intuitively, monotonicity requires that moving an alternative *up* in the agents' preference rankings should not make it *less* likely to be optimal. This condition can be surprisingly difficult to satisfy. For example, if the agents can have any complete and transitive preference relation on A, then any Maskin monotonic SCF must be a constant function (Saijo, 1987). The situation is quite different if we consider *refinements* of Nash equilibrium. For example, there is a sense in which almost any (ordinal) SCR can be fully implemented in *undominated Nash equilibrium* when the agents have complete information (Palfrey and Srivastava, 1991; Jackson, Palfrey and Srivastava, 1994; Sjöström, 1994). Chung and Ely (2003) showed that this possibility result is not robust to small perturbations of the information structure that violate private values (there is a small chance that agent i knows more about agent j's preferences than agent j does). The violation of private values is key. For example, in Sjöström's (1994) mechanism, an agent who knows his own preferences can eliminate his dominated strategies, and a second round of elimination of *strictly* dominated strategies generates the optimal outcome. This construction is robust to small perturbations that respect private values.

A different kind of robustness was studied by McLean and Postlewaite (2002). Consider an economic environment where each agent i observes an independently drawn signal t_i which is correlated with the state θ. The complete information structure is approximated by letting each agent's signal be very accurate. With complete information, any SCF can be truthfully implemented. McLean and Postlewaite (2002) show robustness to perturbations of the information structure: any outcome can be approximated by an incentive-compatible allocation, if the agents' signals are accurate enough. There is no need to assume private values.

The literature on *Bayesian mechanism design* typically assumes each agent i knows only his own type $\theta_i \in \Theta_i$, the agents share a common prior p over $\Theta \equiv \times_{i=1}^{n} \Theta_i$, and the principal knows p. In fact, for truthful implementation with $n \geq 3$, the assumption that the principal knows p is redundant. Suppose for any common prior p on Θ, there is an incentive-compatible revelation mechanism $\Gamma_p = (\times_{i=1}^{n} \Theta_i, h_p)$. By definition, Γ_p truthfully implements the SCF $F_p \equiv h_p$. The mechanism Γ_p is 'parametric', that is, it depends on p. To be specific, consider a quasi-linear public goods environment with independent types, and suppose Γ_p is the well-known mechanism of d'Aspremont and Gérard-Varet (1979). Now consider a nonparametric mechanism Γ, where each agent i

announces p and θ_i. If at least $n-1$ agents report the same p, the outcome is $h_p(\theta_1,\ldots,\theta_n)$. Now, if agent i thinks everyone will announce p truthfully, he may as well do so. If in addition he thinks the other agents report θ_{-i} truthfully, then he should announce θ_i truthfully by incentive compatibility of Γ_p. Therefore, for any common prior p, the nonparametric mechanism Γ truthfully implements F_p. In this sense, the principal can use Γ to extract the agents' shared information about p. Of course, this particular mechanism also has non-truthful equilibria. Choi and Kim (1999) *fully* implemented the d'Aspremont and Gérard-Varet (1979) outcome in *undominated* Bayesian–Nash equilibrium, using a nonparametric mechanism. Naturally, their mechanism is quite complex. Suppose we restrict attention to mechanisms where each agent i only reports θ_i, truthfully in equilibrium. Then the necessary and sufficient condition for full nonparametric Bayesian–Nash implementation for any common prior p is (dominant strategy) incentive compatibility plus the *rectangular property* (Cason et al., 2006).

The d'Aspremont and Gérard-Varet (1979) mechanism is budget balanced and surplus maximizing. The above argument shows that such outcomes can be truthfully implemented by a nonparametric mechanism in quasi-linear environments with independent types. As is well known, this cannot be achieved by any dominant strategy mechanism. Thus, in general, nonparametric truthful implementation is easier than dominant strategy implementation. However, there are circumstances where the two concepts coincide. Bergemann and Morris (2005a) consider a model where each agent i has a *payoff type* $\theta_i \in \Theta_i$ and a *belief type* π_i. The payoff type determines the payoff function $u_i(a, \theta_i)$, while the belief type determines beliefs over other agents' types. The set of socially optimal outcomes $F(\theta)$ depends on payoff types, but not on beliefs. Bergemann and Morris (2005a) show that in quasi-linear environments with no restrictions on side payments (hence no budget-balance requirement), truthful implementation for all possible type spaces with a common prior implies dominant strategy implementation. (For related results, see Section 4.)

Bergemann and Morris (2005b) consider *full* implementation of SCFs in a similar framework. The SCF $F:\Theta\rightarrow A$ is *fully robustly implemented* if there exists a mechanism which fully implements F on all possible type spaces. They make no common prior assumption. Full robust implementation turns out to be equivalent to implementation using iterated elimination of strictly dominated strategies. Although a demanding concept, there are situations where full robust implementation is possible. For example, a Vickrey-Clarke-Groves (VCG) mechanism in a public goods economy with private values and strictly concave valuation functions achieves implementation in strictly dominant strategies. However, Bergemann and Morris (2005b) show the impossibility of full robust implementation when values are sufficiently interdependent.

A generalization of Maskin monotonicity called *Bayesian monotonicity* is necessary for ('parametric') full Bayesian-Nash implementation (Postlewaite and Schmeidler, 1986; Palfrey and Srivastava, 1989a; Jackson, 1991). Again, refinements lead to possibility results (Palfrey and Srivastava, 1989b). Another way to expand the set of implementable SCRs is *virtual* implementation (Abreu and Sen, 1991; Duggan, 1997). Serrano and Vohra (2001) argue that the sufficient conditions for virtual implementation are in fact quite strong.

The work discussed so far is *consequentialist*: only the final outcome matters. The mechanisms are clearly not meant to be descriptive of real-world institutions. For example, they typically require the agents to report 'all they know' before any decision is reached, an extreme form of centralized decision making hardly ever encountered in the real world. (The question of how much information must be transmitted in order to implement a given SCR is addressed by Hurwicz and Reiter, 2006, and Segal, 2004.) Delegating the power to make (verifiable) decisions to the agents would only create additional 'moral hazard' constraints, as discussed in the entry on MECHANISM DESIGN. Since centralization eliminates these moral hazard constraints, it typically strictly dominates decentralization in the basic model. However, as discussed below, by introducing additional aspects such as renegotiation and collusion, we can frequently prove the optimality of more realistic decentralized mechanisms. The implicit assumption is that decentralized decision making is in itself a good thing, which is a mild form of non-consequentialism. (Other non-consequentialist arguments are discussed in Section 4.) We might add that there is, of course, no way to eliminate the moral hazard constraints if the agents take *unverifiable* decisions that cannot be contracted upon. In this case, the issue of centralization versus decentralization of decisions is moot.

2. Renegotiation and credibility

Suppose $n = 2$ and both agents know the true θ. If a revelation mechanism is used and the agents announce different states, then we cannot identify a deviator from a 'consensus', so it may be necessary to punish *both* agents in order to support a truth-telling equilibrium. But this threat is not credible if the agents can avoid punishment by renegotiating the outcome. Maskin and Moore (1999) capture the renegotiation process by an exogenously given function $r : A \times \Theta \rightarrow A$ which maps outcome a in state θ into an efficient outcome $r(a, \theta)$. They derive an incentive-compatibility condition which is necessary for truth-telling when $n = 2$, and show that *renegotiation monotonicity* is necessary for full Nash implementation (see also Segal and Whinston, 2002).

The idea that renegotiation may preclude the implementation of the first-best outcome, even when information is complete, has received attention in models of bilateral trade with relationship-specific investments (the hold-up problem). It is possible to implement the first-best outcome if trade is one-dimensional and investments are 'selfish', in the sense that each agent's investment does not directly influence the other agent's payoff (Nöldeke and Schmidt, 1995; Edlin and Reichelstein, 1996). If investments are not selfish, then the first-best cannot always be achieved, while the second-best can often be implemented without any explicit contract (Che and Hausch, 1999). Segal (1999) found a similar result in a model with k goods and selfish investments, for k large (see also Maskin and Tirole, 1999; Hart and Moore, 1999). It should be noted that the case $n = 2$ is quite special, and adding a third party often alleviates the problem of renegotiation (Baliga and Sjöström, 2006).

Credibility and renegotiation also impact trading with asymmetric information. Suppose the seller can produce goods of different quality, but the buyer's valuation is his private information. It is typically second-best optimal for the seller to offer a contract such that low-valuation buyers consume less than first-best quality ('underproduction'), while high-valuation buyers enjoy 'information rents'. Incentive compatibility guarantees that the buyer reveals his true valuation. Now suppose trading takes place twice, and the buyer's valuation does not change. Suppose the seller cannot credibly commit to a long-run (two-period) contract. If the buyer reveals his true valuation in the first period, then in the second period the seller will leave him no rent. This is typically not the second-best outcome. The seller may prefer a 'pooling' contract which does not fully reveal valuations in the first period, a commitment device which limits his ability to extract second period rents. This idea has important applications. When a regulator cannot commit to a long-run contract, a regulated firm may hide information or exert less effort to cut costs, the *ratchet effect* (Freixas, Guesnerie and Tirole, 1985). A borrower may not exert effort to improve a project knowing that a lender with deep pockets will bail him out, the *soft budget constraint* (Dewatripont and Maskin, 1995a). These problems are exacerbated if the principal is well informed and cannot commit not to use his information. Institutional or organizational design can alleviate the problems. By committing to acquire less information via 'incomplete contracts', or by maintaining an 'arm's-length relationship', the principal can improve efficiency (Dewatripont and Maskin, 1995b; Crémer, 1995). Less frequent regulatory reviews offset the ratchet effect, and a decentralized credit market helps to cut off borrowers from future funding. Long-run contracts can help, but they may be vulnerable to renegotiation (Dewatripont, 1989). In particular, the second-period outcome may be renegotiated if quality levels are known to be different from the first-best. Again, some degree of pooling may be optimal.

If the principal cannot commit even to short-run contracts, then, after receiving the agents' messages, she always chooses an outcome that is optimal given her beliefs. She cannot credibly threaten punishments that she would not want to carry out. Refinements proposed in the cheap-talk literature suggest that a putative pooling equilibrium may be destroyed if an agent can reveal information by 'objecting' in a credible way. This leads to a necessary condition for full implementation with complete information which is reminiscent of Maskin monotonicity, but which involves the principal's preferences (Baliga, Corchón and Sjöström, 1997).

3. Collusion

A large literature on collusion was inspired by Tirole (1986). A key contribution was made by Laffont and Martimort (1997), who assumed an uninformed third party proposes side contracts. This circumvents the signalling problems that might arise if a privately informed agent makes collusive proposals. A side contract for a group of colluding agents is a *collusive mechanism* which must respect incentive compatibility, individual rationality and feasibility constraints. The original mechanism Γ, designed by the principal, is called the *grand mechanism*. The objective is to design an optimal grand mechanism when collusion is possible. Typically, collusion imposes severe limits on what can be achieved.

Baliga and Sjöström (1998) study a model with moral hazard and limited liability. Two agents share information not known to the principal: agent 1's effort is observed by both agents. Agent 2's effort is known only to himself. In the absence of collusion, the optimal grand mechanism specifies a 'message game': agent 2 reports agent 1's effort to the principal. Now suppose the agents can side contract on agent 1's effort, but not on agent 2's effort (which is unobserved). Side contracts can specify side transfers as a function of realized output, but must respect limited liability. This collusion may destroy centralized 'message games', and we obtain a theory of optimal delegation of decision making. For some parameters, it is optimal for the principal to contract only with agent 2, and let agent 2 subcontract with agent 1. This is intuitive, since agent 2 observes agent 1's effort and can contract directly on it. More surprisingly, there are parameter values where it is better for the principal to contract only with agent 1.

Mookherjee and Tsumagari (2004) study a similar model, but with adverse selection: the agents privately observe their own production costs. In this model, delegating to a 'prime supplier' creates 'double marginalization of rents': the prime supplier uses underproduction to minimize the other agent's information rent. A centralized contract avoids this problem. Hence, in this model delegation is always strictly dominated by centralization, even though the agents can collude.

Mookherjee and Tsumagari (2004) assume the agents can side contract before deciding to participate in the grand contract. Che and Kim (2006) assume side contracting occurs only after the decision to participate in the grand mechanism has been made. In this case, collusion does not limit what the principal can achieve. Hence, the timing of side contracting is important. In a complete information environment with $n \geq 3$, Sjöström (1999) showed that neither renegotiation nor collusion limit the possibility of undominated Nash implementation.

4. Other theoretical issues

In quasi-linear environments with uncorrelated types, there exist incentive-compatible mechanisms which maximize the social surplus (for example, d'Aspremont and Gérard-Varet, 1979). But the principal cannot extract all the surplus: the agents must get informational rents. However, Crémer and McLean (1988) showed that the principal can extract all the surplus in auctions with *correlated types*. McAfee and Reny (1992) extended this result to general quasi-linear environments.

Jehiel and Moldovanu (2001) considered a quasi-linear environment with multidimensional (uncorrelated) types and interdependent values. Generically, a standard revelation mechanism cannot be designed to extract information about multidimensional types, and no incentive-compatible and surplus-maximizing mechanism exists. Mezzetti (2004) presents an ingenious *two-stage* mechanism which maximizes the surplus in interdependent values environments, even when types are independent and multidimensional. In the first stage, the mechanism specifies an outcome decision but not transfers. Transfers are determined in the second stage by reports on payoffs realized by the outcome decision. Mezzetti (2007) shows that the principal can sometimes extract all the surplus by this method, even if types are independent. For optimal mechanisms for a profit-maximizing monopolist when consumers have multidimensional types and private values, see Armstrong (1996).

Incentive compatibility does not require that each agent has a dominant strategy. Nevertheless, incentive-compatible outcomes can often be replicated by dominant strategy mechanisms (Mookherjee and Reichelstein, 1992). In quasi-linear environments, all incentive-compatible mechanisms that maximize the social surplus are *payoff-equivalent* to dominant strategy (VCG) mechanisms (Krishna and Perry, 1997; Williams, 1999). However, as pointed out above, dominant strategies (but not incentive compatibility) rules out budget balance.

Bergemann and Välimäki (2002) assume agents can update a common prior by costly information acquisition. Suppose a single-unit auction has two bidders i and j who observe statistically independent private signals θ_i and θ_j. Bidder i's valuation of the good is $u_i(\theta_i, \theta_j) = \alpha\theta_i + \beta\theta_j$, where $\alpha > \beta > 0$. Thus, values are interdependent. Efficiency requires that bidder i gets the good if and only if $\theta_i \geq \theta_j$. Suppose bidders report their signals, the good is allocated efficiently given their reports, and the winning bidder i pays the price $(\alpha + \beta)\theta_j$. This VCG mechanism is incentive compatible (Maskin, 1992). If bidder i acquires negative information which causes him to lose the auction, then he imposes a negative externality on the other bidder (as $\beta > 0$). This implies the bidders have an incentive to collect too much information. Conversely, there is an incentive to collect too little information when $\beta < 0$. Bergemann and Välimäki (2002) provide a general analysis of these externalities. Similar externalities occur when members of a committee must collect information before voting. If the committee is large, each vote is unlikely to be pivotal, and free riding occurs. Persico (2004) shows how the optimal committee is designed to encourage the members to collect information.

Some authors reject consequentialism and instead emphasize agents' *rights*. For example, suppose a mechanism implements *envy-free outcomes*. An agent might still feel unfairly treated if his own bundle is worse than a bundle which another agent *had the right to choose* (but did not). Such agents may demand 'equal rights' (Gaspart, 1995). Unfortunately, once we leave the classical exchange economy, Sen's 'Paretian liberal' paradox (Sen, 1970) suggests that rights are incompatible with efficiency (Deb, Pattanaik and Razzolini, 1997). Sen originally considered rights embodied in SCRs rather than mechanisms. Peleg and Winter (2002) study *constitutional implementation* where the mechanism embodies the same rights as the SCR it implements.

5. Learning from experiments

Cabrales, Charness and Corchón (2003) tested the so-called canonical mechanism for Nash implementation. A Nash equilibrium was played only 13 per cent of the time (20 per cent when monetary fines were used). Remarkably, the optimal outcome was implemented 68 per cent of the time (80 per cent with 'fines'), because deviations from equilibrium strategies frequently did not affect the outcome. This suggests that a desirable property of a mechanism is *fault-solerance*: it should produce optimal outcomes even if some 'faulty' players deviate from the theoretical predictions. Eliaz (2002) showed that, if at most $k<\frac{1}{2}\ n-1$ players are 'faulty' (that is, unpredictable), then full Nash implementation is possible if *no-veto-power* and $(k+1)$*-monotonicity hold.*

Equilibrium play can be justified by epistemic or dynamic theories. According to epistemic theories, common knowledge about various aspects of the game implies equilibrium play even in one-shot games. Experiments provide little support for this. However, there is evidence that players can reach equilibrium through a dynamic adjustment process. If a game is played repeatedly, with no player knowing any other player's payoff function, the outcome frequently converges to a Nash equilibrium of the one-shot complete information game (Smith, 1979). Dynamic theories have been applied to the mechanism design problem (for example, Cabrales and Ponti, 2000). Chen and Tang (1998) and Chen and Gazzale (2004) argue that mechanisms which induce supermodular games produce good long-run outcomes. Unfortunately, these convergence results are irrelevant for decisions that are taken infrequently, or if the principal is too impatient to care only about the long-run outcome.

The idea of dominant strategies is less controversial than Nash equilibrium, and should be more relevant for decisions that are taken infrequently. Unfortunately, experiments on dominant-strategy mechanisms have yielded negative results. Attiyeh, Franciosi and Isaac (2000, p. 112) conclude pessimistically, 'we do not believe that the pivot mechanism warrants further practical consideration .... This is due to the fundamental failure of the mechanism, in our laboratory experiments, to induce truthful value revelation.' However, VCG mechanisms (such as the pivotal mechanism) frequently have a multiplicity of Nash equilibria, some of which produce suboptimal outcomes. Cason et al. (2006) did experiments with *secure* mechanisms, which fully implement an SCR both in dominant strategies and in Nash equilibria. The players were much more likely to use their dominant strategies in secure than in non-secure mechanisms. In the non-secure mechanisms, deviations from dominant strategies tended to correspond to Nash equilibria. However, these deviations typically did not lead to suboptimal *outcomes*. In this sense, the non-secure mechanisms were fault-tolerant. Kawagoe and Mori (2001) report experiments where deviations from dominant strategies typically corresponded to suboptimal Nash equilibria.

In experiments, subjects often violate standard axioms of rational decision making. Alternative theories, such as prospect theory, fit the experimental evidence better. But, if we modify the axioms of individual behaviour, the optimal mechanisms will change. Esteban and Miyagawa (2005) assume the agents have Gul–Pesendorfer preferences (Gul and Pesendorfer, 2001). They suffer from 'temptation', and may prefer a smaller menu (choice set) to a larger one. Suppose each agent first chooses a menu, and then chooses an alternative from this menu. Optimal menus may contain 'tempting' alternatives which are never chosen in equilibrium, because this relaxes the incentive-compatibility constraints pertaining to the choice of menu. Eliaz and Spiegler (2006) assume some agents are 'sophisticated' and some are 'naive'. Sophisticated agents know that they are dynamically inconsistent, and would like to commit to a future decision. Naive agents are unaware that they are dynamically inconsistent. The optimal mechanism screens the agents by providing commitment devices that are chosen only by sophisticated agents.

Experiments reveal the importance of human emotions such as spite or kindness (Andreoni, 1995; Saijo, 2003). In many mechanisms in the theoretical literature, by changing his strategy an agent can have a big impact on another agent's payoff without materially changing his own. Such mechanisms may have little hope of practical success if agents are inclined to manipulate each others' payoffs due to feelings of spite or kindness.

SANDEEP BALIGA AND TOMAS SJÖSTRÖM

See also **incentive compatibility; mechanism design; revelation principle.**

Bibliography

Abreu, D. and Sen, A. 1991. Virtual implementation in Nash equilibria. *Econometrica* 59, 997–1022.

Andreoni, J. 1995. Cooperation in public goods experiments: kindness or confusion? *American Economic Review* 85, 891–904.

Armstrong, M. 1996. Multiproduct nonlinear pricing. *Econometrica* 64, 51–75.

Attiyeh, G., Franciosi, R. and Isaac, R.M. 2000. Experiments with the pivotal process for providing public goods. *Public Choice* 102, 95–114.

Baliga, S., Corchón, L. and Sjöström, T. 1997. The theory of implementation when the planner is a player. *Journal of Economic Theory* 77, 15–33.

Baliga, S. and Sjöström, T. 1998. Decentralization and collusion. *Journal of Economic Theory* 83, 196–232.

Baliga, S. and Sjöström, T. 2006. Contracting with third parties. Working paper No. 75, CSIO, Northwestern University.

Bergemann, D. and Morris, S. 2005a. Robust mechanism design. *Econometrica* 73, 1771–813.

Bergemann, D. and Morris, S. 2005b. Robust implementation: the role of large type spaces. Discussion Paper No. 1519, Cowles Foundation, Yale University.

Bergemann, D. and Välimäki, J. 2002. Information acquisition and efficient mechanism design. *Econometrica* 70, 1007–34.

Cabrales, A., Charness, G. and Corchón, L. 2003. An experiment on Nash implementation. *Journal of Economic Behavior and Organization* 51, 161–93.

Cabrales, A. and Ponti, G. 2000. Implementation, elimination of weakly dominated strategies and evolutionary dynamics. *Review of Economic Dynamics* 3, 247–82.

Cason, T., Saijo, T., Sjöström, T. and Yamato, T. 2006. Secure implementation experiments: do strategy-proof mechanisms really work? *Games and Economic Behavior* 57, 206–35.

Che, Y.K. and Hausch, D. 1999. Cooperative investments and the value of contracting. *American Economic Review* 89, 125–47.

Che, Y.K. and Kim, J. 2006. Robustly collusion-proof implementation. *Econometrica* 74, 1063–107.

Chen, Y. and Gazzale, R. 2004. When does learning in games generate convergence to Nash equilibria? *American Economic Review* 94, 1505–35.

Chen, Y. and Tang, F.F. 1998. Learning and incentive compatible mechanisms for public goods provision: an experimental study. *Journal of Political Economy* 106, 633–62.

Choi, J. and Kim, T. 1999. A nonparametric, efficient public decision mechanism: undominated Bayesian Nash implementation. *Games and Economic Behavior* 27, 64–85.

Chung, K. and Ely, J. 2003. Implementation with near-complete information. *Econometrica* 71, 857–71.

Crémer, J. 1995. Arm's length relationships. *Quarterly Journal of Economics* 110, 275–95.

Crémer, J. and McLean, R. 1988. Full extraction of the surplus in Bayesian and dominant strategy auctions. *Econometrica* 56, 1247–57.

d'Aspremont, C. and Gérard-Varet, L.A. 1979. Incentives and incomplete information. *Journal of Public Economics* 11, 25–45.

Deb, R., Pattanaik, P. and Razzolini, L. 1997. Game forms, rights and the efficiency of social outcomes. *Journal of Economic Theory* 72, 74–95.

Dewatripont, M. 1989. Renegotiation and information revelation over time: the case of optimal labor contracts, *Quarterly Journal of Economics* 104, 589–619.

Dewatripont, M. and Maskin, E. 1995a. Credit and efficiency in centralized and decentralized economies. *Review of Economic Studies* 62, 541–55.

Dewatripont, M. and Maskin, E. 1995b. Contractual contingencies and renegotiation. *RAND Journal of Economics* 26, 704–19.

Duggan, J. 1997. Virtual Bayesian implementation. *Econometrica* 67, 1175–99.

Edlin, A. and Reichelstein, S. 1996. Hold-ups, standard breach remedies and optimal investment. *American Economic Review* 86, 478–501.

Eliaz, K. 2002. Fault tolerant implementation. *Review of Economic Studies* 69, 589–610.

Eliaz, K. and Spiegler, R. 2006. Contracting with diversely naive agents. *Review of Economic Studies* 73, 689–714.

Esteban, S. and Miyagawa, E. 2005. Optimal menu of menus with self-control preferences. Unpublished paper, Penn State University.

Freixas, X., Guesnerie, R. and Tirole, J. 1985. Planning under incomplete information and the ratchet effect. *Review of Economic Studies* 52, 173–92.

Gaspart, F. 1995. Fair implementation in the cooperative production problem: two properties of normal form mechanisms. Unpublished manuscript, FUNDP Namur.

Gul, F. and Pesendorfer, W. 2001. Temptation and self-control. *Econometrica* 69, 1403–35.

Hart, O. and Moore, J. 1999. Foundations of incomplete contracts. *Review of Economic Studies* 66, 115–38.

Hurwicz, L. and Reiter, S. 2006. *Designing Economic Mechanisms*. New York: Cambridge University Press.

Jackson, M. 1991. Bayesian implementation. *Econometrica* 59, 461–77.

Jackson, M., Palfrey, T. and Srivastava, S. 1994. Undominated Nash implementation in bounded mechanisms. *Games and Economic Behavior* 6, 474–501.

Jehiel, P. and Moldovanu, B. 2001. Efficient design with interdependent valuations. *Econometrica* 69, 1237–59.

Kawagoe, T. and Mori, T. 2001. Can the pivotal mechanism induce truthtelling? *Public Choice* 108, 331–54.

Krishna, V. and Perry, M. 1997. Efficient mechanism design. Unpublished manuscript, Penn State University.

Laffont, J.J. and Martimort, D. 1997. Collusion under asymmetric information. *Econometrica* 65, 875–911.

McAfee, P. and Reny, P. 1992. Correlated information and mechanism design. *Econometrica* 60, 395–421.

McLean, R. and Postlewaite, A. 2002. Informational size and incentive compatibility. *Econometrica* 70, 2421–53.

Maskin, E. 1992. Auctions and privatization. In *Privatization*, ed. H. Siebert. Tübingen: J.C.B. Mohr.

Maskin, E. 1999. Nash equilibrium and welfare optimality. *Review of Economic Studies* 66, 23–38.

Maskin, E. and Moore, J. 1999. Implementation and renegotiation. *Review of Economic Studies* 66, 39–56.

Maskin, E. and Tirole, J. 1999. Two remarks on the property rights literature. *Review of Economic Studies* 66, 139–50.

Mezzetti, C. 2004. Mechanism design with interdependent valuations: efficiency. *Econometrica* 72, 1617–26.

Mezzetti, C. 2007. Mechanism design with interdependent valuations: surplus extraction. *Economic Theory* 3, 473–99.

Mookherjee, D. and Reichelstein, S. 1992. Dominant strategy implementation of Bayesian incentive compatible allocation rules. *Journal of Economic Theory* 56, 378–99.

Mookherjee, D. and Tsumagari, M. 2004. The organization of supplier networks: effects of delegation and intermediation. *Econometrica* 72, 1179–219.

Nöldeke, G. and Schmidt, K. 1995. Option contracts and renegotiation: a solution to the hold–up problem. *RAND Journal of Economics* 26, 163–79.

Palfrey, T. and Srivastava, S. 1989a. Implementation with incomplete information in exchange economies. *Econometrica* 57, 115–34.

Palfrey, T. and Srivastava, S. 1989b. Mechanism design with incomplete information: a solution to the implementation problem. *Journal of Political Economy* 97, 668–91.

Palfrey, T. and Srivastava, S. 1991. Nash implementation using undominated strategies. *Econometrica* 59, 479–501.

Peleg, B. and Winter, E. 2002. Constitutional implementation. *Review of Economic Design* 7, 187–204.

Persico, N. 2004. Committee design with endogenous information. *Review of Economic Studies* 71, 165–91.

Postlewaite, A. and Schmeidler, D. 1986. Implementation in differential information economies. *Journal of Economic Theory* 39, 14–33.

Saijo, T. 1987. On constant Maskin monotonic social choice functions. *Journal of Economic Theory* 42, 382–86.

Saijo, T. 2003. Spiteful behavior in voluntary contribution mechanism experiments. In *Handbook of Experimental Economics Results*, ed. C. Plott and V. Smith, Elsevier Science.

Segal, I. 1999. Complexity and renegotiation: a foundation for incomplete contracts. *Review of Economic Studies* 66, 57–82.

Segal, I. 2004. The communication requirements of social choice rules and supporting budget sets. Working Paper No. 39, School of Social Science, Institute for Advanced Study, Princeton.

Segal, I. and Whinston, M. 2002. The Mirrlees approach to mechanism design with renegotiation. *Econometrica* 70, 1–46.

Sen, A. 1970. The impossibility of a Paretian liberal. *Journal of Political Economy* 78, 152–7.

Serrano, R. and Vohra, R. 2001. Some limitations of virtual Bayesian implementation. *Econometrica* 69, 785–92.

Sjöström, T. 1994. Implementation in undominated Nash equilibria without using integer games. *Games and Economic Behavior* 6, 502–11.

Sjöström, T. 1999. Undominated Nash implementation with collusion and renegotiation. *Games and Economic Behavior* 26, 337–52.

Smith, V. 1979. Incentive compatible experimental processes for the provision of public goods. In *Research in Experimental Economics*, vol. 1, ed. V. Smith. Greenwich, CT: JAI Press.

Tirole, J. 1986. Hierarchies and bureaucracies. *Journal of Law, Economics and Organization* 2, 181–214.

Williams, S. 1999. A characterization of efficient, Bayesian incentive compatible mechanisms. *Economic Theory* 14, 155–80.

mixed strategy equilibrium

In many strategic situations a player's success depends upon his actions being unpredictable. Competitive sports are replete with examples. One of the simplest occurs repeatedly in soccer (football): if a kicker knows which side of the goal the goalkeeper has chosen to defend, he will kick to the opposite side; and if the goalkeeper knows to which side the kicker will direct his kick, he will choose that side to defend. In the language of game theory, this is a simple 2×2 game which has no pure strategy equilibrium.

John von Neumann's (1928) theoretical formulation and analysis of such strategic situations is generally regarded as the birth of game theory. Von Neumann introduced the concept of a *mixed strategy*: each player in our soccer example should choose his Left or Right action randomly, but according to some particular binomial process. Every *zero sum* two-person game in which each player's set of available strategies is finite must have a *value* (or *security level*) for each player, and each player must have at least one *minimax* strategy – a strategy that assures him that, no matter how his opponent plays, he will achieve at least his security level for the game, in expected value terms. In many such games the minimax strategies are pure strategies, requiring no mixing; in others, they are mixed strategies.

John Nash (1950) introduced the powerful notion of *equilibrium* in games (including non-zero-sum games and games with an arbitrary number of players): an equilibrium is a combination of strategies (one for each player) in which each player's strategy is a *best* strategy for him against the strategies all the other players are using. An equilibrium is thus a sustainable combination of strategies, in the sense that no player has an incentive to change unilaterally to a different strategy. A *mixed-strategy equilibrium* (MSE) is one in which each player is using a mixed strategy; if a game's only equilibria are mixed, we say it is an MSE game. In two-person zero-sum games there is an equivalence between minimax and equilibrium: it is an equilibrium for each player to use a minimax strategy, and an equilibrium can consist only of minimax strategies.

An example or two will be helpful. First consider the game tic-tac-toe. There are three possible outcomes: Player A wins, Player B wins, or the game ends in a draw. Fully defining the players' possible strategies is somewhat complex, but anyone who has played the game more than a few times knows that each player has a strategy that guarantees him no worse than a draw. These are the players' respective minimax strategies and they constitute an equilibrium. Since they are *pure strategies* (requiring no mixing), tic-tac-toe is not an MSE game.

A second example is the game called 'matching pennies'. Each player places a penny either heads up or tails up; the players reveal their choices to one another simultaneously; if their choices match, Player A gives his penny to Player B, otherwise

Player B gives his penny to Player A. This game has only two possible outcomes and it is obviously zero-sum. Neither of a player's pure strategies (heads or tails) ensures that he won't lose. But by choosing heads or tails randomly, each with probability one-half (for example, by 'flipping' the coin), he ensures that in expected value his payoff will be zero *no matter how his opponent plays*. This 50–50 mixture of heads and tails is thus a minimax strategy for each player, and it is an MSE of the game for each player to choose his minimax strategy.

Figure 1 provides a matrix representation of matching pennies. Player A, when choosing heads or tails, is effectively choosing one of the matrix's two rows; Player B chooses one of the columns; the cell at the resulting row-and-column intersection indicates Player A's *payoff*. Player B's payoff need not be shown, since it is the negative of Player A's (as always in a zero-sum game). Matching pennies is an example of a 2×2 game: each player has two pure strategies, and the game's matrix is therefore 2×2.

Figure 2 depicts our soccer example, another 2×2 MSE game. The kicker and the goalie simultaneously choose either Left or Right; the number in the resulting cell (at the row-and-column intersection) is the probability a goal will be scored, given the players' choices. The probabilities capture the fact that for each combination of choices by kicker and goalie the outcome is still random – a goal is less likely (but not impossible) when their choices match and is more likely (while not certain) when they don't. The specific probabilities will depend upon the abilities of the specific kicker and goalie: the probabilities in Figure 2 might represent, for example, a situation in which the kicker is more effective kicking to the left half of the goal than to the right half. For the specific game in Figure 2 it can be shown that the kicker's minimax strategy is a 50–50 mix between Left and Right and the goalie's minimax strategy is to defend Left 3/5 of the time and Right 2/5. The reader can easily see that the value of the game is therefore 3/5, that is, in the MSE the kicker will succeed in scoring a goal 60 per cent of the time.

Non-zero-sum games and games with more than two players often have mixed strategy equilibria as well. Important examples are decisions whether to enter a competition (such as an industry, a tournament, or an auction), 'wars of attrition' (decisions about whether and when to exit a competition), and models of price

		Player B	
		H	T
Player A	H	–1	1
	T	1	–1

Figure 1

Kicker \ Goalkeeper	L	R
L	0.4	0.9
R	0.8	0.3

Figure 2

dispersion (which explain how the same good may sell at different prices), as well as many others.

How do people actually behave in strategic situations that have mixed strategy equilibria? Does the MSE provide an accurate description of people's behaviour? Virtually from the moment Nash's 1950 paper was distributed in preprint, researchers began to devise experiments in which human subjects play games that have mixed strategy equilibria. The theory has not fared well in these experiments. The behaviour observed in experiments typically departs from the MSE in two ways: participants do not generally play their strategies in the proportions dictated by the game's particular MSE probability distribution; and their choices typically exhibit negative serial correlation – a player's mixed strategy in an MSE requires that his choices be independent across multiple plays, but experimental subjects tend instead to switch from one action to another more often than chance would dictate. Experimental psychologists have reported similar 'switching too often' in many experiments designed to determine people's ability to intentionally behave randomly. The evidence suggests that humans are not very good at behaving randomly.

The results from experiments were so consistently at variance with the theory that empirical analysis of the concept of MSE became all but moribund for nearly two decades, until interest was revived by Barry O'Neill's (1987) seminal paper. O'Neill pointed out that there were features of previous experiments that subtly invalidated them as tests of the theory of mixed strategy equilibrium, and he devised a clever but simple experiment that avoided these flaws. Although James Brown and Robert Rosenthal (1990) subsequently demonstrated that the behaviour of O'Neill's subjects was still inconsistent with the theory, the correspondence between theory and observation was nevertheless closer in his experiment than in prior experiments.

Mark Walker and John Wooders (2001) were the first to use field data instead of experiments to evaluate the theory of mixed strategy equilibrium. They contended that, while the rules and mechanics of a simple MSE game may be easy to learn quickly, as required in a laboratory experiment, substantial experience is nevertheless required in order to develop an understanding of the strategic subtleties of playing even simple MSE games. In short, an MSE game may be easy to play but not easy to play *well*. This fact alone may account for much of the theory's failure in laboratory experiments.

Instead of using experiments, Walker and Wooders applied the MSE theory to data from professional tennis matches. The 'serve' in tennis can be described as a 2×2 MSE game exactly like the soccer example in Figure 2: the server chooses which direction to serve, the receiver chooses which direction to defend, and the resulting payoff is the probability the server wins the point. Walker and Wooders obtained data from matches between the best players in the world, players who have devoted their lives to the sport and should therefore be expert in the strategic subtleties of this MSE game. Play by these world-class tennis players was found to correspond quite closely to the MSE predictions. Subsequent research by others, with data from professional tennis and soccer matches, has shown a similar correspondence between theory and observed behaviour.

Thus, the empirical evidence to date indicates that MSE is effective for explaining and predicting behaviour in strategic situations at which the competitors are experts and that it is less effective when the competitors are novices, as experimental subjects typically are. This leaves several obvious open questions. In view of the enormous disparity in expertise between world-class athletes and novice experimental subjects, how can we determine, for specific players, whether the MSE yields an appropriate prediction or explanation of their play? And when MSE is not appropriate, what *is* a good theory of play? We clearly need a generalization of current theory, one that includes MSE, that tells us in addition when MSE is 'correct', and that explains behavior when MSE is not correct. Moreover, the need for such a theory extends beyond MSE games to the theory of games more generally.

A more general theory will likely comprise either an alternative, more general notion of equilibrium or a theory of out-of-equilibrium behaviour in which some players may, with enough experience, come to play as the equilibrium theory predicts. Recent years have seen research along both lines. Among the most promising developments are the notion of quantal response equilibrium introduced by Richard McKelvey and Thomas Palfrey (1995), the theory of level-*n* thinking introduced by Dale Stahl and Paul Wilson (1994), and the idea of reinforcement learning developed by Ido Erev and Alvin Roth (1998).

MARK WALKER AND JOHN WOODERS

See also **purification.**

Bibliography

Brown, J. and Rosenthal, R. 1990. Testing the minimax hypothesis: a re-examination of O'Neill's game experiment. *Econometrica* 58, 1065–81.

Erev, I. and Roth, A.E. 1998. Predicting how people play games: reinforcement learning in experimental games with unique, mixed strategy equilibria. *American Economic Review* 88, 848–81.

McKelvey, R. and Palfrey, T. 1995. Quantal response equilibria for normal-form games. *Games and Economic Behavior* 10, 6–38.

Nash, J.F. 1950. Equilibrium points in N person games. *Proceedings of the National Academy of Sciences* 36, 48–9.

O'Neill, B. 1987. Nonmetric test of the minimax theory of two-person zerosum games. *Proceedings of the National Academy of Sciences* 84, 2106–9.
Stahl, D. and Wilson, P. 1994. Experimental evidence on players' models of other players. *Journal of Economic Behavior & Organization* 25, 309–27.
von Neumann, J. 1928. Zur Theorie der Gesellschaftsspiele. *Mathematische Annalen* 100, 295–320.
Walker, M. and Wooders, J. 2001. Minimax play at Wimbledon. *American Economic Review* 91, 1521–38.

Nash equilibrium, refinements of

Game theory studies decisions by several persons in situations with significant interactions. Two features distinguish it from other theories of multi-person decisions. One is explicit consideration of each person's available strategies and the outcomes resulting from combinations of their choices; that is, a complete and detailed specification of the 'game'. Here a person's strategy is a complete plan specifying his action in each contingency that might arise. In non-cooperative contexts, the other is a focus on optimal choices by each person separately. John Nash (1950; 1951) proposed that a combination of mutually optimal strategies can be characterized mathematically as an *equilibrium*. According to Nash's definition, a combination is an equilibrium if each person's choice is an optimal response to others' choices. His definition assumes that a choice is optimal if it maximizes the person's expected utility of outcomes, conditional on knowing or correctly anticipating the choices of others. In some applications, knowledge of others' choices might stem from prior agreement or communication, or accurate prediction of others' choices might derive from 'common knowledge' of strategies and outcomes and of optimizing behaviour. Because many games have multiple equilibria, the predictions obtained are incomplete. However, equilibrium is a weak criterion in some respects, and therefore one can refine the criterion to obtain sharper predictions (Harsanyi and Selten, 1988; Hillas and Kohlberg, 2002; Kohlberg, 1990; Kreps, 1990).

Here we describe the main refinements of Nash equilibrium used in the social sciences. Refinements were developed incrementally, often relying on ad hoc criteria, which makes it difficult for a non-specialist to appreciate what has been accomplished. Many refinements have been proposed but we describe only the most prominent ones. First we describe briefly those refinements that select equilibria with simple features, and then we focus mainly on those that invoke basic principles adapted from single-person decision theory.

Equilibria with simple features

Nash's construction allows each person to choose randomly among his strategies. But randomization is not always plausible, so in practice there is a natural focus on equilibria in 'pure' strategies, those that do not use randomization. There is a similar focus on strict equilibria, those for which each person has a unique optimal strategy in response to others' strategies. In games with some symmetries among the players, the symmetric equilibria are those that reflect these symmetries. In applications to dynamic interactions the most useful equilibria are those that, at each stage, depend only on that portion of prior history that is relevant for outcomes in the future. In particular, when the dynamics of the game are stationary one selects equilibria that are stationary or that are Markovian in that they depend only on state variables that summarize the history relevant for the future. Applications to computer science select equilibria or, more often, approximate equilibria, using strategies that can be implemented by simple algorithms. Particularly useful are equilibria that rely only on limited recall of past events and actions and thus economize on memory or computation.

Refinements that require strategies to be admissible

One strategy is strictly dominated by another if it yields strictly inferior outcomes for that person regardless of others' choices. Because an equilibrium never uses a strictly dominated strategy, the same equilibria persist when strictly dominated strategies are deleted, but after deletion it can be that some remaining strategies become strictly dominated. A refinement that exploits this feature deletes strictly dominated strategies until none remain, and then selects those equilibria that remain in the reduced game. If a single equilibrium survives then the game is called 'dominance solvable'. An equilibrium can, however, use a strategy that is weakly dominated in that it would be strictly dominated were it not for ties – in decision theory such a strategy is said to be inadmissible. A prominent criterion selects equilibria that use only admissible strategies, and sometimes this is strengthened by iterative deletion of strictly dominated strategies after deleting the inadmissible strategies. A stronger refinement uses *iterative deletion of* (both strictly and weakly) *dominated strategies* until none remain; however, this procedure is ambiguous because the end result can depend on the order in which weakly dominated strategies are deleted.

A particular order is used for dynamic games that decompose into a succession of subgames as time progresses. In this case, those strategies that are weakly dominated because they are strictly dominated in final subgames are deleted first, then those in penultimate subgames, and so on. In games with 'perfect information' as defined below this procedure implements the criterion called 'backward induction' and the equilibria that survive are among those that are 'subgame-perfect' (Selten, 1965). In general a subgame-perfect equilibrium is one that induces an equilibrium in each subgame. Figure 1 depicts an example in which there are two Nash equilibria, one in which A moves down because she anticipates that B will move down, and a second that is subgame-perfect because in the subgame after A moves across, B also moves across, which yields him a higher payoff than down.

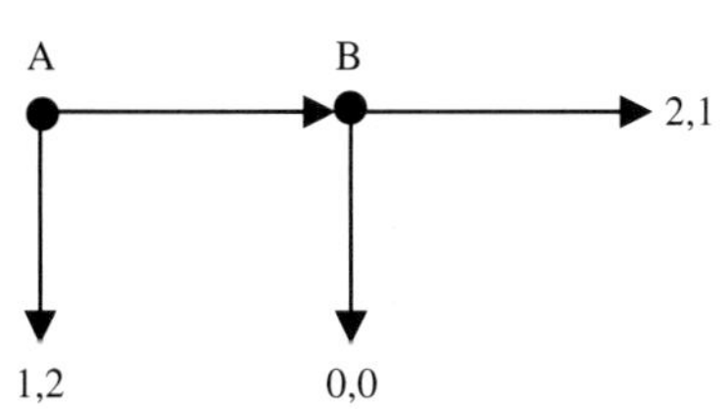

Figure 1 Player A moves down or across, in which case player B moves down or across. Payoffs for A and B are shown at the end of each sequence of moves

The informal criterion of 'forward induction' has several formulations. Kohlberg and Mertens (1986) require that a refined set of equilibria contains a subset that survives deletion of strategies that are not optimal responses at any equilibrium in the set. Van Damme (1989; 1991) requires that if player A rejects a choice X in favour of Y or Z then another player who knows only that Y or Z was chosen should consider Z unlikely if it is chosen only in equilibria that yield player A outcomes worse than choosing X, whereas Y is chosen in an equilibrium whose outcome is better. A typical application mimics backward induction but in reverse – if a person previously rejected a choice with an outcome that would have been superior to the outcomes from all but one equilibrium of the ensuing subgame, then presumably the person is anticipating that favourable equilibrium and intends to use his strategy in that equilibrium of the subgame. In Figure 2, if A rejects the payoff 5 from Down then B can infer that A intends to play Top in the ensuing subgame, yielding payoff 6 for both players.

Dynamic games

Before proceeding further we describe briefly some relevant features of dynamic games, that is, games in which a player acts repeatedly, and can draw inferences about others' strategies, preferences, or private information as the game progresses. A dynamic game is said to have 'perfect information' if each person knows initially all the data of the game, and the prior history of his and others' actions whenever he acts, and they do not act simultaneously. In such a game each action initiates a subgame; hence backward induction yields a unique subgame-perfect equilibrium if there are no ties. But in many dynamic games there are no subgames. This is so whenever some person acts without knowing all data of the game relevant for the future. In Figure 3 player C acts without knowing whether player A or B chose down.

The source of this deficiency is typically that some participant has private information – for example, about his own preferences or about outcomes – or because his actions are observed imperfectly by some others. Among parlour games, chess is a game with perfect information (if players remember whether each king has been castled). Bridge and poker are games with imperfect information because the cards in one player's hand are not known to others when they bet. In practical settings,

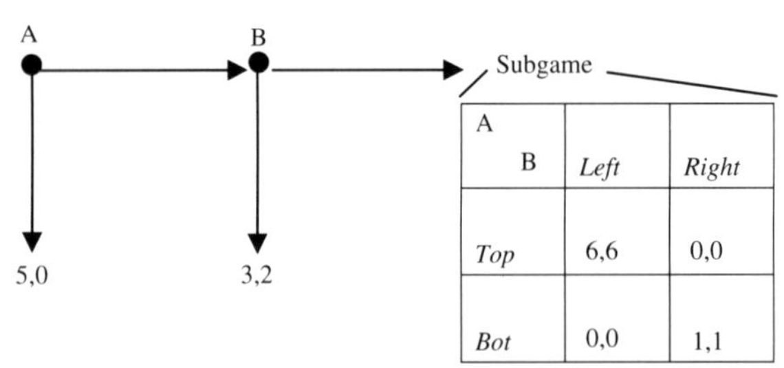

Figure 2 First A and then B can avoid playing the subgame in which simultaneously each chooses between two options

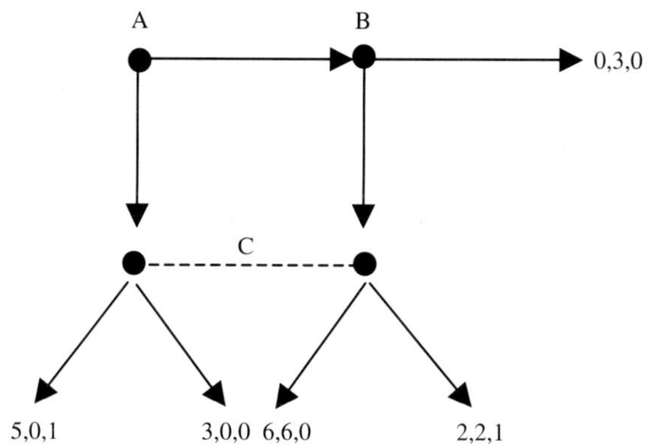

Figure 3 Player A moves down or across, in which case player B moves down or across. Player C does not observe whether it was A or B who moved down when she chooses to move left or right

auctions and negotiations resemble poker because each party acts (bids, offers, and so on) without knowing others' valuations of the transaction. Analyses of practical economic games usually assume (as we do here) 'perfect recall' in the sense that each player always remembers what he knew and did previously. If bridge is treated as a two-player game between teams, then it has imperfect recall because each team alternately remembers and forgets the cards in one member's hand as the bidding goes round the table, but bridge has perfect recall if it is treated as a four-player game. In card games like bridge and poker each player can derive the probability distribution of others' cards from the assumption that the deck of cards was thoroughly shuffled. Models of economic games impose analogous assumptions; for example, a model of an auction assumes that each bidder initially assesses a probability distribution of others' valuations of the item for sale, and then updates this assessment as he observes their bids. More realism is obtained from more complicated scenarios; for example, it could be that player A is uncertain about player B's assessment of player A's valuation. In principle the model could allow a hierarchy of beliefs – A's probability assessment of B's assessment of A's assessment of .... To adopt a proposal by John Harsanyi (1967–1968) developed by Mertens and Zamir (1985), such situations are modelled by assuming that each player is one of several types. The initial joint distribution of types is commonly known among the players, but each player knows his own type, which includes a specification of his available strategies, his preferences over outcomes, and, most importantly, his assessment of the conditional probabilities of others' types given his own type. In poker, for instance, a player's type includes the hand of cards he is dealt, and his hand affects his beliefs about others' hands.

Refinements of Nash equilibrium are especially useful in dynamic games. Nash equilibria do not distinguish between the case in which each player commits initially and irrevocably to his strategy throughout the game, and the case in which a player continually re-optimizes as the game progresses. The distinction is lost because the definition of Nash equilibrium presumes that players will surely adhere to their strategies chosen initially. Most refinements of Nash equilibrium are intended to resurrect this important distinction. Ideally one would like each Nash equilibrium to bear a label telling whether it assumes implicit commitment or relies on incredible threats or promises. Such features are usually evident in the equilibria of trivially simple games, but in more complicated games they must be identified by augmenting the definition of Nash equilibrium with additional criteria.

In the sequel we describe two classes of refinements in detail, but first we summarize their main features, identify the main selection criteria they use, and mention the names of some specific refinements. Both classes are generalizations of backward induction and subgame perfection, and they obtain similar results, but their motivation and implementation differ.

1. The criterion of sequential rationality

The presumption that commitment is irrevocable is flawed if other participants in the game do not view commitment to a strategy as credible. Commitment can be advantageous, of course, but if commitment is possible (for example, via enforceable contractual arrangements) then it should properly be treated as a distinct strategy. Absent commitment, some Nash equilibria are suspect because they rely implicitly on promises or threats that are not credible. For example, one Nash equilibrium might enable an incumbent firm to deter another firm from entering its market by threatening a price war. If such a threat succeeds in deterring entry then it is costless to the incumbent because it is never challenged; indeed, it can be that this equilibrium is sustained only by the presumption that the incumbent will never need to carry out the threat. But this threat is not credible if, after entry occurs, the incumbent would recognize that accommodation is more profitable than a price war. In such contexts, the purpose of a refinement is to select an alternative Nash equilibrium that anticipates correctly that entry will be followed by accommodation. For instance, the subgame-perfect equilibrium in Figure 1 satisfies this criterion.

Refinements in the first class exclude strategies that are not credible by requiring explicitly that a strategy is optimal in each contingency, even if it comes as a surprise. (We use the term 'contingency' rather than the technical term 'information set' used in game theory – it refers to any situation in which the player chooses an action.) These generally require that a player's strategy is optimal initially (as in the case of commitment), *and* that in each subsequent contingency in which the player might act his strategy remains optimal for the remainder of the game, even if the equilibrium predicts that the contingency should not occur. This criterion is called 'sequential rationality'. As described later, three such refinements are *perfect Bayes*, *sequential*, and *lexicographic* equilibria, each of which can be strengthened further by imposing additional criteria such as *invariance*, the *intuitive criterion* and *divinity*.

2. The criterion of perfection or stability

The presumption that commitment is irrevocable is also flawed if there is some chance of deviations. If a player might 'tremble' or err in carrying out his intended strategy, or his valuation of outcomes might be slightly different from others anticipated, then other players can be surprised to find themselves in unexpected situations. Refinements that exploit this feature are implemented in two stages. In the first stage one identifies the Nash equilibria of a perturbation of the original game, usually obtained by restricting each player to randomized strategies that assign positive probabilities to all his original pure strategies. In the second stage one identifies those equilibria of the original game that are limits of equilibria of the perturbed game as this restriction is relaxed to allow inferior strategies to have zero probabilities.

Refinements in the second class also exclude strategies that are not credible, but refinements in this class implement sequential rationality indirectly. The general criterion that is invoked is called ‘perfection’ or ‘stability’, depending on the context. In each case a refinement is obtained from analyses of perturbed games. This second class of refinements is typically more restrictive than the first class due to the stronger effects of perturbations. As described later, two such refinements are *perfect* and *proper* equilibria. These are equilibria that are perturbed slightly by *some* perturbation of the players’ strategies. A more stringent refinement selects a subset of equilibria that is *truly perfect* or *stable* in the sense that it is perturbed only slightly by *every* perturbation of players’ strategies. This refinement selects a subset of equilibria rather than a single equilibrium because there need not exist a single equilibrium that is *essential* in that it is perturbed slightly by every perturbation of strategies. A stringent refinement selects a subset that is *hyperstable* in that it is stable against perturbations of both players’ strategies and their valuations of outcomes, or against perturbations of their optimal responses; and further, it is *invariant* in that it is unaffected by addition or deletion of redundant strategies.

The crucial role of perturbations in the second class of refinements makes them more difficult for non-specialists to understand and appreciate, but they have a prominent role in game theory because of their desirable properties. For example, in a two-player game a perfect equilibrium is equivalent to an equilibrium that uses only admissible strategies. In general, refinements in the second class have the advantage that they satisfy several selection criteria simultaneously.

After this overview, we now turn to detailed descriptions of the various refinements.

Refinements that require sequential rationality

In dynamic games with perfect information, the implementation of backward induction is unambiguous because in each contingency the player taking an action there knows exactly the subgame that follows. In chess, for example, the current positions of the pieces determine how the game can evolve subsequently. Moreover, if he anticipates his opponent’s strategy then he can predict how the opponent will respond to each possible continuation of his own strategy. Using this prediction he can choose an optimal strategy for the remainder of the game by applying the *principle of optimality* – his optimal strategy in the current subgame consists of his initial action that, when followed by his optimal strategies in subsequent subgames, yields his best outcome. Thus, in principle (although not in practice, since chess is too complicated) his optimal strategy can be found by working backward from final positions through all possible positions in the game.

In contrast, in a game with imperfect information a player's current information may be insufficient to identify the prior history that led to this situation, and therefore insufficient to identify how others will respond in the future, even if he anticipates their strategies. In poker, for example, knowledge of his own cards and anticipation of others' strategies are insufficient to predict how they will respond to his bets. Their strategies specify how they will respond conditional on their cards but, since he does not know their cards, he remains uncertain what bets they will make in response to his bets. In this case, it is his assessment of the probability distribution of their cards that enables construction of his optimal strategy. That is, this probability distribution can be combined with their strategies to provide him with a probabilistic prediction of how they will bet in response to each bet he might make. Using this prediction he can again apply the principle of optimality to construct an optimal strategy by working backward from the various possible conclusions of the game.

Those refinements that select equilibria satisfying sequential rationality use an analogous procedure. The analogue of the probability distribution of others' cards is a system of 'beliefs', one for each contingency in which the player might find himself. Each belief is a conditional probability distribution on the prior history of the game given the contingency at which he has arrived. Thus, to whatever extent he is currently uncertain about others' preferences over final outcomes or their prior actions, his current belief provides him with a probability distribution over the various possibilities. As in poker, this probability distribution can be combined with his anticipation of their strategies to provide him with a probabilistic prediction of how they will act in response to each action he might take – and again, using this prediction he can apply the principle of optimality to construct an optimal strategy by working backward from the various possible conclusions of the game.

There is an important proviso, however. These refinements require that, whenever one contingency follows another with positive probability, the belief at the later one must be obtained from the belief at the earlier one by Bayes' rule. This ensures consistency with the rules of conditional probability. But, importantly, it does not restrict a player's belief at a contingency that was unexpected, that is, had zero probability according to his previous belief and the other players' strategies.

In Figure 3, in one Nash equilibrium A chooses down, B chooses across, and C chooses left. This is evidently not sequential because if A were to deviate then B could gain by choosing down. In a sequential equilibrium B chooses down and each of A and C randomizes equally between his two strategies. The strategies of A and B imply that C places equal probabilities on which of A and B chose down.

The weakest refinement selects a *perfect-Bayes* equilibrium (Fudenberg and Tirole, 1991). This requires that each player's strategy is consistent with some system of beliefs such that (*a*) his strategy is optimal given his beliefs and others' strategies, and (*b*) his beliefs satisfy Bayes' rule (wherever it applies) given others' strategies. A stronger refinement selects *sequential* equilibria (Kreps and Wilson, 1982). A sequential equilibrium requires that each player's system of beliefs is consistent with the structure of the game. Consistency is defined formally as the requirement that each player's system of beliefs is the limit of the conditional probabilities induced by players' strategies in some perturbed game, as described previously. A further refinement selects *quasi-perfect* equilibria (van Damme, 1984), which requires admissibility of a player's strategy in continuation from each contingency, excluding any chance that he himself might deviate from his intended strategy. And even stronger are *proper* equilibria (Myerson, 1978), described later. This sequence of progressively stronger refinements is typical. Because proper implies quasi-perfect implies sequential implies perfect-Bayes, one might think that it is sufficient to always use properness as the refinement. However, the prevailing practice in the social sciences is to invoke the weakest refinement that suffices for the game being studied. This reflects a conservative attitude about using unnecessarily restrictive refinements. If, say, there is a unique sequential equilibrium that uses only admissible strategies, then one refrains from imposing stronger criteria.

Additional criteria can be invoked to select among sequential equilibria. In Figure 4 there is a sequential equilibrium in which both types of A move left and B randomizes equally between middle and bottom, and another in which both types of A move right and B chooses middle. An alternative justification for the second, due to Hillas (1998), is shown in Figure 5, where the game is restructured so that A either commits initially to left or they play the subgame with simultaneous choices of strategies. The criterion of subgame perfection selects the second equilibrium in Figure 4 because in Figure 5 the subgame has a unique equilibrium with payoff 6 for A that is superior to his payoff 4 from committing to left.

These refinements can be supplemented with additional criteria that restrict a player's beliefs in unexpected contingencies. The most widely used criteria apply to contexts in which one player B could interpret the action of another player A as revealing private information; that is, A's action might signal something about A's

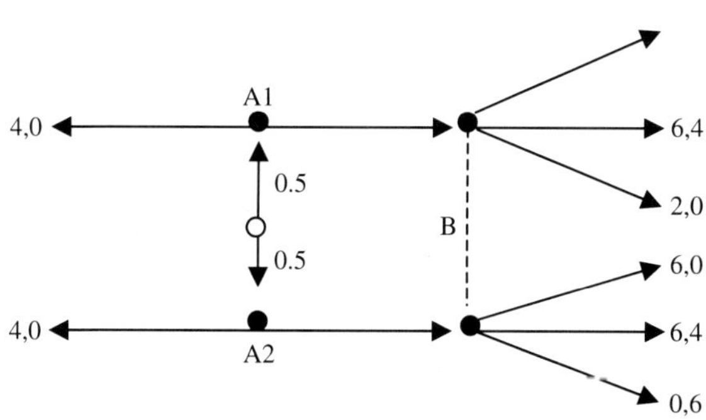

Figure 4 Nature chooses whether player A's type is A1 or A2 with equal probabilities. Then A chooses Left or Right, in which case player B, without knowing A's type, chooses one of three options

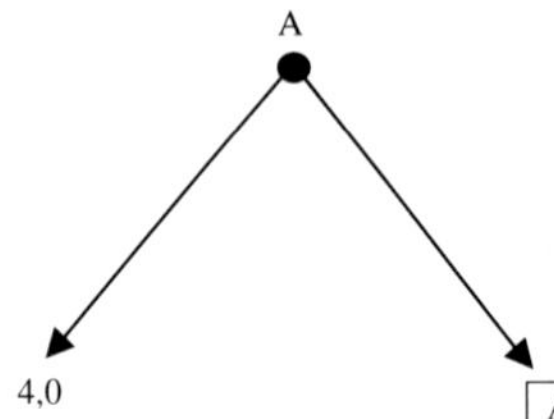

A / B	*Top*	*Middle*	*Bottom*
1.*Left* 2.*Right*	5,0	5,2	2,3
1.*Right* 2.*Left*	3,2	5,2	3,0
1.*Right* 2.*Right*	4,3	**6,4 ***	1,3

Figure 5 The game in Figure 4 restructured so that either A commits to Left regardless of his type, or plays a subgame with simultaneous moves in which he chooses one of his other three type-contingent strategies. The payoffs 6,4 to A and B from the unique Nash equilibrium of the subgame are shown with an asterisk

type. These criteria restrict B's belief (after B observes A deviating from the equilibrium) to one that assigns positive probability only to A's types that might possibly gain from the deviation, provided it were interpreted by B as a credible signal about A's type. The purpose of these criteria is to exclude beliefs that are blind to A's attempts to signal what his type is when it would be to A's advantage for B to recognize the signal. In effect, these criteria reject equilibria that commit a player to unrealistic beliefs. Another interpretation is that these criteria reject equilibria in which A is 'threatened by B's beliefs' because B stubbornly retains these beliefs in spite of plausible evidence to the contrary.

The simplest version requires that B's belief assigns zero probability to those types of A that cannot possibly gain by deviating, regardless of how B responds. The *intuitive* criterion (Cho and Kreps, 1987) requires that there cannot be some type of A that surely gains from deviating in every continuation for which B responds with a strategy that is optimal based on a belief that assigns zero probability to those types of A that cannot gain from the deviation. That is, an equilibrium fails the intuitive criterion if B's belief fails to recognize that A's deviation is a credible signal about his type. They apply this criterion to the game in Figure 6, which has two sequential equilibria. In one both types of A choose left and B chooses down or up contingent on left or right. In another both types choose right and B chooses up or down contingent on left or right. In both equilibria B's belief in the unexpected event (right or left respectively) assigns probability greater than 0.5 to A's type A1. The intuitive criterion rejects the second equilibrium because if A2 were to deviate by choosing left, and then B recognizes that this deviation credibly signals A's type A2 (because type A1 cannot gain by deviating regardless of B's response) and therefore B chooses down, then type A2 obtains payoff 3 rather than his equilibrium payoff 2.

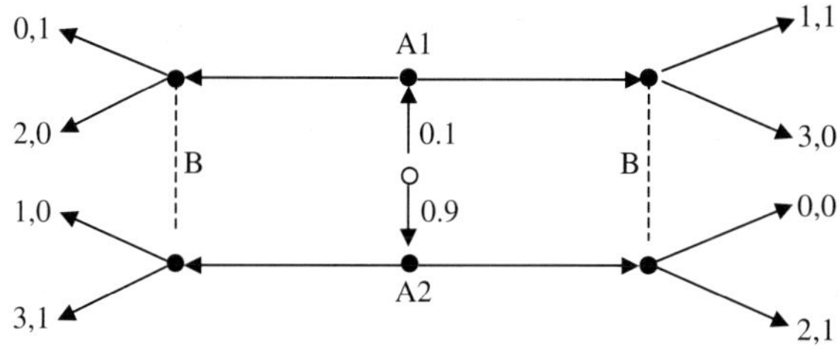

Figure 6 A signalling game in which Nature chooses A's type A1 or A2, then A chooses left or right, and then B, without knowing A's type, chooses up or down

Cho and Kreps also define an alternative version, called the 'equilibrium domination' criterion. This criterion requires that, for each continuation in which B responds with a strategy that is optimal based on a belief that assigns zero probability to those types of A that cannot gain from deviating, there cannot be some type of A that gains from deviating. More restrictive is the criterion **D1** (Banks and Sobel, 1987), also called 'divinity' when it is applied iteratively, which requires that, if the set of B's responses for which one type of A gains from deviating is larger than the set for which a second type gains, then B's beliefs must assign zero probability to the second type. The criterion **D2** is similar except that some (rather than just one) types of A gain. All these criteria are weaker than the *never weak best reply* criterion that requires an equilibrium to survive deletion of a player's strategy that is not an optimal reply to any equilibrium with the same outcome. In Figure 6 this criterion is applied by observing that the second equilibrium does not survive deletion of those strategies of A in which type A2 chooses left.

A *lexicographic* equilibrium (Blume, Brandenburger and Dekel, 1991a; 1991b) uses a different construction. Each player is supposed to rely on a sequence of 'theories' about others' strategies. He starts the game by assuming that his first theory of others' strategies is true, and uses his optimal strategy according to that theory. He continues doing so until he finds himself in a situation that cannot be explained by his first theory. In this case, he abandons the first theory and assumes instead that the second theory is true – or if it too cannot explain what has happened then he proceeds to the next theory in the sequence. This provides a refinement of Nash equilibrium because each player anticipates that deviation from his optimal strategy for any theory will provoke others to abandon their current theories and strategies and thus respond with their optimal strategies for their next theories consistent with his deviant action. Lexicographic equilibria can be used to represent nearly any refinement. The hierarchy of a player's theories serves basically the same role as his system of beliefs, but the focus is on predictions of other players' strategies in the future rather than probabilities of what they know or have done in the past. The lexicographic specification has the same effect as considering small perturbations of strategies; for example, the sequence of strategies approximating a perfect or proper equilibrium can be used to construct the hierarchy of theories.

Refinements derived from perturbed games

The other major class of refinements relies on perturbations to select among the Nash equilibria. The motive for this approach stems from a basic principle of decision theory –the *equivalence* of alternative methods of deriving optimal strategies. This principle posits that constructing a player's optimal strategy in a dynamic game by invoking auxiliary systems of beliefs and the iterative application of the principle of optimality (as in perfect-Bayes and sequential equilibria) is a useful computational procedure, but the same result should be obtainable from an initial choice of a strategy, that is, an optimal plan for the entire game of actions taken in each contingency. Indeed, the definition of Nash equilibrium embodies this principle. Proponents therefore argue that whatever improvements come from dynamic analysis can and should be replicated by static analysis of initial choices among strategies, supplemented by additional criteria. (We use the terms 'static' and 'dynamic' analysis rather than the technical terms 'normal-form' and 'extensive-form' analysis used in game theory.) The validity of this argument is evident in the case of subgame-perfect equilibria of games with perfect information, which can be derived either from the principle of optimality using backward induction, or by iterative elimination of weakly dominated strategies in a prescribed order. The argument is reinforced by major deficiencies of dynamic analysis; for example, we mentioned above that a sequential equilibrium can use inadmissible strategies. Another deficiency is failure to satisfy the criterion of *invariance*, namely, the set of sequential equilibria can depend on which of many equivalent descriptions of the dynamics of the game is used (in particular, on the addition or deletion of redundant strategies).

On this view one should address directly the basic motive for refinement, which is to exclude equilibria that assume implicitly that each player commits initially to his strategy – since Nash equilibria do not distinguish between cases with and without commitment. Thus one considers explicitly that during the game any player might deviate from his equilibrium strategy for some exogenous reason that was not represented in the initial description of the game. Recognition of the possibility of deviations, however improbable they might be, then ensures that a player's strategy includes a specification of his optimal response to others' deviations from the equilibrium. The objective is therefore to characterize those equilibria that are affected only slightly by small probabilities of deviant behaviours or variations in preferences. This programme is implemented by considering perturbations of the game. These can be perturbations of strategies or payoffs, but actually the net effect of a perturbation of others' strategies is to perturb a player's payoffs.

In the following we focus on the perturbations of the static (that is, the normal form) of the game but similar perturbations can also be applied to the dynamic version (that is, the extensive form) by applying them to each contingency separately. This is done by invoking the principle that a dynamic game can also be analysed in a static framework by treating the player acting in each contingency as a new player (interpreted as the player's agent who acts solely in that contingency) in the 'agent-normal-form' of the game, where the new player's payoffs agree with those of the original player.

The construction of a *perfect* equilibrium (Selten, 1975) illustrates the basic method, which uses two steps.

1. For each small positive number ε one finds an ε-*perfect* equilibrium, defined by the requirement that each player's strategy has the following property: every one of his pure strategies is used with positive probability, but any pure strategy that is an inferior response to the others' strategies has probability no more than ε. Thus an ε-perfect equilibrium supposes that every strategy, and therefore every action during the game, might occur, even if it is suboptimal.
2. One then obtains a perfect equilibrium as the limit of a convergent subsequence of ε-perfect equilibria.

One method of constructing an ε-perfect equilibrium starts by specifying for each player i a small probability $\delta_i < \varepsilon$ and a randomized strategy σ_i that uses every pure strategy with positive probability – that is, the strategy combination σ is 'completely mixed'. One then finds an ordinary Nash equilibrium of the perturbed game in which each player's payoffs are as follows: his payoff from each combination of all players' pure strategies is replaced by his expected payoff when each player i's pure strategy is implemented only with probability $1-\delta_i$ and with probability δ_i that player uses his randomized strategy σ_i instead. In this context one says that the game is perturbed by less than ε toward σ – we use this phrase again later when we describe stable sets of equilibria. An equilibrium of this perturbed game induces an ε-perfect equilibrium of the original game.

An alternative definition of perfect equilibrium requires that each player's strategy is an optimal response to a convergent sequence of others' strategies for which all their pure strategies have positive probability – this reveals explicitly that optimality against small probabilities of deviations is achieved, and that a perfect equilibrium uses only admissible strategies. In fact, a perfect equilibrium of the agent-normal-form induces a sequential equilibrium of the dynamic version of the game. Moreover, if the payoffs of the dynamic game are generic (that is, not related to each other by polynomial equations) then every sequential equilibrium is also perfect.

A stronger refinement selects *proper* equilibria (Myerson, 1978). This refinement supposes that the more inferior the expected payoff from a strategy is, the less likely it is to be used. The construction differs only in step 1: if one pure strategy S is inferior to another T in response to the others' strategies then S has probability no more than ε times the probability of T. A proper equilibrium induces a sequential equilibrium in every one of the equivalent descriptions of the dynamic game.

A perfect or proper equilibrium depends on the particular perturbation used to construct an ε-perfect or ε-proper equilibrium. Sometimes a game has an equilibrium that is *essential* or *truly perfect* in that any σ can be used when perturbing the game by less than ε toward σ, as above. This is usual for a static game with generic payoffs because in this case its equilibria are isolated and vary continuously with perturbations. However, such equilibria rarely exist in the important case that the static game represents a dynamic game, since in this case some strategies have the same equilibrium payoffs. This occurs because there is usually considerable freedom about how a player acts in contingencies off the predicted path of the equilibrium; in effect, the same outcome results whether the player 'punishes' others only barely enough to deter deviations, or more than enough. Indeed, for a dynamic game with generic payoffs, all the equilibria in a connected set yield the same equilibrium outcome because they differ only off the predicted path of equilibrium play. One must therefore consider sets of equilibria when invoking stringent refinements like truly perfect. One applies a somewhat different test to sets of equilibria. When considering a set of equilibria one requires that every sufficiently small perturbation (within a specified class) of the game has an equilibrium near some equilibrium in the set. Some refinements insist on a minimal closed set of equilibria with this property, but here we ignore minimality.

The chief refinement of this kind uses strategy perturbations to generate perturbed games. Kohlberg and Mertens (1986) say that a set of equilibria is *stable* if for each neighbourhood of the set there exists a positive probability ε such that, for every completely mixed strategy combination σ, each perturbation of the game by less than ε toward σ has an equilibrium within the neighbourhood. Stability can be interpreted as truly perfect applied to sets of equilibria and using the class of payoff perturbations generated by strategy perturbations. Besides the fact that a stable set always exists, it satisfies several criteria: it uses only *admissible* strategies, it contains a stable set of the reduced game after deleting a strategy that is weakly dominated or an inferior response to all equilibria in the set (these assure *iterative elimination of weakly dominated strategies* and a version of *forward induction*), and it is *invariant* to addition or deletion of redundant strategies. However, examples are known in which a stable set of a static game does not include a sequential equilibrium of the dynamic game it represents. This failure to satisfy the backward induction criterion can be remedied in various ways that we describe next.

One approach considers the larger class of all payoff perturbations. In this case, invariance to redundant strategies is not assured so it is imposed explicitly. For this, say that two games are equivalent if deletion of all redundant strategies results in the same reduced game. Similarly, randomized strategies in these two games are equivalent if they yield the same randomization over pure strategies of the reduced game. Informally, a set of equilibria is hyperstable if, for every payoff perturbation of every equivalent game, there is an equilibrium equivalent to one near the set. Two formal versions are the following. Kohlberg and Mertens (1986) say that a set S of equilibria is *hyperstable* if, for each neighbourhood N of those strategies in an equivalent game that are equivalent to ones in S, there is a sufficiently small neighbourhood P of payoff perturbations for the equivalent game such that every game in P has an equilibrium in N. A somewhat stronger version is the following. A set S of equilibria of a game G is *uniformly hyperstable* if, for each neighbourhood N of S, there is a $\delta > 0$ such that every game in the δ-neighbourhood of any game equivalent to G has an equilibrium equivalent to one in N. This version emphasizes that uniform hyperstability is closely akin to a kind of continuity with respect to payoff perturbations of equivalent games. Unfortunately, both of these definitions are complex, but the second actually allows a succinct statement in the case that the set S is a 'component' of equilibria, namely, a maximal connected set of the Nash equilibria. In this case the component is uniformly hyperstable if and only if its topological index is non-zero, and thus *essential* in the sense used in algebraic topology to characterize a set of fixed points of a function that is slightly affected by every perturbation of the function. This provides a simply computed test of whether a component is uniformly hyperstable.

Hyperstable sets tend to be larger than stable sets of equilibria because they must be robust against a larger class of perturbations, but for this same reason the criterion is actually stronger. Within a hyperstable component there is always a stable set satisfying the criteria listed previously. There is also a proper equilibrium that induces a sequential equilibrium in every dynamic game with the same static representation – thus, the criterion of *backward induction* is also satisfied. Selecting a stable subset or a proper equilibrium inside a hyperstable component may be necessary because there can be other equilibria within a hyperstable component that use inadmissible strategies. Nevertheless, for a dynamic game with generic payoffs, all the equilibria within a single component yield the same outcome, since they differ only off the path of equilibrium play, so for the purpose of predicting the outcome rather than players' strategies it is immaterial which equilibrium is considered. However, examples are known in which an inessential hyperstable component contains two stable sets with opposite indices with respect to perturbations of strategies.

The most restrictive refinement is the revised definition of stability proposed by Mertens (1989). Although this definition is highly technical, it can be summarized briefly as follows for the mathematically expert reader. Roughly, a closed set of equilibria is (Mertens-) *stable* if the projection map (from its neighbourhood in the graph of the Nash equilibria into the space of games with perturbed strategies) is essential. Such a set satisfies all the criteria listed previously, and several more. For instance, it satisfies the *small-worlds* criterion (Mertens, 1992), which requires that adding other players whose strategies have no effect on the payoffs for the original players has no effect on the selected strategies of the original players. The persistent mystery in the study of refinements is why such sophisticated constructions seem to be necessary if a single definition is to satisfy all the criteria simultaneously. The clue seems to be that, because Nash equilibria are the solutions of a fixed-point problem, a fully adequate refinement must ensure that fixed points exist for every perturbation of this problem.

The state of the art of refinements

The development of increasingly stronger refinements by imposing ad hoc criteria incrementally was a preliminary to more systematic development. Eventually, one wants to identify decision-theoretic criteria that suffice as axioms to characterize refinements. The two groups of refinements described above approach this problem differently. Those that consider perturbations seek to verify whether there exist refinements that satisfy many or (in the case of Mertens-stability) most criteria. From its beginning in the work of Selten (1975), Myerson (1978), and Kohlberg and Mertens (1986), this has been a productive exercise, showing that refinements can enforce more stringent criteria than Nash (1950; 1951) requires. However, the results obtained depend ultimately on the class of perturbations considered, since Fudenberg, Kreps and Levine (1988) show that each Nash equilibrium of a game is the limit of strict equilibria of perturbed games in a very general class. Perturbations are mathematical artefacts used to identify refinements with desirable properties, but they are not intrinsic to a fundamental theory of rational decision making in multi-person situations. Those in the other group directly impose decision-theoretic criteria – admissibility, iterative elimination of dominated or inferior strategies, backward induction, invariance, small worlds, and so on. Their ultimate aim is to characterize refinements axiomatically. But so far none has obtained an ideal refinement of the Nash equilibria.

SRIHARI GOVINDAN AND ROBERT B. WILSON

See also **behavioural game theory; epistemic game theory: incomplete information; Nash program.**

Bibliography

Banks, J. and Sobel, J. 1987. Equilibrium selection in signaling games. *Econometrica* 55, 647–61.

Blume, L., Brandenburger, A. and Dekel, E. 1991a. Lexicographic probabilities and choice under uncertainty. *Econometrica* 59, 61–79.

Blume, L., Brandenburger, A. and Dekel, E. 1991b. Lexicographic probabilities and equilibrium refinements. *Econometrica* 59, 81–98.

Cho, I. and Kreps, D. 1987. Signaling games and stable equilibria. *Quarterly Journal of Economics* 102, 179–221.

Fudenberg, D., Kreps, D. and Levine, D. 1988. On the robustness of equilibrium refinements. *Journal of Economic Theory* 44, 351–80.

Fudenberg, D. and Tirole, J. 1991. Perfect Bayesian equilibrium and sequential equilibrium. *Journal of Economic Theory* 53, 236–60.

Harsanyi, J. 1967–1968. Games with incomplete information played by 'Bayesian' players, I–III. *Management Science* 14, 159–82, 320–34, 486–502.

Harsanyi, J. and Selten, R. 1988. *A General Theory of Equilibrium Selection in Games.* Cambridge, MA: MIT Press.

Hillas, J. 1998. How much of 'forward induction' is implied by 'backward induction' and 'ordinality'? Mimeo. Department of Economics, University of Auckland.

Hillas, J. and Kohlberg, E. 2002. The foundations of strategic equilibrium. In *Handbook of Game Theory*, vol. 3, ed. R. Aumann and S. Hart. Amsterdam: North-Holland/Elsevier Science Publishers.

Kohlberg, E. 1990. Refinement of Nash equilibrium: the main ideas. In *Game Theory and Applications*, ed. T. Ichiishi, A. Neyman and Y. Tauman. San Diego: Academic Press.

Kohlberg, E. and Mertens, J.-F. 1986. On the strategic stability of equilibria. *Econometrica* 54, 1003–38.

Kreps, D. 1990. *Game Theory and Economic Modeling*. New York: Oxford University Press.

Kreps, D. and Wilson, R. 1982. Sequential equilibria. *Econometrica* 50, 863–94.

Mertens, J.-F. 1989. Stable equilibria – a reformulation, Part I: definition and basic properties. *Mathematics of Operations Research* 14, 575–624.

Mertens, J.-F. 1992. The small worlds axiom for stable equilibria. *Games and Economic Behavior* 4, 553–64.

Mertens, J.-F. and Zamir, S. 1985. Formulation of Bayesian analysis for games with incomplete information. *International Journal of Game Theory* 14, 1–29.

Myerson, R. 1978. Refinement of the Nash equilibrium concept. *International Journal of Game Theory* 7, 73–80.

Nash, J. 1950. Equilibrium points in n-person games. *Proceedings of the National Academy of Sciences USA* 36, 48–9.

Nash, J. 1951. Non-cooperative games. *Annals of Mathematics* 54, 286–95.

Selten, R. 1965. Spieltheoretische Behandlung eines Oligopolmodells mit Nachfragetragheit. *Zeitschrift fur die gesamte Staatswissenschaft* 121, 301–24, 667–89.

Selten, R. 1975. Reexamination of the perfectness concept for equilibrium points in extensive games. *International Journal of Game Theory* 4, 25–55.

van Damme, E. 1984. A relation between perfect equilibria in extensive form games and proper equilibria in normal form games. *International Journal of Game Theory* 13, 1–13.

van Damme, E. 1989. Stable equilibria and forward induction. *Journal of Economic Theory* 48, 476–96.

van Damme, E. 1991. *Stability and Perfection of Nash Equilibria.* Berlin: Springer-Verlag.

Nash program

In game theory, 'Nash program' is the name given to a research agenda, initiated in Nash (1953), intended to bridge the gap between the cooperative and non-cooperative approaches to the discipline.

Many authors have contributed to the program since its beginnings (see Serrano, 2005, for a comprehensive survey). The current article concentrates on a few salient contributions. One should begin by introducing some preliminaries and providing definitions of some basic concepts.

Preliminaries

The non-cooperative approach to game theory provides a rich language and develops useful tools to analyse strategic situations. One clear advantage of the approach is that it is able to model how specific details of the interaction may affect the final outcome. One limitation, however, is that its predictions may be highly sensitive to those details. For this reason it is worth also analysing more abstract approaches that attempt to obtain conclusions that are independent of such details. The cooperative approach is one such attempt.

Here are the primitives of the basic model in cooperative game theory. Let $N = \{1, \ldots, n\}$ be a finite set of players. For each S, a non-empty subset of N, we shall specify a set $V(S)$ containing $|S|$-dimensional payoff vectors that are feasible for coalition S. Thus, a reduced form approach is taken because one does not explain what strategic choices are behind each of the payoff vectors in $V(S)$. In addition, in this formulation, referred to as the characteristic function, it is implicitly assumed that the actions taken by the complement coalition (those players not in S) cannot prevent S from achieving each of the payoff vectors in $V(S)$. There are more general models in which these sorts of externalities are considered, but for the most part the contributions to the Nash program have been confined to the characteristic function model. Given a collection of sets $V(S)$, one for each S, the theory formulates its predictions on the basis of solution concepts.

A solution is a mapping that assigns a set of payoff vectors in $V(N)$ to each characteristic function $(V(S))_{S \subseteq N}$. Thus, a solution in general prescribes a set, although it can be single-valued (when it assigns a unique payoff vector as a function of the fundamentals of the problem). The leading set-valued cooperative solution concept is the core, while the most used single-valued ones are the Nash bargaining solution and the Shapley value.

There are several criteria to evaluate the reasonableness or appeal of a cooperative solution. One could start by defending it on the basis of its definition alone. In the case of the core, this will be especially relevant: in a context in which players can freely get together in groups, the prediction should be payoff vectors that cannot be improved upon by any coalition. Alternatively, one can propose axioms, abstract principles, that one would like the solution to have, and the next step is to pursue their logical consequences. Historically, this was the first argument to justify the Nash solution and the Shapley value. However, some may think that the definition may be somewhat arbitrary, or one may object that the axiomatic approach is 'too abstract'. By proposing non-cooperative games that specify the details of negotiation, the Nash program may help to counter these criticisms. First, the procedure will tell a story about how coalitions form and what sort of interaction among players is happening. In that process, because the tools of non-cooperative game theory are used for the analysis, the cooperative solution will be understood as the outcome of a series of strategic problems facing individual players. Second, novel connections and differences among solutions may now be uncovered from the distinct negotiation procedures that lead to each of them. Therefore, a result in the Nash program, referred to as a 'non-cooperative foundation' or 'non-cooperative implementation' of a cooperative solution, enhances its significance, being looked at now from a new perspective. Focusing on the features of the rules of negotiation that lead to different cooperative solutions takes one a long way in opening the 'black box' of how a coalition came about, and contributes to a deeper understanding of the circumstances under which one solution versus another may be more appropriate to use.

The Nash bargaining solution

A particular case of a characteristic function is a two-player bargaining problem. In it, $N = \{1, 2\}$ is the set of players. The set $V(\{1, 2\})$, a compact and convex subset of $\mathbb{R}$, is the set of feasible payoffs if the two players reach an agreement. Compactness may follow from the existence of a bounded physical pie that the parties are dividing, and convexity is a consequence of expected utility and the potential use of lotteries. The sets $(V(\{i\}))_{i \in N}$ are subsets of $\mathbb{R}$, and let $d_i = \max V(\{i\})$ be the disagreement payoff for player i, that is, the payoff that i will receive if the parties fail to reach an agreement. It is assumed that $V(\{1, 2\})$ contains payoff vectors that Pareto dominate the disagreement payoffs. A solution assigns a feasible payoff pair to each bargaining problem.

This is the framework introduced in Nash (1950), where he proposes four axioms that a solution to bargaining problems should have. First, expected utility implies that, if payoff functions are rescaled via positive affine transformations, so must be the solution (scale invariance). Second, the solution must prescribe a Pareto efficient payoff pair (efficiency). Third, if the set $V(\{1, 2\})$ is symmetric with respect to the 45 degree line and $d_1 = d_2$, the solution must lie on that line (symmetry). Fourth, the solution must be independent of 'irrelevant' alternatives, that is, it must pick the same point if it is still feasible after one eliminates other points from the feasible set (IIA). Because of scale invariance, there is no loss of generality in normalizing the disagreement payoff to 0. We call the resulting problem a normalized problem.

Nash (1950) shows that there exists a unique solution satisfying scale invariance, efficiency, symmetry and IIA, and it is the one that assigns to each normalized bargaining problem the point (u_1, u_2) that maximizes the product $v_1 v_2$ over all $(v_1, v_2) \in V(\{1, 2\})$. Today we refer to this as the 'Nash solution'. The use of the Nash solution is pervasive in applications and, following the axioms in Nash (1950), it is usually viewed as a normatively appealing resolution to bargaining problems.

In the first paper of the Nash program, Nash (1953) provides a non-cooperative approach to his axiomatically derived solution. This is done by means of a simple demand game. The two players are asked to demand simultaneously a payoff: player 1 demands v_1 and player 2 demands v_2. If the pair (v_1, v_2) is feasible, so that $(v_1, v_2) \in V(\{1, 2\})$, the corresponding agreement and split of the pie takes place to implement these payoffs. Otherwise, there is disagreement and payoffs are 0. To fix ideas, let us think of the existence of a physical pie of size 1 that is created if agreement is reached, while no pie is produced otherwise. Thus, player i's demand v_i corresponds to demanding a share x_i of the pie, $0 \leq x_i \leq 1$, such that player i's utility or payoff from receiving x_i is v_i.

The Nash demand game admits a continuum of Nash equilibria. Indeed, every point on the Pareto frontier of $V(\{1, 2\})$ is a Nash equilibrium outcome, as is the disagreement payoff point if each player demands the payoff corresponding to having the entire pie. However, Nash (1953) introduces uncertainty concerning the exact size of the pie. Now players, when formulating their demands, must have to take into account the fact that with some probability the pair of demands may lead to disagreement, even if they add up to less than 1. Then, it can be shown that the optimal choice of demands at a Nash equilibrium of the demand game with uncertain pie converges to the Nash solution payoffs as uncertainty becomes negligible. Hence, the Nash solution arises as the rule that equates marginal gain (through the increase in one's demanded share) and marginal loss (via the increase in the probability of disagreement) for each player when the problem is subject to a small degree of noise and demands/commitments are made simultaneously.

Rubinstein (1982) proposes a different non-cooperative procedure. In it, time preferences – impatience – and credibility of threats are the main forces that drive the equilibrium. The game is a potentially infinite sequence of alternating offers. In period 0, player 1 begins by making the first proposal. If player 2 accepts it, the game ends; otherwise, one period elapses and the rejector will make a counter-proposal in period 1, and so on. Let $\delta \in [0, 1)$ be the common per period discount factor, and let $v_i(\cdot)$ be player i's utility function over shares of the pie, assumed to be concave and strictly monotone. Thus, if player i receives a share x_i in an agreement reached in period t, his payoff is $\delta^{t-1} v_i(x_i)$. Perpetual disagreement has a payoff of 0.

Using subgame perfect equilibrium as the solution concept (the standard tool to rule out non-credible threats in dynamic games of complete information), Rubinstein (1982) shows that there exists a unique prediction in his game. Specifically, the unique subgame perfect equilibrium prescribes an immediate agreement on the splits $(x, 1-x)$ – offered by player 1 – and $(y, 1-y)$ – by player 2 – which are described by the following equations:

$$v_1(y) = \delta v_1(x)$$
$$v_2(1-x) = \delta v_2(1-y).$$

That is, at the unique equilibrium, the player acting as a responder in a period is offered a share that makes him exactly indifferent between accepting and rejecting it to play the continuation: the bulk of the proof is to show that any other behaviour relies on non-credible threats.

As demonstrated in Binmore, Rubinstein and Wolinsky (1986), the unique equilibrium payoffs of the Rubinstein game, regardless of who is the first proposer, converge to the Nash solution payoffs as $\delta \to 1$. First, note that the above equations imply that, for any value of δ, the product of payoffs $v_1(x)v_2(1-x)$ is the same as the product $v_1(y)v_2(1-y)$. Thus, both points, $(v_1(x), v_2(1-x))$ and $(v_1(y), v_2(1-y))$, lie on the same hyperbola of equation $v_1 v_2 = K$ and, in addition, since they correspond to efficient agreements, both points also lie on the Pareto frontier of $V(\{1,2\})$. Finally, as $\delta \to 1$, one has that $x \to y$ so that the two proposals (the one made by player 1 and the other by player 2) converge to one and the same, the one that yields the Nash solution payoffs. Thus, credible threats in dynamic negotiations in which both players are equally and almost completely patient also lead to the Nash solution.

The Shapley value

Now consider an n-player coalitional game where payoffs are transferable in a one-to-one rate among different players (for instance, because utility is money for all of them). This means that $V(S)$, the feasible set for coalition S, is the set of payoffs $(x_i)_{i \in S}$ satisfying the inequality $\sum_{i \in S} x_i \leq v(S)$ for some real number $v(S)$. This is called a transferable utility or TU game in characteristic function form. The number $v(S)$ is referred to as the 'worth of S', and it expresses S's initial position (for example, the maximum total utility that the group S of agents can achieve in an exchange economy by redistributing their endowments when utility is quasi-linear).

Therefore, without loss of generality, we can describe a TU game as a collection of real numbers $(v(S))_{S \subseteq N}$. A solution is then a mapping that assigns to each TU game a set of payoffs in the set $V(N)$, that is, vectors $(x_1, \ldots, x_n)$ such that $\sum_{i \in N} x_i \leq v(N)$. In this section, as in the previous one, we shall require that the solution be single-valued. Shapley (1953) is interested in solving in a fair way the problem of distribution of surplus among the players, when taking into account the worth of each coalition. To do this, he resorts to the axiomatic method. First, the payoffs must add up to $v(N)$, which means that the entire surplus is allocated (efficiency). Second, if two players are substitutes because they contribute the same to each coalition, the solution should treat them equally (symmetry). Third, the solution to the sum of two TU games must be the sum of what it awards to each of the two games (additivity). Fourth, if a player contributes nothing to every coalition, the solution should pay him nothing (dummy).

The result in Shapley (1953) is that there is a unique single-valued solution to TU games satisfying efficiency, symmetry, additivity and dummy. It is what today we call the Shapley value, the function that assigns to each player i the payoff

$$\mathrm{Sh}_i(N, v) = \sum_{S, i \in S} \frac{(|S| - 1)!(|N| - |S|)!}{|N|!} \times [v(S) - v(S \backslash \{i\})].$$

That is, the Shapley value awards to each player the average of his marginal contributions to each coalition. In taking this average, all orders of the players are considered to be equally likely. Let us assume, also without loss of generality, that $v(\{i\}) = 0$ for each player i.

Hart and Mas-Colell (1996) propose the following non-cooperative procedure. With equal probability, each player $i \in N$ is chosen to publicly make a feasible proposal to the others: $(x_1, \ldots, x_n)$ is such that the sum of its components cannot exceed $v(N)$. The other players get to respond to it in sequence, following a pre-specified order. If all accept, the proposal is implemented; otherwise, a random device is triggered. With probability $0 \leq \delta < 1$, the same game continues being played among the same n players (thus, a new proposer will be chosen again at random among them), but with probability $1-\delta$ the proposer leaves the game. He is paid 0 and his resources are removed so that, in the next period, proposals to the remaining $n-1$ players cannot add up to more than $v(N \backslash \{i\})$. A new proposer is chosen at random among the set $N \backslash \{i\}$, and so on.

As shown in Hart and Mas-Colell (1996), there exists a unique stationary subgame perfect equilibrium payoff profile of this procedure, and it actually coincides with the Shapley value payoffs for any value of δ . (Stationarity means that strategies cannot be history dependent.) As $\delta \to 1$, the Shapley value payoffs are also obtained not only in expectation but independently of who the proposer is. One way to understand this result, as done in Hart and Mas-Colell (1996), is to check that the rules of the procedure and stationary behaviour in it are in agreement with Shapley's axioms. That is, the equilibrium relies on immediate acceptances of proposals, stationary strategies treat substitute players similarly, the equations describing the equilibrium have an additive structure, and dummy players will have to receive 0 because no resources are destroyed if they are asked to leave. It is also worth stressing the important role in the procedure of players' marginal contributions to coalitions: following a rejection, a proposer incurs the risk of being thrown out and the others of losing his resources, which seem to suggest a 'price' for them.

The core

The idea of agreements that are immune to coalitional deviations was first introduced to economic theory in Edgeworth (1881), who defined the set of coalitionally stable allocations of an economy under the name 'final settlements'. Edgeworth envisioned this concept as an alternative to Walrasian equilibrium (Walras, 1874), and was also the first to investigate the connections between the two concepts. Edgeworth's notion, which today we refer to as 'the core', was rediscovered and introduced to game theory in Gillies (1959). Therefore, the origins of the core were not axiomatic. Rather, its simple definition appropriately describes stable outcomes in a context of unfettered coalitional interaction. (The axiomatizations of the core came much later: see, for example, Peleg, 1985; 1986; Serrano and Volij, 1998).

For simplicity, let us continue to assume that we are studying a TU game. In this context, the core is the set of payoff vectors $x = (x_1, \ldots, x_n)$ that are feasible, that is, $\sum_{i \in N} x_i \leq v(N)$, and such that there does not exist any coalition $S \subseteq N$ for which $\sum_{i \in S} x_i < v(S)$. If such a coalition S exists, we shall say that S can improve upon or block x, and x is deemed unstable. The core usually prescribes a set of payoffs instead of a single one, and it can also prescribe the empty set in some games.

To obtain a non-cooperative implementation of the core, the procedure must embody some feature of anonymity, since the core is usually a large set and it contains payoffs where different players are treated very differently. Perry and Reny (1994) build in this anonymity by assuming that negotiations take place in continuous time, so that anyone can speak at the beginning of the game instead of having a fixed order. The player that gets to speak first makes a proposal consisting of naming a coalition that contains him and a feasible payoff for that coalition. Next, the players in that coalition get to respond. If they all accept the proposal, the coalition leaves and the game continues among the other players. Otherwise, a new proposal may come from any player in N. It is shown that, if the TU game has a non-empty core (as well as any of its subgames), the stationary subgame perfect equilibrium outcomes of this procedure coincide with the core. If a core payoff is proposed to the grand coalition, there are no incentives for individual players to reject it. Conversely, a non-core payoff cannot be sustained because any player in a blocking coalition has an incentive to make a proposal to that coalition, who will accept it (knowing that the alternative, given stationarity, would be to go back to the non-core status quo). Moldovanu and Winter (1995) offer a discrete-time version of the mechanism: in their work, the anonymity required is imposed on the solution concept by looking at order-independent equilibria.

Serrano (1995) sets up a market to implement the core. The anonymity of the procedure stems from the random choice of broker. The broker announces a vector $(x_1, \ldots, x_n)$, where the components add up to $v(N)$. One can interpret x_i as the price for the productive asset held by player i. Following an arbitrary order, the remaining players either accept or reject these prices. If player i accepts, he sells his asset to the broker for the price x_i and leaves the game. Those who reject get to buy from the broker, at the called out prices, the portfolio of assets of their choice if the broker still has them. If a player rejects but does not get to buy the portfolio of assets he would like because someone else took them before, he can always leave the market with his own asset. The broker's payoff is the worth of the final portfolio of assets that he holds, plus the net monetary transfers that he has received. Serrano (1995) shows that the prices announced by the broker will always be his top-ranked vectors in the core. If the

TU game is such that gains from cooperation increase with the size of coalitions, the set of all subgame perfect equilibrium payoffs of this procedure will coincide with the core. Core payoffs are here understood as those price vectors where all arbitrage opportunities in the market have been wiped out. Finally, yet another way to build anonymity in the procedure is by allowing the proposal to be made by brokers outside of the set *N*, as done in Pérez-Castrillo (1994).

ROBERTO SERRANO

See also **bargaining; non-cooperative games (equilibrium existence); Shapley value.**

Bibliography

Binmore, K., Rubinstein, A. and Wolinsky, A. 1986. The Nash bargaining solution in economic modelling. *RAND Journal of Economics* 17, 176–88.

Edgeworth, F. 1881. Mathematical psychics. In *F. Y. Edgeworth's Mathematical Psychics and Further Papers on Political Economy*, ed. P. Newman. Oxford: Oxford University Press, 2003.

Gillies, D. 1959. Solutions to general non-zero-sum games. In *Contributions to the Theory of Games IV*, ed. A. Tucker and R. Luce. Princeton, NJ: Princeton University Press.

Hart, S. and Mas-Colell, A. 1996. Bargaining and value. *Econometrica* 64, 357–80.

Moldovanu, B. and Winter, E. 1995. Order independent equilibria. *Games and Economic Behavior* 9, 21–34.

Nash, J. 1950. The bargaining problem. *Econometrica* 18, 155–62.

Nash, J. 1953. Two person cooperative games. *Econometrica* 21, 128–40.

Peleg, B. 1985. An axiomatizationof the core of cooperative games without side payments. *Journal of Mathematical Economics* 14, 203–14.

Peleg, B. 1986. On the reduced game property and its converse. *International Journal of Game Theory* 15, 187–200.

Pérez-Castrillo, D. 1994. Cooperative outcomes through non-cooperative games. *Games and Economic Behavior* 7, 428–40.

Perry, M. and Reny, P. 1994. A non-cooperative view of coalition formation and the core. *Econometrica* 62, 795–817.

Rubinstein, A. 1982. Perfect equilibrium in a bargaining model. *Econometrica* 50, 97–109.

Serrano, R. 1995. A market to implement the core. *Journal of Economic Theory* 67, 285–94.

Serrano, R. 2005. Fifty years of the Nash program, 1953–2003. *Investigaciones Ecońomicas* 29, 219–58.

Serrano, R. and Volij, O. 1998. Axiomatizations of neoclassical concepts for economies. *Journal of Mathematical Economics* 30, 87–108.

Shapley, L. 1953. A value for *n*-person games. In *Contributions to the Theory of Games II*, ed. A. Tucker and R. Luce. Princeton, NJ: Princeton University Press.

Walras, L. 1874. *Elements of Pure Economics, or the Theory of Social Wealth*, trans. W. Jaffé. Philadelphia: Orion Editions, 1984.

non-cooperative games (equilibrium existence)

1. Introduction

Nash equilibrium is *the* central notion of rational behaviour in non-cooperative game theory (see Osborne and Rubinstein, 1994, for a discussion of Nash equilibrium, including motivation and interpretation). Our purpose here is to discuss various conditions under which a strategic form game possesses at least one Nash equilibrium.

Strategic settings arising in economics are often naturally modelled as games with infinite strategy spaces. For example, models of price and spatial competition (Bertrand, 1883; Hotelling, 1929), quantity competition (Cournot, 1838), auctions (Milgrom and Weber, 1982), patent races (Fudenberg et al., 1983), and so on, typically allow players to choose any one of a continuum of actions. The analytic convenience of the continuum from both an equilibrium characterization and a comparative statics point of view is perhaps the central reason for the prevalence and usefulness of infinite-action games. Because of this, our treatment will permit both finite-action and infinite-action games.

Games with possibly infinite strategy spaces can be divided into two categories: those with continuous payoffs and those with discontinuous payoffs. Cournot oligopoly models and Bertrand price-competition models with differentiated products, as well as all finite-action games, are important examples of continuous games, while Bertrand price-competition with homogeneous products, auctions, and Hotelling spatial competition are important examples in which payoffs are discontinuous. Equilibrium existence results for both continuous and discontinuous games will be reviewed here. We begin with some notation.

A strategic form game, $G = (S_i, u_i)_{i=1}^N$, consists of a positive finite number, N, of players, and for each player $i \in \{1, \ldots, N\}$, a non-empty set of pure strategies, S_i, and a payoff function $u_i : S \to \mathbb{R}$, where $S = \times_{i=1}^N S_i$. The notations s_{-i} and S_{-i} have their conventional meanings: $s_{-i} = (s_1, \ldots, s_{i-1}, s_{i+1}, \ldots, s_N)$ and $S_{-i} = \times_{j \neq i} S_j$. Throughout, we assume that each S_i is a subset of some metric space and that, if any finite number of sets are each endowed with a topology, then the product of those sets is endowed with the product topology.

2. Continuous games

2.1 Pure strategy Nash equilibria

Pure strategy equilibria are more basic than their mixed strategy counterparts for at least two reasons. First, pure strategies do not require the players to possess preferences over lotteries. Second, mixed strategy equilibrium existence results often follow as corollaries of the pure strategy results. It is therefore natural to consider first the case of pure strategies.

Definition. $s^* \in S$ is a pure strategy Nash equilibrium of $G = (S_i, u_i)_{i=1}^N$ if for every player i, $u_i(s^*) \geq u_i(s_i, s^*_{-i})$ for every $s_i \in S_i$.

An important and very useful result is the following.

Theorem 1 If each S_i is a non-empty, compact, convex subset of a metric space, and each $u_i(s_1, \ldots, s_N)$ is continuous in $(s_1, \ldots, s_N)$ and quasi-concave in s_i, then $G = (S_i, u_i)_{i=1}^N$ possesses at least one pure strategy Nash equilibrium.
Proof. For each player i, and each $s_{-i} \in S_{-i}$, let $B_i(s_{-i})$ denote the set of maximizers in S_i of $u_i(\cdot, s_{-i})$. The continuity of u_i and the compactness of S_i ensure that $B_i(s_{-i})$ is non-empty and also ensure, given the compactness of S_{-i}, that the correspondence, $B_i : S_{-i} \twoheadrightarrow S_i$ is upper hemi-continuous. The quasi-concavity of u_i in s_i implies that $B_i(s_{-i})$ is convex. Consequently, each B_i is upper hemicontinuous, non empty-valued and convex-valued. All three of these properties are therefore inherited by the correspondence $B : S \twoheadrightarrow S$ defined by $B(s) = \times_{i=1}^N B_i(s_{-i})$ for each $s \in S$. Consequently, we may apply Glicksberg's (1952) fixed point theorem to B and conclude that there exists $\hat{s} \in S$ such that $\hat{s} \in B(\hat{s})$. This $\hat{s}$ is therefore a pure strategy Nash equilibrium. Q.E.D.

Remark. Theorem 1 remains valid when 'metric space' is replaced by 'locally convex Hausdorff topological vector space'. See Glicksberg (1952).

Remark. The convexity property of strategy sets and the quasi-concavity of payoffs in own action cannot be dispensed with. For example, strategy sets are not convex in matching pennies, and, even though the continuity and compactness assumptions hold there, no pure strategy equilibrium exists. On the other hand, in the two-person zero-sum game in which both players' compact convex pure strategy set is $[-1,1]$ and player 1's payoff function is $u_1(s_1, s_2) = |s_1 + s_2|$, all of the assumptions of Theorem 1 hold except the quasi-concavity of u_1 in s_1. But this is enough to preclude the existence of a pure strategy equilibrium because in any such equilibrium player 2's payoff would have to be zero (given s_1, 2 can choose $s_2 = -s_1$) and 1's payoff would have to be positive (given s_2, 1 can choose $s_1 \neq -s_2$).

Remark. More general results for continuous games can be found in Debreu (1952) and Schafer and Sonnenschein (1975). Existence results for games with strategic complements on lattices can be found in Milgrom and Roberts (1990) and Vives (1990).

2.2 Mixed strategy Nash equilibria

A *mixed strategy* for player i is a probability measure, m_i, over S_i. If S_i is finite, then $m_i(s_i)$ denotes the probability assigned to $s_i \in S_i$ by the mixed strategy m_i, and i's set of mixed strategies is the compact convex subset of Euclidean space $M_i = \{m_i \in [0, 1]^{\#S_i} : \sum_{s_i \in S_i} m_i(s_i) = 1\}$.

In general, we shall not require S_i to be finite. Rather, we shall suppose only that it is a subset of some metric space. In this more general case, a mixed strategy for player i is a (regular, countably additive) probability measure, m_i, over the Borel subsets of S_i; for any Borel subset A of S_i, $m_i(A)$ denotes the probability assigned to A by the mixed strategy m_i. Player i's set of such mixed strategies, M_i, is then convex. Further, if S_i is compact, the convex set M_i is compact in the weak-∗ topology (see, for example, Billingsley, 1968).

Extend $u_i : S \to \mathbb{R}$ to $M = \times_{i=1}^{N} M_i$ by an expected utility calculation (hence, the $u_i(s)$ are assumed to be von Neumann–Morgenstern utilities). That is, define $u_i(m_1, \ldots, m_N) = \int_{S_1} \cdots \int_{S_N} u_i(s_1, \ldots, s_N) dm_1 \ldots dm_N$ for all $m = (m_1, \ldots, m_N) \in M$. (This is an extension because we view S as a subset of M; each $s \in S$ is identified with the $m \in M$ that assigns probability one to s.) Finally, let $G = (M_i, u_i)_{i=1}^{N}$ denote the *mixed extension* of $G = (S_i, u_i)_{i=1}^{N}$.

Definition. $m^* \in M$ is a mixed strategy Nash equilibrium of $G = (S_i, u_i)_{i=1}^{N}$ if m^* is a pure strategy Nash equilibrium of the mixed extension, G, of G. That is, if for every player i, $u_i(m^*) \geq u_i(m_i, m_{-i}^*)$ for every $m_i \in M_i$.

Because $u_i(m_i, m_{-i})$ is linear and therefore quasi-concave, in $m_i \in M_i$ for each $m_{-i} \in M_{-i}$, and because continuity of $u_i(\cdot)$ on S implies continuity of $u_i(\cdot)$ on M (in the weak-∗ topology), Theorem 1 applied to the mixed extension of G yields the following basic mixed strategy Nash equilibrium existence result:

Corollary 1. If each S_i is a non-empty compact subset of a metric space, and each $u_i(s)$ is continuous in $s \in S$, then $G = (S_i, u_i)_{i=1}^{N}$ possesses at least one mixed strategy Nash equilibrium, $m^* \in M$.

Remark. Note that Corollary 1 does not require $u_i(s_i, s_{-i})$ to be quasi-concave in S_i, nor does it require the S_i to be convex.

Remark. Corollary 1 yields von Neumann's (1928) classic result for two-person zero-sum games as well as Nash's (1950; 1951) seminal result for Enite games as special cases. To obtain Nash's result, note that if each S_i is Enite, then each u_i is continuous on S in the discrete metric. Hence, the corollary applies and we conclude that every finite game possesses at least one mixed strategy Nash equilibrium.

Remark. To see how Theorem 1 can be applied to obtain the existence of mixed strategy equilibria in Bayesian games, see Milgrom and Weber (1985).

Remark. See Glicksberg (1952) for a generalization to non-metrizable strategy spaces.

3. Discontinuous games

The basic challenge one must overcome in extending equilibrium existence results from continuous games to discontinuous games is the failure of the best reply correspondence to satisfy the properties required for application of a fixed point theorem. For example, in auction or Bertrand price-competition settings, discontinuities in payoffs sometimes preclude the existence of best replies. The best reply correspondence then fails to be non-empty valued, and Glicksberg's theorem, for example, cannot be applied.

A natural technique for overcoming such difficulties is to approximate the infinite strategy spaces by a sequence of finer and finer *finite* approximations. Each of the approximating finite games is guaranteed to possess a mixed strategy equilibrium (by Corollary 1) and the resulting sequence of equilibria is guaranteed, by compactness, to possess at least one limit point. Under appropriate assumptions, the limit point is a Nash equilibrium of the original game. This technique has been cleverly employed in Dasgupta and Maskin's (1986) pioneering work, and also by Simon (1987). However, while this finite approximation technique can yield results on the existence of *mixed* strategy Nash equilibria, it is unable to produce equally general existence results for pure strategy Nash equilibria. The reason, of course, is that the approximating games, being finite, are guaranteed to possess mixed strategy, but not necessarily pure strategy, Nash equilibria. Consequently, the sequence of equilibria, and so also the limit point, cannot be guaranteed to be pure.

One might be tempted to conclude that, unlike the continuous game case where the mixed strategy result is a special case of the pure strategy result, discontinuous games require a separate treatment of pure and mixed strategy equilibria. But such a conclusion would be premature. A connection between pure and mixed strategy equilibrium existence results similar to that for continuous games can be obtained for discontinuous games by considering a different kind of approximation. Rather than approximating the infinite strategy spaces by a sequence of finite approximations, one can instead approximate the discontinuous payoff functions by a sequence of continuous payoff functions. This payoff-approximation technique is employed in Reny (1999), whose main result we now proceed to describe. All of the definitions, notation, and conventions of the previous sections remain in effect. In particular, each S_i is a subset of some metric space. (This is for simplicity of presentation only. The results to follow hold in non metrizable settings as well. See Reny, 1999.)

3.1 Better-reply security

Definition. Player i can *secure* a payoff of $\alpha \in \mathbb{R}$ at $s \in S$ if there exists $s_i \in S_i$, such that $u_i(s_i, s'_{-i}) \geq \alpha$ a for all s'_{-i} close enough to s_{-i}.

Thus, a payoff can be secured by i at s if i has a strategy that guarantees at least that payoff even if the other players deviate slightly from s.

A pair $(s, u) \in S \times \mathbb{R}^N$ is in the closure of the graph of the vector payoff function if $u \in \mathbb{R}^N$ is the limit of the vector of player payoffs for some sequence of strategies converging to s. That is, if $u = \lim_n(u_1(s^n), \ldots, u_N(s^n))$ for some $s^n \to s$.

Definition. A game $G = (S_i, u_i)_{i=1}^N$ is *better-reply secure* if whenever (s^*, u^*) is in the closure of the graph of its vector payoff function and s^* is not a Nash equilibrium, some player i can secure a payoff strictly above u_i^* at s^*.

All games with continuous payoff functions are better-reply secure. This is because if (s^*, u^*) is in the closure of the graph of the vector payoff function of a continuous game, we must have $u^* = (u_1(s^*), \ldots, u_N(s^*))$. Also, if s^* is not a Nash equilibrium then some player i has a strategy s_i such that $u_i(s_i, s^*_{-i}) > u_i(s^*)$, and continuity ensures that this inequality is maintained even if the others deviate slightly from s^*. Consequently, player i can secure a payoff strictly above $u_i^* = u_i(s^*)$.

The import of better-reply security is that it is also satisfied in many discontinuous games. For example, Bertrand's price-competition game, many auction games, and many games of timing are better-reply secure.

3.2 Pure strategy Nash equilibria

The following theorem provides a pure strategy Nash equilibrium existence result for discontinuous games.

Theorem 2. (Reny, 1999). If each S_i is a non-empty, compact, convex subset of a metric space, and each $u_i(s_1, \ldots, s_N)$ is quasi-concave in s_i, then $G = (S_i, u_i)_{i=1}^N$ possesses at least one pure strategy Nash equilibrium if in addition G is better-reply secure.

Remark. Theorem 1 is a special case of Theorem 2 because every continuous game is better-reply secure.

Remark. A classic result due to Sion (1958) states that every two-person zero-sum game with compact strategy spaces in which player 1's payoff is upper-semi-continuous and quasi-concave in his own strategy, and lower-semi-continuous and quasi-convex in the opponent's strategy, has a value and each player has an optimal pure strategy. (Sion does not actually prove the existence of optimal strategies, but this follows rather easily from his compactness assumptions and his result that the game has a value, that is, that infsup = supinf.) It is not difficult to show that Sion's result is a special case of Theorem 2.

Remark. A related result that weakens quasi-concavity but adds conditions to the sum of the players' payoffs can be found in Baye, Tian and Zhou (1993). Dasgupta and Maskin (1986) provide two interesting pure strategy equilibrium existence results, both of which require each player's payoff function to upper semi-continuous in the vector of all players' strategies.

3.3 Mixed strategy Nash equilibria

One easily obtains from Theorem 2 a mixed strategy equilibrium existence result (the analogue of Corollary 1) by treating each M_i as if it were player i's pure strategy set and by applying the definition of better-reply security to the mixed extension $G = (M_i, u_i)$ of G. This observation yields the following result.

Corollary 2 (Reny, 1999). If each S_i is a non-empty, compact, convex subset of a metric space, then $G = (S_i, u_i)_{i=1}^N$ possesses at least one mixed strategy Nash equilibrium if in addition its mixed extension, $G = (M_i, u_i)$, is better-reply secure.

Remark. Better-reply security of G neither implies nor is implied by better-reply security of G. (See Reny, 1999, for sufficient conditions for better-reply security.)

Remark. Corollary 1 is a special case of Corollary 2 because continuity of each $u_i(s)$ in $s \in S$ implies (weak-$*$) continuity of $u_i(m)$ in $m \in M$, which implies that the mixed extension, G, is better-reply secure.

Remark. Corollary 2 has as special cases the mixed strategy equilibrium existence results of Dasgupta and Maskin (1986), Simon (1987) and Robson (1994).

Remark. Theorem 2 can similarly be used to obtain a result on the existence of mixed strategy equilibria in discontinuous Bayesian games by following Milgrom and Weber's (1985) seminal distributional strategy approach. One simply replaces Milgrom and Weber's payoff continuity assumption with the assumption that the Bayesian game is better-reply secure in distributional strategies. An example of this technique is provided in the next subsection.

3.4 An application to auctions

Auctions are an important class of economic games in which payoffs are discontinuous. Furthermore, when bidders are asymmetric, in general one cannot prove existence of equilibrium by construction, as in the symmetric case. Consequently, an existence theorem applicable to discontinuous games is called for. Let us very briefly sketch how Theorem 2 can be applied in this case.

Consider a first-price single-object auction with N bidders. Each bidder i receives a private value $v_i \in [0, 1]$ prior to submitting a sealed bid, $b_i \geq 0$. Bidder i's value is drawn independently according to the continuous and positive density f_i. The highest bidder wins the object and pays his bid. Ties are broken randomly and equiprobably. Losers pay nothing.

Because payoffs are not quasi-concave in own bids, one cannot appeal directly to Theorem 2 to establish the existence of an equilibrium in pure strategy bidding functions. On the other hand, it is not difficult to show that all mixed strategy equilibria are pure and non-decreasing. Hence, to obtain an existence result for pure strategies, it suffices to show that there is an equilibrium in mixed, or equivalently in distributional, strategies. (In this context, a distributional strategy for bidder i is a joint probability distribution over his values and bids with the property that the marginal density over his values is f_i; see Milgrom and Weber, 1985.)

Because the set of distributional strategies for each bidder is a non-empty compact convex metric space and each bidder's payoff is linear in his own distributional strategy, Theorem 2 can be applied so long as a first-price auction game in distributional strategies is better-reply secure. Better-reply security can be shown to hold by using the facts that payoff discontinuities occur only when there are ties in bids and that bidders can always break a tie in their favour by increasing their bid slightly. Consequently, a Nash equilibrium in distributional strategies exists and, as mentioned above, this equilibrium is pure and non-decreasing.

3.5 Endogenous sharing rules

Discontinuities in payoffs sometimes arise endogenously. For example, consider a political game in which candidates first choose a policy from the interval [0, 1] and each voter among a continuum then decides for whom to vote. Voters vote for the candidate whose policy they most prefer, and if there is more than one such candidate it is conventional to assume that voters randomize equiprobably over them. The behaviour of voters in the second stage can induce discontinuities in the payoffs of the candidates in the first stage since a candidate can discontinuously gain or lose a positive fraction of votes by choosing a policy that, instead of being identical to another candidate's policy, is just slightly different from it.

Simon and Zame (1990) suggest an elegant way to handle such discontinuities. In particular, for the political game example above, they would not insist that voters, when indifferent, randomize equiprobably. Indeed, applying subgame perfection to the two-stage game would permit voters to randomize in any manner whatsoever over those candidates whose policies they most prefer. With this in mind, if s is a joint pure strategy for the N candidates specifying a location for each, let us denote by $U(s)$ the resulting *set* of payoff vectors for the N candidates when all best replies of the voters are considered. If no voter is indifferent, then $U(s)$ contains a single payoff vector. On the other hand, if some voters are indifferent (as would be the case if two or more candidates chose the same location) and $U(s)$ is not a singleton, then distinct payoff vectors in $U(s)$ correspond to different ways the indifferent voters can randomize between the candidates among whom they are indifferent.

The significance of the correspondence $U(\cdot)$ is this. Suppose that we are able to select, for each s, a payoff vector $u(s) \in U(s)$ in such a way that some joint mixed strategy m^* for the N candidates is a Nash equilibrium of the induced policy-choice game between them when their vector payoff function is $u(\cdot)$. Then m^* together with the voter behaviour that is implicit in the definition of $u(s)$ for each s, constitutes a subgame perfect equilibrium of the original two-stage game. Thus, solving the original problem with potentially endogenous discontinuities boils down to obtaining an appropriate selection from $U(\cdot)$. Simon and Zame (1990) provide a general result concerning the existence of such selections, which they refer to as 'endogenous sharing rules'. This method therefore provides an additional tool for obtaining equilibrium existence when discontinuities are present. Simon and Zame's main result is as follows.

Theorem 3. (Simon and Zame, 1990). Suppose that each S_i is a compact subset of a metric space and that $U : S \twoheadrightarrow \mathbb{R}^N$ is a bounded, upper hemi-continuous, non-empty-valued, convex-valued correspondence. Then for each player i, there is a measurable payoff function, $u_i : S \to \mathbb{R}$, *such that* $(u_1(s), \ldots, u_N(s)) \in U(s)$ for every $s \in S$ and such that the game $(S_i, u_i)_{i=1}^N$ possesses at least one mixed strategy Nash equilibrium.

Remark. Theorem 3 applies to the political game example above because for any policy choice s of the N candidates, the resulting set of payoff vectors $U(s)$ is convex, a fact that follows from the presence of a continuum of voters. It can also be shown that, as a correspondence, $U(\cdot)$ is upper hemi-continuous.

Remark. In the context of Bayesian games, an even more subtle endogenous-sharing rule result can be found in Jackson et al. (2002). This result, too, can be very helpful in dealing with discontinuous games. Indeed, Jackson and Swinkels (2005) have shown how it can be used to obtain equilibrium existence results in a variety of auction settings, including double auctions.

PHILIP J. RENY

See also **epistemic game theory: incomplete information; strategic and extensive form games.**

Bibliography

Baye, M., Tian, G. and Zhou, J. 1993. Characterizations of the existence of equilibria in games with discontinuous and non-quasiconcave payoffs. *Review of Economic Studies* 60, 935–48.

Bertrand, J. 1883. Théorie mathématique de la richesse sociale. *Journal des Savants* 67, 499–508.

Billingsley, P. 1968. *Convergence of Probability Measures.* New York: John Wiley and Sons.

Cournot, A. 1838. *Researches into the Mathematical Principles of the Theory of Wealth*, ed. N. Bacon. New York: Macmillan, 1897.

Dasgupta, P. and Maskin, E. 1986. The existence of equilibrium in discontinuous economic games, I: theory. *Review of Economic Studies* 53, 1–26.

Debreu, G. 1952. A social equilibrium existence theorem. *Proceedings of the National Academy of Sciences* 38, 386–93.

Fudenberg, D., Gilbert, R., Stiglitz, J. and Tirole, J. 1983. Preemption, leapfrogging, and competition in patent races. *European Economic Review* 22, 3–31.

Glicksberg, I. 1952. A further generalization of the Kakutani fixed point theorem. *Proceedings of the American Mathematical Society* 3, 170–4.

Hotelling, H. 1929. The stability of competition. *Economic Journal* 39, 41–57.

Jackson, M. and Swinkels, J. 2005. Existence of equilibrium in single and double private value auctions. *Econometrica* 73, 93–139.

Jackson, M., Simon, L., Swinkels, J. and Zame, W. 2002. Communication and equilibrium in discontinuous games of incomplete information. *Econometrica* 70, 1711–40.

Milgrom, P. and Roberts, J. 1990. Rationalizability, learning, and equilibrium in games with strategic complementarities. *Econometrica* 58, 1255–77.

Milgrom, P. and Weber, R. 1982. A theory of auctions and competitive bidding. *Econometrica* 50, 1089–122.

Milgrom, P. and Weber, R. 1985. Distributional strategies for games with incomplete information. *Mathematics of Operations Research* 10, 619–32.

Nash, J. 1950. Equilibrium points in n-person games. *Proceedings of the National Academy of Sciences* 36, 48–9.

Nash, J. 1951. Non-cooperative games. *Annals of Mathematics* 54, 286–95.

Osborne, M. and Rubinstein, A. 1994. *A Course in Game Theory.* Cambridge, MA: MIT Press.

Reny, P. 1999. On the existence of pure and mixed strategy Nash equilibria in discontinuous games. *Econometrica* 67, 1029–56.

Robson, A. 1994. An 'informationally robust' equilibrium in two-person nonzero-sum games. *Games and Economic Behavior* 2, 233–45.

Schafer, W. and Sonnenschein, H. 1975. Equilibrium in Abstract economies without ordered preferences. *Journal of Mathematical Economics* 2, 345–8.

Simon, L. 1987. Games with discontinuous payoffs. *Review of Economic Studies* 54, 569–97.

Simon, L. and Zame, W. 1990. Discontinuous games and endogenous sharing rules. *Econometrica* 58, 861–72.

Sion, M. 1958. On general minimax theorems. *Pacific Journal of Mathematics* 8, 171–6.
Vives, X. 1990. Nash equilibrium with strategic complementarities. *Journal of Mathematical Economics* 19, 305–21.
von Neumann, J. 1928. Zur Theorie der Gesellshaftspiele. *Mathematische Annalen* 100, 295–320. Trans. S. Bargmann [On the theory of games of strategy] in *Contributions to the Theory of Games*, vol. 4, ed. R. Luce and A. Tucker. Princeton: Princeton University Press, 1959.

psychological games

Traditional game-theoretic models presume that utilities depend on actions. While this framework is quite general (it can, for example, accommodate profit-maximization, altruism, inequity aversion and Rawlsian maximin preferences) it is not rich enough to adequately describe several psychological or social aspects of motivation which depend directly on beliefs (about beliefs) in addition to which actions are chosen. The following example illustrates.

Karen feels guilty if she lets others down. When paying her landscaper (Jim), this influences her tipping. The more she believes Jim believes he will receive as a tip, the more she gives. More precisely, she gives just as much as she believes Jim believes he will get, in order to avoid the feelings of guilt that will plague her if she gives less.

Beyond depicting something arguably realistic, the example illustrates in the simplest possible way how one may have to transcend traditional game theory to model a belief-dependent motivation. Consider a standard game form where Karen chooses a tip t such that $0 \leq t \leq w$, where w is the number of dollars in her wallet, and where the landscaper has no choice (his strategy set is modelled as a singleton $\{x\}$). Karen's choice of tip thus pins down a strategy profile (t,x). In traditional game theory, payoffs are defined on strategy profiles (or on endnodes induced by strategy profiles), so Karen's best choice (or choices) would be independent of her belief about Jim's belief about her choice of tip. This runs counter to the example.

Gilboa and Schmeidler (1988) and Geanakoplos, Pearce and Stacchetti (1989) present several examples that illustrate the inadequacy of traditional methods of representing preferences that reflect various forms of belief-dependent motivation. Geanakoplos, Pearce and Stacchetti develop a new analytical framework, in which the centrepiece is the notion of a *psychological game*, which may be seen as a generalization of a traditional game and which can model some of the desired effects. A psychological game differs from a traditional game in that utilities are defined on beliefs (about actions and beliefs), as well as on which actions are chosen. (The term 'game with belief-dependent motivation' would be more descriptive than the term 'psychological game', but I stick with the latter, which has become established.)

Reciprocity

The best-known example of a psychological games-based application is Rabin's (1993) highly influential model of reciprocity, according to which players wish to act kindly (unkindly) in response to kind (unkind) actions. The key notion of kindness depends on beliefs in such a way that reciprocal motivation can be described only by using psychological games. To see why, suppose that I jump out in front of your car, blocking your way, so that you can't cross a bridge and therefore arrive late to an important meeting. Am I kind? Clearly one cannot say without knowing what my beliefs are. If I believe the bridge is as sturdy as bridges usually are and I am just goofing around, then I am unkind. However, if I believe the bridge is about to collapse, then I am kind. Arguably, I would be kind even if I mistook a sturdy bridge for a dangerous one. So should you be kind or unkind in return? The answer depends on your beliefs about my kindness, and hence on your beliefs about my beliefs. It takes a psychological game to model that. (The example given here is similar in spirit to another example given in Rabin, 1998, p. 23. Rabin's model is normal-form based. See Dufwenberg and Kirchsteiger, 2004, for an extension to extensive game forms. See Fehr and Gächter, 2000, and Sobel, 2005, for general discussions of why reciprocity has important economic consequences.)

Emotions

Reciprocity is but one form of motivation that can be modelled by means of psychological games. Many emotions are good candidates. In his article 'Emotions and Economic Theory', Elster (1998) argues that a variety of emotions have important economic consequences, and he laments how little attention economists have paid to this. He argues that a key characteristic of emotions is that 'they are triggered by beliefs' (1998, p. 49). He discusses anger, hatred, guilt, shame, pride, admiration, regret, rejoicing, disappointment, elation, fear, hope, joy, grief, envy, malice, indignation, jealousy, surprise, boredom, sexual desire, enjoyment, worry, and frustration. He asks (1998, p. 48): '[H]ow can emotions help us explain behavior for which good explanations seem to be lacking?' Psychological games may be useful for providing answers.

But little work has been done. One exception is Caplin and Leahy's (2004) health care model in which a physician is concerned with a patient's belief-dependent anxiety (compare also Caplin and Leahy, 2001). Another exception is the emotion of guilt for which a string of results, both theoretical and experimental, have been established for the specific context of trust games (see Huang and Wu (1994), Dufwenberg (1995; 2002), Dufwenberg and Gneezy (2000), Bacharach, Guerra and Zizzo (2007), and Charness and Dufwenberg (2006). I shall elaborate in some detail on these latter findings (borrowing eclectically from the cited works), since they may be suggestive of the importance of psychological games more generally in a variety of ways.

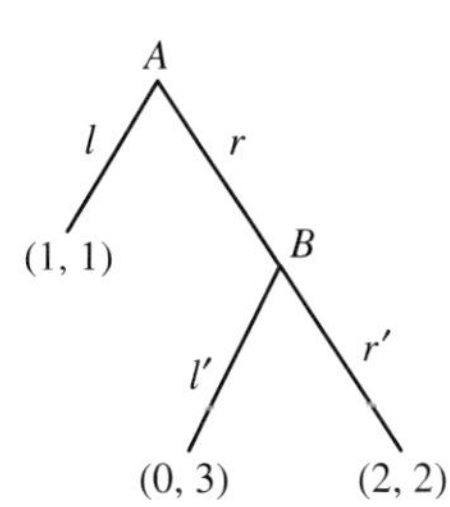

Figure 1

Consider the game in Figure 1, where payoffs reflect money income (first for player *A*, then for player *B*) but not the players' preferences which may depend also on guilt as will be indicated.

Assume that the more strongly player *B* believes that player *A* believes that *B* will make choice *r*′, the more guilt *B* would feel making choice *l*′ and the more likely *B* is to make choice *r*′. Specifically, the players' utilities at the various end nodes in the game form of Figure 1 coincide with the monetary payoffs, *except* following the choice sequence (*r*, *l*′) where *B*'s utility is $3 \cdot (1 - \beta)$ rather than 3, and where β is a measure of *B*'s belief (with range from 0 to 1) about *A*'s belief that *B* will choose *r*′. (More specifically, *B* has a probability measure describing her beliefs about which probability *A* assigns to the choice *r*′ conditional on *A* choosing *r*; β is the mean of that measure.) Say that *B* is *guilt averse*. This is all modelled in the psychological game in Figure 2:

I wish to make several points. First, the guilt aversion modelled in Figure 2 is similar to that involved in the above example featuring Karen and Jim. In fact, the idea that people feel guilty in proportion to the degree to which they do not live up to another's expectations can be extended to any game. (See Battigalli and Dufwenberg, 2007, for a recent attempt at doing this.)

Second, one can test for guilt aversion experimentally, but this requires one to measure *B*'s belief β. This can be done by inviting subjects to make guesses about one another's choices and guesses, rewarding accuracy in the guesswork. Such experimental tests have indicated that the prediction of guilt aversion is empirically supported in trust games. (The involved form of belief elicitation could conceivably be usefully complemented by two other forms of measurement: emotional self-reports and neurological methods such as functional magnetic resonance imaging.)

Third, guilt aversion may provide the seeds of a theory why communication can help foster trust and cooperation. To illustrate with reference to Figure 2, suppose that, before play, *A* and *B* meet and talk. Player *B* looks player *A* in the eye and *promises* to choose *r*′. If *A* believes this, and if *B* believes that *A* believes this, then guilt aversion would make *B* live up to her promise. A promise by *B* can thus feed a self-fulfilling circle of beliefs about beliefs that *r*′ will be chosen. In combination with guilt aversion, words may be tools that create commitment power, which may in turn foster trust and cooperation.

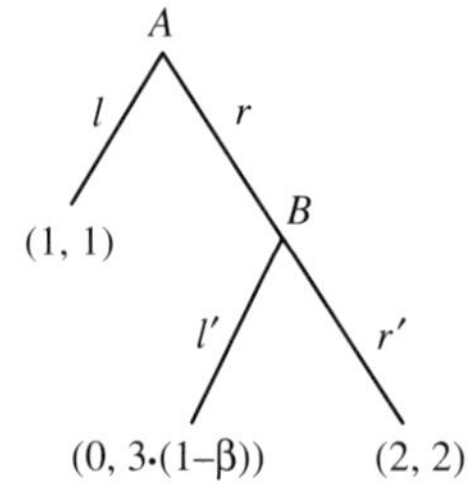

Figure 2

Fourth, even without communication between A and B, one may argue that if B is guilt averse (as described above) then trust and cooperation will ensue. If A is rational and maximizes his expected monetary income (recall that we have assumed that A is selfish in this way) then by choosing r he *signals* a certain strength of belief in B choosing r'; if A did not assign a probability of at least 1/2 to B choosing r' then he would rather chose l. If B figures this out, it puts a lower bound of 1/2 on β. So B is forced to hold a belief such that she would feel so guilty if she choose l' that she prefers r'; in numbers, with $\beta \geq 1/2$ we get $3 \cdot (1-\beta) < 2$. If A figures this out, he should of course choose r. The illustrated phenomenon has been labelled *psychological forward induction*.

To sum up: the idea of guilt aversion is intuitively plausible, experimentally testable, empirically supported, relevant for explaining why communication matters to economic behaviour, and suggestive of intriguing signalling issues that may shape emotions and behaviour. These insights concern a very special emotion and a very special psychological game, but seem profound given that limited scope. One may reasonably suspect that exciting conclusions are in store also for other emotions and other strategic settings, and that these conclusions may in part concern communication or belief signalling.

Social rewards

The discussion so far may have been misleading with its rather heavy emphasis on reciprocity and emotions. Psychological game theory may be relevant also for describing certain social rewards (norms, respect and status), where decision makers somehow care about the opinions or views of others. Bernheim (1994) and Dufwenberg and Lundholm (2001) present models that bear this out. These authors do not make explicit mention of psychological games, but if one takes a close look at the mathematical details one can discover connections.

Developing the theory

One might hope that Geanakoplos, Pearce and Stacchetti's framework is appropriate for tackling all the interesting problems to which psychological games may be relevant. However, this is not the case. A careful scrutiny reveals that their approach is too restrictive to handle many plausible forms of belief-dependent motivation (as they acknowledge themselves; see 1989, pp. 70, 78). There are several reasons, including the following:

(1) Geanakoplos, Pearce and Stacchetti only allow initial beliefs to enter the domain of a player's utility, while many forms of belief-dependent motivation require *updated* beliefs to play a role.

(2) Geanakoplos, Pearce and Stacchetti only allow a player's own beliefs to enter the domain of his utility, while there are conceptual and technical reasons to let *others'* beliefs matter.

(3) Geanakoplos, Pearce and Stacchetti follow the traditional extensive-games approach of letting strategies influence utilities only in so far as they influence

which end node is reached, but many forms of belief-dependent motivation become compelling in conjunction with preferences that depend on strategies in ways not captured by end nodes.

(4) Geanakoplos, Pearce and Stacchetti restrict attention to equilibrium analysis, but in many strategic situations there is little compelling reason to expect players to coordinate on an equilibrium, and one may wish to explore alternative assumptions.

(1) is manifest, for example, in the above psychological forward induction argument which hinges crucially on B's motivation depending on an updated belief. (2) is relevant, for example, for modelling social rewards (compare the above comments on Bernheim's and Dufwenberg and Lundholm's models). As regards (3), one can show that the issue comes up if one wants to model, for example, regret, disappointment or guilt. (4) echoes considerations relevant also for traditional games; equilibrium play is not a self-evident proposition in many contexts, for example if one assumes (only) that there is common belief in rationality or in learning scenarios.

The list (1)–(4) is adapted from Battigalli and Dufwenberg (2005), who elaborate in more detail on each issue and take first steps towards developing psychological game theory in the indicated directions. Their approach draws crucially on Battigalli and Siniscalchi's (1999) work on how to represent hierarchies of conditional beliefs.

Decision-theoretic foundations

The decision-theoretic foundations of psychological game theory are not well understood. Classical decision theory (say, von Neumann and Morgenstern) does not apply straightforwardly. Too see this, take the emotion of disappointment as an example. It is plausible that disappointment is a belief-dependent emotion. To exemplify, I have I just failed to win a million dollars and I am not at all disappointed, which, however, I clearly would be if I were playing poker and knew I would win a million dollars unless my opponent got extremely lucky drawing to an inside straight, and then he hit his card. Another example could be based on the lotteries used in the so-called Allais paradox. In both cases the level of disappointment, which if anticipated might affect choice behaviour, may plausibly depend on the strength of the prior belief that a decision maker will win a lot of money. It follows that, unless consequences are described so as to include a specification of disappointment, the so-called 'independence axiom' will not make much sense for decision makers who are prone to disappointment.

Decision theorists have given related matters some attention, but not a lot. Machina (1981, pp. 172–3; 1989, p. 1662) presents examples in spirit related to the one million dollar example above. Bell (1985), Loomes and Sugden (1986), Karni (1992), and Karni and Schlee (1995) go on to develop models in which utility may depend directly on beliefs; the latter two references take axiomatic approaches. Robin Pope has written extensively, over many years, about how conventional decision theory excludes various forms of belief-dependent motivation; Pope (2004) expounds her programme and gives further references. Caplin and Leahy (2001) develop a model of 'psychological expected utility' that admits belief-dependent motivation. However, these contributions mainly develop perspectives for settings with single decision-makers, and more will be needed to address games more generally.

Conclusion

Research on psychological games is still in its infancy. This is true for all facets of investigation: the development of basic classes of games and solution concepts, the investigation of decision-theoretic underpinning, tests of empirical (most likely experimental) validity, and finally applied theoretical work which uses psychological game theory to analyse various economic models. In each of these domains some work has been done which is indicative of the viability of the line of research, and there is good reason to be thrilled about the prospects for future research.

MARTIN DUFWENBERG

See also **behavioural economics and game theory; behavioural game theory.**

Bibliography

Bacharach, M., Guerra, G. and Zizzo, D. 2007. The self-fulfilling property of trust: An experimental study. *Theory and Decision* 63, 349–88.

Battigalli, P. and Dufwenberg, M. 2005. Dynamic psychological games. Working Paper No. 287, IGIER, Bocconi University.

Battigalli, P. and Dufwenberg, M. 2007. Guilt in games. American Economics Review, 97, 170–6.

Battigalli, P. and Siniscalchi, M. 1999. Hierarchies of conditional beliefs and interactive epistemology in dynamic games. *Journal of Economic Theory* 88, 188–230.

Bell, D. 1985. Disappointment in decision making under uncertainty. *Operations Research* 33, 1–27.

Bernheim, B.D. 1994. A theory of conformity. *Journal of Political Economy* 102, 841–77.

Caplin, A. and Leahy, J. 2001. Psychological expected utility and anticipatory feelings. *Quarterly Journal of Economics* 116, 55–79.

Caplin, A. and Leahy, J. 2004. The supply of information by a concerned expert. *Economic Journal* 114, 487–505.

Charness, G. and Dufwenberg, M. 2006. Promises and Partnership. *Econometrica* 74, 1579–601.

Dufwenberg, M. 1995. Time consistent wedlock with endogenous trust. In doctoral dissertation, *Economic Studies* 22, Uppsala University.

Dufwenberg, M. 2002. Marital investment, time consistency, and emotions. *Journal of Economic Behavior and Organization* 48, 57–69.

Dufwenberg, M. and Gneezy, U. 2000. Measuring beliefs in an experimental lost wallet game. *Games and Economic Behavior* 30, 163–82.

Dufwenberg, M. and Kirchsteiger, G. 2004. A theory of sequential reciprocity. *Games and Economic Behavior* 47, 268–98.

Dufwenberg, M. and Lundholm, M. 2001. Social norms and moral hazard. *Economic Journal* 111, 506–25.

Elster, J. 1998. Emotions and economic theory. *Journal of Economic Literature* 36, 47–74.

Fehr, E. and Gächter, S. 2000. Fairness and retaliation: the economics of reciprocity. *Journal of Economic Perspectives* 14, 159–81.

Geanakoplos, J., Pearce, D. and Stacchetti, E. 1989. Psychological games and sequential rationality. *Games and Economic Behavior* 1, 60–79.

Gilboa, Y. and Schmeidler, D. 1988. Information dependent games: can common sense be common knowledge? *Economics Letters* 27, 215–21.

Huang, P. and Wu, H.-M. 1994. More order without more law: a theory of social norms and organizational cultures. *Journal of Law, Economics & Organization* 10, 390–406.

Karni, E. 1992. Utility theory with probability dependent outcome valuation. *Journal of Economic Theory* 57, 111–24.

Karni, E. and Schlee, E. 1995. Utility theory with probability dependent outcome valuation: extensions and applications. *Journal of Risk and Uncertainty* 10, 127–42.

Loomes, G. and Sugden, R. 1986. Disappointment and dynamic consistency in choice under uncertainty. *Review of Economic Studies* 53, 271–82.

Machina, M. 1981. 'Rational' decision making versus 'rational' decision modeling. *Journal of Mathematical Psychology* 24, 163–75.

Machina, M. 1989. Dynamic consistency and non-expected utility models of choice under uncertainty. *Journal of Economic Literature* 27, 1622–68.

Pope, R. 2004. Biases from omitted risk effects in standard gamble utilities. *Journal of Health Economics* 23, 1029–50.

Rabin, M. 1993. Incorporating fairness into game theory and economics. *American Economic Review* 83, 1281–302.

Rabin, M. 1998. Psychology and economics. *Journal of Economic Literature* 36, 11–46.

Sobel, J. 2005. Interdependent preferences and reciprocity. *Journal of Economic Literature* 43, 392–436.

purification

In a mixed strategy equilibrium of a complete information game, players randomize between their actions according to a particular probability distribution, even though they are indifferent between their actions. Two criticisms of such mixed strategy equilibria are (*a*) that players do not seem to randomize in practice, and (*b*), if a player were to randomize, why would he or she choose to do so according to probabilities that make other players indifferent between their strategies?

Since many games have no pure strategy equilibria, it is important to provide a more compelling rationale for the play of mixed strategy equilibria.

Harsanyi (1973) gave a 'purification' interpretation of mixed strategy equilibrium that resolves these criticisms. The complete information-game payoffs are intended as an approximate description of the strategic situation, but surely do not capture every consideration in the minds of the players. In particular, suppose that a player has some small private inclination to choose one action or another independent of the specified payoffs, but this information is not known to the other players. Then the behaviour of such players will look – to their opponents and to outside observers – as if they are randomizing between his actions, even though they do not experience the choice as randomization. Because of the private payoff perturbation, they will not in fact be indifferent between their actions, but will almost always be choosing a strict best response. Harsanyi's remarkable purification theorem showed that all equilibria (pure or mixed) of almost all complete information games are the limit of pure strategy equilibria of perturbed games where players have independent small shocks to payoffs.

There are other interpretations of mixed strategy play: Reny and Robson (2004) present an analysis that unifies the purification interpretation with the 'classical' interpretation that players randomize because they think that there is a small chance that their mixed strategy may be observed in advance by other players. But Harsanyi's purification theorem justly provides the leading interpretation of mixed strategy equilibria among game theorists today.

I first review Harsanyi's theorem. Harsanyi's result applies to regular equilibria of complete information games with independent payoff shocks; since many equilibria of interest (especially in dynamic games) are not regular, Harsanyi's result cannot be relied upon in many economic settings of interest; I therefore briefly review what little is known about such extensions.

Harsanyi's theorem has two parts: (*a*) pure strategy equilibria always exist in suitably perturbed versions of a complete information game; and (*b*) for any regular equilibrium of a complete information game and any sequence of such perturbed games converging to the complete information game, there is a sequence of pure strategy equilibria converging to the regular equilibrium. An important literature has ignored the latter approachability question and focused on the former pure strategy existence question, identifying conditions on an information structure – much weaker than Harsanyi's – to establish the existence of pure strategy equilibria. I conclude by reviewing these papers.

Harsanyi's theorem

Consider two players engaging in the symmetric coordination game below.

	A	B
A	2,2	0,0
B	0,0	1,1

As well as the pure strategy Nash equilibria (A, A) and (B, B), this game has a symmetric mixed strategy Nash equilibrium where each player chooses A with probability $\frac{1}{3}$ and B with probability $\frac{2}{3}$.

But suppose that, in addition to these common knowledge payoffs, each player i observes a payoff shock depending on the action he or she chooses. Thus,

	A	B
A	$2+\varepsilon.\eta_{1A}, 2+\varepsilon.\eta_{2A}$	$\varepsilon.\eta_{1A}, \varepsilon.\eta_{2B}$
B	$\varepsilon.\eta_{1B}, \varepsilon.\eta_{2A}$	$1+\varepsilon.\eta_{1B}, 1+\varepsilon.\eta_{2B}$

where $\varepsilon > 0$ is a commonly known parameter measuring the size of payoff shocks and (η_{1A}, η_{1B}) and (η_{2A}, η_{2B}) are distributed independently of each other, and player i observes only (η_{iA}, η_{iB}). Finally, assume that, for each player i, $\eta_i = \eta_{iA} - \eta_{iB}$ is distributed according to a continuous density f on the real line with corresponding c.d.f. F.

This perturbed game is one with incomplete information, where a player's strategy is a function $s_i : \mathbb{R} \to \{A, B\}$. In equilibrium, each player will follow a threshold strategy of the form

$$s_i(\eta_i) = \begin{cases} A, & \text{if } \eta_i \geq z_i \\ B, & \text{if } \eta_i < z_i \end{cases}.$$

Under such a strategy, the *ex ante* probability that player i will choose action B is $\pi_I = F(z_i)$, and the probability he or she will choose A is $1 - \pi_i$. Thus we can re-parameterize the strategy as

$$s_i(\eta_i) = \begin{cases} A, & \text{if } \eta_i \geq F^{-1}(\pi_i) \\ B, & \text{if } \eta_i < F^{-1}(\pi_i) \end{cases}.$$

Let us look for a strategy profile (s_1, s_2) of the incomplete information game, parameterized by (π_1, π_2), that forms an equilibrium of the incomplete information game. Since player 1 thinks that player 2 will choose action A with probability $1-\pi_2$ and action B with probability π_2, player 1's expected payoff gain from choosing action A over action B is then

$$2(1-\pi_2)+\varepsilon.\eta_1-\pi_2.$$

Thus player 1's best response must be to follow a threshold strategy with threshold

$$F^{-1}(\pi_1)=\frac{3\pi_2-2}{\varepsilon}$$

or

$$\varepsilon F^{-1}(\pi_1)=3\pi_2-2.$$

Symmetrically, we have

$$\varepsilon F^{-1}(\pi_2)=3\pi_1-2.$$

Thus, there will be a symmetric equilibrium where both players choose action B with probability π if and only if

$$\varepsilon F^{-1}(\pi)=3\pi-2.$$

For small ε, this equation has three solutions tending 0, $\frac{2}{3}$ and 1 as $\varepsilon \to 0$. These solutions correspond to the three symmetric Nash equilibria of the above complete information game, respectively: (a) both always choose A, (b) both choose B with probability $\frac{2}{3}$, and (c) both always choose B.

Harsanyi's purification theorem generalizes the logic of this example. If we add small independent noise to each player's payoffs, then each player will almost always have a unique best response and thus the perturbed game will have a pure strategy equilibrium. There is a system of equations that describes equilibria of the unperturbed game. If that system of equations is regular, then a small perturbation of the system of equations will result in a nearby equilibrium.

I will report a statement of Harsanyi's result due to Govindan, Reny and Robson (2003), which weakens a number of the technical conditions in the original theorem.

Consider an I player complete information game where each player i has a finite set of possible actions A_i and a payoff function $g_i : A \to \mathbb{R}$ where $A = A_1 \times \cdots \times A_I$. An equilibrium $\alpha \in \Delta(A_1) \times \cdots \times \Delta(A_I)$ is a *regular Nash equilibrium* of the complete information game if the Jacobian determinant of a continuously differentiable map characterizing equilibrium is non-zero at α (see van Damme, 1991, Definition 1.5.1, p. 39).

The μ-perturbed game is then an incomplete information game where each player i privately observes a vector $\eta_i \in \mathbb{R}^{|A|}$. Player i's payoff in the incomplete information game if action profile a is chosen is then $g_i(a)+\eta_{ia}$; thus η_i is a private value shock. Each η_i is independently drawn according to a measure μ_i, where each μ_i assigns probability 0 to i's expectation of η_i being equal under any pair of i's pure strategies a_i and a_i', given any mixed strategy profile of the other players; Govindan, Reny and Robson (2003) note that this weak condition is implied by μ_i being absolutely continuous with respect to Lebesgue measure on $\mathbb{R}^{|A|}$. A pure strategy for player i in the μ-perturbation is a function $s_i : \mathbb{R}^{|A|} \to A_i$. A pure strategy profile s induces a probability distribution over actions $V_s \in \Delta(A)$, where

$$\nu_s(a) = \Pr_{\mu}\{\eta : s_i(\eta_i) = a_i \text{ for each } i\}$$

Theorem (Harsanyi, 1973; Govindan, Reny and Robson, 2003). Suppose that α is a regular Nash equilibrium of the complete information game and that, for each i, μ_i^n converges to a point mass at $0 \in \mathbb{R}^{|A|}$. Then for all $\varepsilon > 0$ and all large enough n, the μ-perturbed game has a pure strategy equilibrium inducing a distribution on A that is within ε of α, that is,

$$|\nu_s(a) - \prod_{i=1}^{I} \alpha_i(a_i)| \leq \varepsilon$$

for all $a \in A$.

The pure strategy equilibria are 'essentially strict', that is, almost all types have a strict best response. An elegant proof in Govindan, Reny and Robson (2003) simplifies Harsanyi's original proof.

Dynamic games

Harsanyi's theorem applies only to regular equilibria of a complete information game. Harsanyi noted that all equilibria of almost all finite complete information games are regular, where 'almost all' means with probability one under Lebesgue measure on the set of payoffs. Of course, normal form games derived from general extensive form games are not generic in this sense. This raises the question of whether mixed strategy equilibria of extensive form games are purifiable in Harsanyi's sense.

Here is an economic example suggesting why this is an important question. Consider an infinite overlapping generations economy where agents live for two periods; the young are endowed with two units of an indivisible and perishable consumption good, and the old have no endowment. Each young agent has the option of transferring one unit of consumption to the current old agent. Each agent's utility function is strictly increasing in own consumption when young and old, and values smoothed consumption (one when young, one when old) strictly more highly than consuming the endowment (two when young, none when old). Under perfect information, this game has a 'social security' subgame perfect Nash equilibrium where each young agent transfers one unit to the old agent if and only if no young agent failed to do so in the past. But suppose instead that each young agent observes only whether the previous young agent made a transfer, and restrict attention to subgame perfect Nash equilibria. Then Bhaskar (1998) has shown that there is no pure strategy equilibrium with a positive probability of transfers (in fact, this conclusion remains true if all agents only observe history of any commonly known finite length). To see why, suppose there was such an equilibrium: if the young agent at date t does not transfer, then the young agent at date $t + 1$ must punish by not making a transfer; but the young agent at date $t + 2$ did not observe the date t outcome, and so will think that the young agent at date $t + 1$ deviated, and will therefore not make transfers; so the young agent at date $t + 1$ would have an incentive to make transfers, and not to punish as required by the equilibrium strategy.

However, Bhaskar shows that there are mixed strategy equilibria with positive transfers. In particular, there is an equilibrium where the young always transfers in the first period or if he or she observed transfers in the previous period, and randomizes between making transfers or not if he or she did not observe transfers. This strategy profile attains the efficient outcome and involves mixing off the equilibrium path only. It is natural to ask whether this equilibrium can be 'purified': suppose that each young agent obtains a small 'altruism' payoff shock that makes transfers to the old slightly attractive. The mixed strategy might then be the limit of pure strategy equilibria where the more altruistic agents make the transfers and the less altruistic agents do not. However, Bhaskar shows that the mixed strategy equilibria cannot be purified. The logic of Harsanyi's purification result breaks down because the equilibrium is not regular.

Very little is known in general about purifiability of mixed strategy equilibria in extensive form games. Results will presumably depend on the regularity of the equations characterizing equilibria and the modelling of payoff choices in the extensive form (for example, do shocks occur at the beginning of the game or at each decision node?). The best hope of a positive purification result would presumably be in finite dynamic games, where Harsanyi's regularity techniques might be applied. But Bhaskar (2000) gives an example of a simple finite extensive form game where mixed strategy equilibria are not purifiable because of the non-regularity of equilibria even for generic assignment of payoffs to terminal nodes. Mixed strategy equilibria play an important role in recent developments of the theory of repeated games. Bhaskar, Mailath and Morris (2006) report some positive and negative purification results in that context.

Purification without approachability

Harsanyi's purification theorem has two parts. First, the 'purification' part: all equilibria of the perturbed game are essentially pure; second, the 'approachability' part: every equilibrium of a complete information game is the limit of equilibria of such perturbed games. For the first part, Harsanyi's theorem uses the assumption of sufficiently diffuse independent payoff shocks. Only the second part required the strong regularity properties of the complete information game equilibria.

Radner and Rosenthal (1982) addressed a weaker version of the purification part of Harsanyi's theorem, asking what conditions on the information system of an incomplete information game will ensure that for every equilibrium (perhaps mixed) there exists an outcome equivalent pure strategy equilibrium. Thus they did not ask that every equilibrium be essentially pure and they did not seek to approximate mixed strategy equilibria of any unperturbed game. Each agent observing a signal with an atomless independent distribution is clearly sufficient for such a 'purification existence' result (whether or not the signal is payoff relevant). But what if there is correlation?

A simple example from Radner and Rosenthal (1982) illustrates the difficulty. Suppose that two players are playing matching pennies and each player i observes a payoff-irrelevant signal x_i, where (x_1, x_2) are uniformly distributed on $\{(x_1, x_2) \in \mathbb{R}^2_+ | 0 \leq x_1 \leq x_2 \leq 1\}$. In any equilibrium, almost all types of each player must assign probability $\frac{1}{2}$ to his or her opponent choosing each action (otherwise, that player would be able to obtain a payoff greater than his or her value in the zero sum game). Yet it is impossible to generate pure strategies of the players that make this property hold true. Another illustration of the importance of correlation for purification occurs in Carlsson and van Damme (1993), where it is shown that, while small independent noise leads to Harsanyi's purification result, small highly correlated noise leads to the selection of a unique equilibrium (the comparison is made explicitly in their Appendix B).

Radner and Rosenthal (1982) show the existence of a pure strategy equilibrium if each player observes a payoff-irrelevant signal with an atomless distribution and each player i's payoff-irrelevant signal and payoff-relevant information (which may be correlated) are independent of each other player's payoff-irrelevant signal. This result extends if players observe additional finite private signals which are also uncorrelated with others' atomless payoff-irrelevant signals. Their method of proof builds on the argument of Schmeidler (1973) showing the existence of a pure strategy equilibrium in a game with a continuum of players. Radner and Rosenthal (1982) also present a number of counter-examples – in addition to the matching pennies example above – with non-existence of pure strategy equilibrium. Milgrom and Weber (1985) show the existence of a pure strategy equilibrium if type spaces are atomless and independent conditional on a finite valued common state variable with payoff interdependence occurring only via the common state variable. Their result – which neither implies nor is implied by the Radner and Rosenthal (1982) conditions – has been used in many applications. Aumann et al. (1983) show that, if every player has a conditionally atomless distribution over others' types (that is, his or her conditional distribution has no atoms for almost every type), there exists a pure strategy ε-equilibrium. Their theorem thus covers the matching pennies example described above.

The existence of such purifications deals with one of the two criticisms of mixed strategy equilibria raised above: people do not appear to randomize. In particular, in any such purification the 'randomization' represents the uncertainty in a player's mind about how other players will act, rather than deliberate randomization. This interpretation of mixed strategies was originally emphasized by Aumann (1974).

STEPHEN MORRIS

See also **global games; mixed strategy equilibrium.**

Bibliography

Aumann, R. 1974. Subjectivity and correlation in randomized strategies. *Journal of Mathematical Economics* 1, 67–96.

Aumann, R., Katznelson, Y., Radner, R., Rosenthal, R. and Weiss, B. 1983. Approximate purification of mixed strategies. *Mathematics of Operations Research* 8, 327–41.

Bhaskar, V. 1998. Informational constraints and overlapping generations model: folk and anti-folk theorems. *Review of Economic Studies* 65, 135–49.

Bhaskar, V. 2000. The robustness of repeated game equilibria to incomplete payoff information. Mimeo, University College London.

Bhaskar, V., Mailath, G. and Morris, S. 2006. Purification in the infinitely-repeated Prisoners' Dilemma. Discussion Paper No. 1571, Cowles Foundation.

Carlsson, H. and van Damme, E. 1993. Global games and equilibrium selection. *Econometrica* 61, 989–1018.

Govindan, S., Reny, P. and Robson, A. 2003. A short proof of Harsanyi's purification theorem. *Games and Economic Behavior* 45, 369–74.

Harsanyi, J. 1973. Games with randomly disturbed payoffs: a new rationale for mixed-strategy equilibrium points. *International Journal of Game Theory* 2, 1–23.

Milgrom, P. and Weber, R. 1985. Distributional strategies for games with incomplete information. *Mathematics of Operations Research* 10, 619–32.

Radner, R. and Rosenthal, R. 1982. Private information and pure-strategy equilibria. *Mathematics of Operations Research* 7, 401–9.

Reny, P. and Robson, A. 2004. Reinterpreting mixed strategy equilibria: a unification of the classical and Bayesian views. *Games and Economic Behavior* 48, 355–84.

Schmeidler, D. 1973. Equilibrium points of non-atomic games. *Journal of Statistical Physics* 7, 295–301.

van Damme, E. 1991. *The Stability and Perfection of Nash Equilibria*. New York: Springer-Verlag.

repeated games

Repeated games provide a formal and quite general framework to examine why self-interested agents manage to cooperate in a long-term relationship.

Formally, repeated games refer to a class of models where the same set of agents repeatedly play the same game, called the 'stage game', over a long (typically, infinite) time horizon. In contrast to the situation where agents interact only once, *any* mutually beneficial outcome can be sustained as an equilibrium when agents interact repeatedly and frequently. A formal statement of this fact is known as the folk theorem.

Repeated games and the general theories of efficiency

Thanks to the developments since the mid-1970s, economics now recognizes three general ways to achieve efficiency: (*a*) competition; (*b*) contracts; and (*c*) long-term relationships. For standardized goods and services, with a large number of potential buyers and sellers, promoting market competition is an effective way to achieve efficiency. This is formulated as the classic First and Second Welfare Theorems in general equilibrium theory. There are, however, other important resource allocation problems which do not involve standardized goods and services. Resource allocation within a firm or an organization is a prime example, as pointed out by Ronald Coase (1937), and examples abound in social and political interactions. In such cases, aligning individual incentives with social goals is essential for efficiency, and this can be achieved by means of *incentive schemes* (penalties or rewards). The incentive schemes, in turn, can be provided in two distinct ways: by a formal contract or by a long-term relationship. The penalties and rewards specified by a formal contract are enforced by the court, while in a long-term relationship the value of future interaction serves as the reward and penalty to discipline the agents' current behaviour. The theory of contracts and mechanism design concern the former case, and the theory of repeated games deals with the latter. These theories provide general methods to achieve efficiency, and have become important building blocks of modern economic theory.

An example: collusion of gas stations and the trigger strategy

Consider two gas stations located right next to each other. They have identical and constant marginal cost c (the wholesale price of gasoline) and compete by publicly posting their prices. Suppose their joint profit is maximized when they both charge $p = 10$, whereby each receives a large profit π. Although this is the best outcome for them, they have an incentive to deviate. By slightly undercutting its price, each can steal all the customers from its opponent, and its profit (almost) doubles. The only price free from such profitable deviation is $p = c$, where their profit is equal to zero. In other words, the only Nash equilibrium in the price competition game is an *inefficient* (for the gas stations) outcome where both charge $p = c$. This situation is the rule rather than the exception: the Nash equilibrium in the stage game, the only outcome that agents can credibly achieve in a one-shot interaction, is quite often inefficient for them. This is because agents seek only their private benefits, ignoring the benefits or costs of their actions for their rivals.

In reality, however, gas stations enjoy positive profits, even when there is another station nearby. An important reason may well be that their interaction is not one-shot. Formally, the situation is captured by a *repeated game*, where the two gas stations play the price competition game (the stage game) over an infinite time horizon $t = 0, 1, 2, \ldots$. Consider the following repeated game strategy:

1. Start with the optimal price $p = 10$.
2. Stick to $p = 10$ as long as no player (including oneself) has ever deviated from $p = 10$.
3. Once anyone (including oneself) deviated, charge $p = c$ for ever.

This can be interpreted as an explicit or implicit agreement of the gas stations: charge the monopoly price $p = 10$, and any deviation triggers cut-throat price competition ($p = c$ with zero profit). Let us now check whether each player has any incentive to deviate from this strategy. Note that, if neither station deviates, each enjoys profit π every day. As we saw above, a player can (almost) double its stage payoff by slightly undercutting the agreed price $p = 10$. Hence the short-term gain from deviation is at most π. If one deviates, however, its future payoff is reduced from π to zero in each and every period in the future. Now assume that the players discount future profits by the *discount factor* $\delta \in (0, 1)$. The number δ measures the value of a dollar in the next period. The discounted future loss is $\delta\pi + \delta^2\pi + \cdots = \frac{\delta}{1-\delta}\pi$. If this is larger than the short-term gain from defection (π), no one wants to deviate from the collusive price $p = 10$. The condition is $\pi \leq \delta/(1-\delta)\pi$, or equivalently, $1/2 \leq \delta$.

Next let us check whether the players have an incentive to carry out the threat (the cut-throat price competition $p = c$). Since $p = c$ is the Nash equilibrium of the stage game, charging $p = c$ in each period is a best reply if the opponent always does so. Hence, the players are choosing mutual best replies. In this sense, the threat of $p = c$ is credible or self-enforcing.

In summary, under the strategy defined above, players are choosing mutual best replies *after any history*, as long as $1/2 \leq \delta$. In other words, the strategy constitutes a *subgame perfect equilibrium* in the repeated game. Similarly, in a general game, any outcome which Pareto dominates the Nash equilibrium can be sustained by a strategy which reverts to the Nash equilibrium after a deviation. Such a strategy is called a *trigger strategy*.

Three remarks: multiple equilibria, credibility of threat and renegotiation, and finite versus infinite horizon

A couple of remarks are in order about the example. First, the trigger strategy profile is not the only equilibrium of the repeated game. The repetition of the stage game Nash equilibrium ($p = c$ for ever) is also a subgame perfect equilibrium. Are there any other equilibria? Can we characterize *all* equilibria in a repeated game? The latter question appears to be formidable at first sight, because there are an infinite number of repeated game strategies, and they can potentially be quite complex. We do have, however, some complete characterizations of all equilibria of a repeated game, such as folk theorems and self-generation conditions as will be discussed subsequently.

Second, one may question the credibility of the threat ($p = c$ for ever). In the above example, credibility was formalized as the subgame perfect equilibrium condition. According to this criterion, the threat $p = c$ is credible because a *unilateral* deviation by a *single* player is never profitable. The threat $p = c$, however, may be upset by *renegotiation*. When players are called upon to carry out this grim threat after a deviation, they may well get together and agree to 'let bygones be bygones'. After all, when there is a better equilibrium in the repeated game (for example, the trigger strategy equilibrium), why do we expect the players to stick to the inefficient one ($p = c$)? This is the problem of *renegotiation proofness* in repeated games. The problem is trickier than it appears, however, and economists have not yet agreed on what is the right notion of renegotiation proofness for repeated games. The reader may get a sense of difficulty from the following observation. Suppose the players have successfully renegotiated away $p = c$ to play the trigger strategy equilibrium again. This is self-defeating, however, because the players now have an incentive to deviate, as they may well anticipate that the threat $p = c$ will be again subject to renegotiation and will not be carried out. For a comprehensive discussion of this topic (and also of a number of major technical results on repeated games), see an excellent survey by D. Pearce (1990).

Third, let me comment on the assumption of an *infinite* time horizon. Suppose that the gas stations are to be closed by the end of next year (due to a new zoning plan, for example). This situation can be formulated as a *finitely* repeated game. On the last day of their business, the gas stations just play the stage game, and therefore they have no other choice but to play the stage game equilibrium $p = c$. In the penultimate day, they rationally anticipate that they will play $p = c$ *irrespective of* their current action. Hence they are effectively playing the stage game in the penultimate day, and again they choose $p = c$. By induction, the *only* equilibrium of the finitely repeated price competition is to charge $p = c$ in *every* period. The impossibility of cooperation holds no matter how long the time horizon is, and it is in sharp contrast to the infinite horizon case.

Although one may argue that players do not really live infinitely long (so that the finite horizon case is more realistic), there are some good reasons to consider the infinite horizon models. First, even though the time horizon is finite, if players do not know in advance exactly *when* the game ends, the situation can be formulated as an infinitely repeated game. Suppose that, with probability $r > 0$, the game ends at the end of any given period. This implies that, *with probability 1*, the game ends in a finite horizon. Note, however, that the expected discounted profit is equal to $\pi(0) + (1 - r)\delta\pi(1) + (1 - r)^2\delta^2\pi(2) + \ldots$, where $\pi(t)$ is the stage payoff in period t.

This is identical to the payoff in an infinitely repeated game with discount factor $\delta' = (1-r)\delta$. Second, the drastic 'discontinuity' between the finite and infinite horizon cases in the price competition example hinges on the uniqueness of equilibrium in the stage game. Benoit and Krishna (1985) show that, if each player has multiple equilibrium payoffs in the stage game, the long but finite horizon case enjoys the same scope for cooperation as the infinite horizon case (the folk theorem, discussed below, approximately holds for T-period repeated game, when $T \to \infty$).

The repeated game model

Now let me present a general formulation of a repeated game. Consider an infinitely repeated game, where players $i = 1, 2, \ldots, N$ repeatedly play the same stage game over an infinite time horizon $t = 0, 1, 2, \ldots$. In each period, player i takes some action $a_i \in A_i$, and her payoff in that period is given by a stage game payoff function $g_i(a)$, where $a = (a_1, \ldots, a_N)$ is the action profile in that period. The repeated game payoff is given by

$$\Pi_i = \sum_{t=0}^{\infty} g_i(a(t))\delta^t,$$

where $a(t)$ denotes the action profile in period t and $\delta \in (0, 1)$ is the discount factor. It is often quite useful to look at the *average payoff* of the repeated game, which is defined to be $(1-\delta)\Pi_i$. Note that, if one receives the same payoff x in each period, the repeated game payoff is $\Pi_i = x + \delta x + \delta^2 x + \cdots = x/(1-\delta)$. This example helps to understand the definition of average payoff: in this case $(1-\delta)\Pi_i$ is indeed equal to x, the payoff per period.

A *history* up to time t is the sequence of realized action profiles before $t : h^t = (a(0), a(1), \ldots, a(t-1))$. A *repeated game strategy* for player i, denoted by s_i, is a complete contingent action plan, which specifies a current action after any history: $a_i(t) = s_i(h^t)$ (a minor note: to determine $a_i(0)$, we introduce a dummy history h^0 such that $a_i(0) = s_i(h^0)$). A repeated game strategy profile $s = (s_1, \ldots, s_N)$ is a *subgame perfect equilibrium* if it specifies mutual best replies after any history.

The folk theorem

Despite the fact that a repeated game has an infinite number of strategies, which can be arbitrarily complicated, we do have a *complete* characterization of equilibrium payoffs. The folk theorem shows exactly which payoff points can be achieved in a repeated game.

Before stating the theorem, we need to introduce a couple of concepts. First, let us determine the set of physically achievable average payoffs in a repeated game. Note that, by alternating between two pure strategy outcomes, say u and v, one may achieve any point between u and v as the average payoff profile. Hence, an average payoff profile can be a weighted average (in other words, a convex combination) of pure strategy payoff profiles in the stage game. Let us denote the set of all such points by V.

Formally, the set of *feasible average payoff profiles* V is the smallest convex set that contains the pure strategy payoff profiles of the stage game.

Second, let us determine the points in V that cannot possibly be an equilibrium outcome. For example, if a player has an option to stay out to enjoy zero profit in each period, it is a priori clear that her equilibrium average payoff cannot be less than zero. In general, there is a payoff level that a player can guarantee herself in any equilibrium, and this is formulated as the *minimax* payoff. Formally, the minimax payoff for player i is defined as $\underline{v}_i = \min_{\alpha_{-i}} \max_{\alpha_i} g_i(\alpha)$, where $\alpha = (\alpha_1, \ldots, \alpha_N)$ is a mixed action profile (α_i is a probability distribution over player i's pure actions) and $g_i(\alpha)$ is the associated expected payoff. To understand why min and max are taken in that particular order, consider the situation where player i always *correctly anticipates what others do.* If player i knows that others choose $\alpha_{-i} = (\alpha_1, \ldots \alpha_{i-1}, \alpha_{i+1}, \ldots, \alpha_N)$, he can play a best reply against α_{-i} to obtain $\max_{\alpha_i} g_i(\alpha)$. Note well that $\max_{\alpha_i} g_i(\alpha)$ is a function of α_{-i}. In the worst case, where others take the most damaging actions α_{-i}, player i obtains the minimax payoff (this is exactly what the definition says). From this definition it is clear that, in any equilibrium of the repeated game, *the average payoff to each player is at least her minimax payoff.* In any equilibrium, each player correctly anticipates what others do, and simply by playing the stage game best reply in each period, any player can make sure that her average payoff is more than her minimax payoff. (A comment: we consider mixed strategies in the definition of the minimax payoff because in many games the minimax payoff is smaller when we consider mixed strategies.)

From what we saw, now it is clear that the set of equilibrium average payoff profiles of a repeated game is *at most* $V^* = \{v \in V | \forall i \ v_i > \underline{v}_i\}$. (The points with $v_i = \underline{v}_i$ are excluded to avoid minor technical complications.) The set V^* is called the *feasible and individually rational payoff set.* This is the set of physically achievable average payoff profiles in the repeated game where each player receives more than her minimax payoff. The folk theorem shows that any point in this 'maximum possible region' can indeed be an equilibrium outcome of the repeated game. (Throughout this article, I maintain a minor technical assumption that each player has a finite number of actions in the stage game.)

Folk theorem In an N-player infinitely repeated game, any feasible and individually rational payoff profile $v \in V^*$ can be achieved as the average payoff profile of a subgame perfect equilibrium when the discount factor δ is close enough to 1, provided that either $N = 2$, or $N \geq 3$ and no two players have identical interests.

Formally, no two players have identical interests if there are *no* players i and j ($i \neq j$) whose payoffs satisfy $g_i(a) = bg_j(a) + c$, $b > 0$ (that is, no two players have the same preferences over the stage game outcomes). This is a 'generic' condition that is almost always satisfied: the case where players have identical interests is very special in the sense that the equality $g_i(a) = bg_j(a) + c$ fails by even a slight change of the payoff functions. Hence, the folk theorem provides a general theory of efficiency: it shows that, for virtually any game, any mutually beneficial outcome can be achieved in a long term relationship, if the discount factor is close to 1. *Although game-theoretic predictions quite often depend on the fine details of the model, this result is a notable exception for its generality.*

The crucial condition in the folk theorem is a high discount factor. The discount factor δ may measure the (subjective) patience of a player, or, it may be equal to $1/(1+r)$, where r is the interest rate per period. Although the discount factor may not be directly observable (in particular, in the former case), it should be high when one period is short. Hence, an empirically testable implication is that players who have daily interaction (such as the gas stations in our example) have a better scope for cooperation than those who interact only once a year. An important message of the folk theorem is that a high *frequency of interaction* is essential for the success of a long term relationship.

The name 'folk theorem' comes from the fact that game theorists had anticipated that something like it should be true long before it was precisely formulated and proved. In this sense, the assertion had been folklore in the game theorist community. The proof is, however, by no means obvious, and there is a body of literature to prove the theorem in various degrees of generality. Early contributions include Aumann (1959), Friedman (1971) and Rubinstein (1979). The statement above is based on Fudenberg and Maskin (1986) and its generalization by Abreu, Dutta and Smith (1994). The proof is constructive: a clever strategy, which has a rather simple structure, is constructed to support any point in V^*.

Repeated games versus formal contracts

To discuss the scope of applications, I now compare a long-term relationship (repeated game) and a formal contract as a means to enforce efficient outcomes. As our gas station example shows, quite often an agent has an incentive to deviate from an efficient outcome, because it increases her private returns at the expense of the social benefit. Such a deviation can be deterred if we impose a sufficiently high penalty so that the *incentive constraint*

$$\text{gain from deviation} \leq \text{penalty}$$

is satisfied. This is the basic and common feature of repeated games and contracts. A formal contract explicitly specifies the penalty and it is enforced by the court. In repeated games, the penalty is indirectly imposed through future interaction. In this sense the theory of repeated games can be regarded as the theory of *informal or relational contracts.*

When is a long-term relationship a better way to achieve cooperation than a formal contract? First, a long-term relationship is useful when a formal contract is too costly or impractical. For example, it is often quite costly for a third party (the court) to verify whether there was any deviation from an agreement, while defections may be directly observed by the players themselves. In practice, what constitutes 'cooperation' is often so fuzzy or complicated that it is hard to write it down explicitly, although the players have a common and good understanding about what it is. 'Pulling enough weight' in a joint research project may be a good example. In those situations, a long-term relationship is a more practical way to achieve cooperation than a formal contract. In fact, a classic study by Macaulay (1963) indicates that the vast majority of business transactions are executed without writing formal contracts. Second, there are some cases where a court powerful enough to enforce formal contracts simply does not exist. For example, in many problems in development economics and economic history, the legal system is highly imperfect. Even for developed countries in the modern age, there are no legal institutions which have enough binding power to enforce international agreements. Hence, repeated games provide a useful framework to address such problems as the organization of medieval trade, informal mutual insurance in developing countries, international policy coordination, and measures against global warming. Lastly, there is no legal system to enforce cartels or collusion, because the existing legal system refuses to enforce any contract that violates antitrust laws. Hence a long-term relationship is the only way to enforce a cartel or collusive agreement.

Is the folk theorem a negative result?

The theory of repeated games based on the folk theorem is often criticized because it does not, as the criticism goes, have any predictive power. The folk theorem basically says that anything can be an equilibrium in a repeated game. One could argue, however, that this criticism is misplaced if we regard the theory of repeated games as a theory of informal contracts. Just as anything can be enforced when the party agrees to sign a binding contract, in repeated games any (feasible and individually rational) outcome is sustained if the players agree on an equilibrium. Enforceability of a wide range of outcomes is the essential property of effective contracts, formal or informal. The folk theorem correctly captures this essential feature.

This criticism is valid, however, in the sense that the theory of repeated games does not provide a widely accepted criterion for equilibrium selection. When we regard a repeated game as an informal contract, where the players explicitly try to agree on which equilibrium to play, the problem of equilibrium selection boils down to the problem of bargaining. In such a context, it is natural to assume that an efficient point (in the set of equilibria) is played. In the vast majority of applied works of repeated games with symmetric stage games (such as the gas stations example), it is common to look at the best symmetric equilibrium. In contrast, when players try to find an equilibrium through trial and error, the theory of repeated games is rather silent about which equilibrium is likely to be selected. A large body of computer simulation literature on the evolution of cooperation, pioneered by Axelrod (1984), may be regarded as an attempt to address this issue.

Imperfect monitoring

So far we assumed that players can perfectly observe each other's actions. In reality, however, long term relationships are often plagued by *imperfect monitoring*. For example, a country may not verify exactly how much CO_2 is emitted by neighbouring countries. Workers in a joint project may not directly observe each others' effort. Electronic appliance shops often offer secret discounts for their customers, and each shop may not know exactly how much is charged by its rivals. In such situations, however, there are usually some pieces of information, or *signals*, which imperfectly reveal what actions have been taken. Published meteorological data indicates the amount of CO_2 emission, the success of the project is more likely with higher effort, and a shop's sales level is related (although not perfectly) to its rivals' prices.

According to the nature of the signals, repeated games with imperfect monitoring are classified into two categories: the case of *public monitoring*, where players commonly observe a public signal, and the case of *private monitoring*, where each player observes a signal that is not observable to others. Hence, the CO_2 emission game and the joint-project game are examples with imperfect public monitoring (published meteorological data and the success of the project are publicly observed), while the secret price-cutting game by electronic shops is a good example with imperfect private monitoring (one's sales level is private information).

This difference may appear to be a minor one, but, somewhat surprisingly, it is not. The imperfect *public* monitoring case shares many features with the *perfect* monitoring case, and we now have a good understanding of how it works. In contrast, the imperfect private monitoring case is not fully understood, and we have only some partial characterizations of equilibria. In what follows, I sketch the main results in the imperfect public and private monitoring cases.

Imperfect public monitoring

At first sight, this case might look much more complicated than the perfect monitoring case, but those two cases are similar in the sense that they share a *recursive structure*. Consider the set W^* of all average payoff profiles associated with the subgame perfect equilibria of a perfect monitoring repeated game. Any point $w \in W^*$ is a weighted average of the current payoff g and the continuation payoff $w' : (1 - \delta)g + \delta w'$. The continuation payoff typically changes when a player deviates from g, in such a way that the short-term gain from deviation is wiped out. Subgame perfection requires that all continuation payoffs are chosen from the equilibrium set W^*. In this sense, W^* is generated by itself, and this stationary or recursive structure turns out to be quite useful in characterizing the set of equilibria.

The set of equilibria in an imperfect public monitoring game also shares the same structure. Consider the equilibria where the public signal determines which continuation equilibrium to play. When a player deviates from the current equilibrium action, it affects both her current payoff and (through the public signal) her continuation payoff. The equilibrium action should be enforceable in the sense that any gain in the former should be wiped out in the latter, and this is easier when the continuation payoff admits large variations. Formally, given the range of continuation payoffs W, we can determine the set $B(W)$ of enforceable average payoffs. The larger the set W is, the more actions can be enforced in the current period (and therefore the larger the set $B(W)$ is). As in the perfect monitoring case, the equilibrium payoff set $W = W^*$ generates itself: it satisfies the *self-generation condition* of Abreu, Pearce and Stacchetti (1990) $W \subseteq B(W)$. W^* is the largest (bounded) set satisfying this condition, and the condition is in fact satisfied with equality. Conversely, it is easy to show that any (bounded) set satisfying the self-generation condition is contained in the equilibrium payoff set W^*.

This provides a simple and powerful characterization of equilibria, which is an essential tool to prove the folk theorem in the imperfect public monitoring case. The folk theorem shows that, despite the imperfection of monitoring, we can achieve any feasible and individually rational payoff profile under a certain set of conditions.

Before presenting a formal statement, let me sketch the basic ideas behind the folk theorem. When monitoring is imperfect, players have to be punished when a 'bad' signal outcome ω is observed, and this may happen with a positive probability even if no one defects. For example, in the joint project game, the project may fail even though everyone works hard. A crucial difference between the perfect and imperfect monitoring cases is that, in the latter, punishment occurs *on the equilibrium path*. The resulting welfare loss, however, can be negligible under certain conditions.

Consider a two-player game, where the probability distribution of the signal $\omega \in \Omega = \{\omega^1, \ldots, \omega^K\}$, when no one defects, is given by $P^* = (p^*(\omega^1), \ldots, p^*(\omega^K))$ in Figure 1. Suppose that each player's defection changes the probability distribution to exactly the same point P'. Then, there is absolutely no way to tell which player deviates, so that the only way to deter a defection is to *punish all players simultaneously*, when a 'bad' outcome emerges. This means that surplus is thrown away, and we are bound to have substantial welfare loss. Now consider a case where different players' actions affect the signal asymmetrically: player 1's defection leads to point P', while the defection by player 2 leads to P''. In this asymmetric case, one can *transfer* future payoff from player 1 to 2 when player 1's defection is suspected. Under such a transfer, surplus is never thrown away, and this enables us to achieve efficiency.

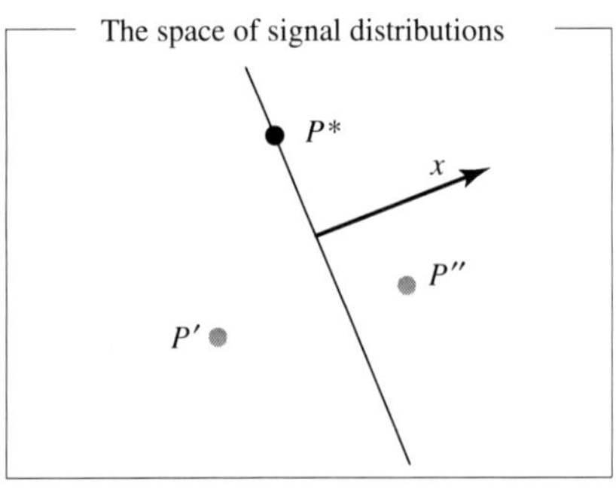

Figure 1

More precisely, consider the normal vector x of the hyperplane separating P' and P'' in the figure, and let $w_1 = x$ and $w_2 = -x$ be the continuation payoffs of player 1 and player 2 respectively. Figure 1 indicates that player 1's expected continuation payoff $P \cdot w_1 = P \cdot x$ is reduced by her own defection ($P' \cdot x < P^* \cdot x$). Similarly, player 2's defection reduces her expected continuation payoffs ($P^* \cdot (-x) > P''(-x)$). Note that this asymmetric punishment scheme does not reduce the joint payoff, because by construction $w_1 + w_2$ is identically equal to 0. This is an essential idea behind the folk theorem under imperfect public monitoring: *When different players' deviations are statistically distinguished, asymmetric punishment deters defections without welfare loss.*

When can we say that different players' deviations are statistically *distinguished*? Note well that the above construction is impossible when P'' is exactly in between P^* and P' (that is, when P'' is a convex combination of P^* and P'). Such a case can be avoided if P^*, P' and P'' are linearly independent. The linear independence of the equilibrium signal distribution (P^*) and the distributions associated with the players' unilateral deviations (P' and P''), is a precise formulation of what it means when the signal 'statistically distinguishes different players' deviations'.

Let us now generalize this observation. Given an action profile (for simplicity of exposition, assume it is pure) to be sustained, there is an associated signal distribution P^*. Consider any pair of players i and j, and let $|A_k|$ be the number of player k's actions ($k = i, j$) in the stage game. Since each player $k = i, j$ has $|A_k| - 1$ ways to deviate, we have $|A_i| + |A_j| - 2$ signal distributions associated with their unilateral deviations. If those distributions and the equilibrium distribution P^*, altogether $|A_i| + |A_j| - 1$ vectors, are linearly independent, we say that the signal can discriminate between deviations by i and deviations by j. This is called the *pairwise full rank condition*. This holds only when the dimension of the signal space ($|\Omega|$, the number of signal outcomes) is larger than the number of those vectors (that is, $|\Omega| \geq |A_i| + |A_j| - 1$). Conversely, if this inequality is satisfied, the pairwise full rank condition holds 'generically' (that is, it holds unless the signal distributions have a very special structure, such as exact symmetry). This leads us to the folk theorem under imperfect public monitoring (this is a restatement of Fudenberg, Levine and Maskin, 1994, in terms of genericity):

Folk theorem under imperfect public monitoring

Suppose that the signal space is large enough in the sense that $|\Omega| \geq |A_i| + |A_j| - 1$ holds for each pair of players i and j. Then, for a generic choice of the signal distributions and the stage game, any feasible and individually rational payoff profile $v \in V^*$ can be asymptotically achieved by a sequential equilibrium as the discount factor δ tends to 1.

In contrast to the perfect monitoring case, the proof is non-constructive. Rather than explicitly constructing equilibrium strategies, the theorem is proved by showing that any smooth subset of V^* is self-generating. In fact, the exact structure of the equilibrium strategy profile to sustain, for example, an efficient point is not so well understood. Sannikov (2005) shows that detailed structure of equilibrium strategies can be obtained if the model is formulated in continuous time.

Imperfect private monitoring

Now consider the case where all players receive a private signal about their opponents' actions. Although this has a number of important applications (a leading example is the secret price cutting model), this part of research is still in its infancy. Hence, rather than just summarizing definitive results as in the previous subsections, I explain in somewhat more technical detail the source of difficulties and the nature of existing approaches.

The difficulties come from a subtle but crucial difference from the perfect or public monitoring case. I explain below the difference from three viewpoints, in the increasing order of technicality.

1. In the perfect or public monitoring case, players share a mutual understanding about when and whom to punish. They can cooperate to implement a specific punishment, and, more importantly, they can mutually provide the incentives to carry out the punishment. This convenient feature is lost when players have diverse private information about each others' actions.
2. In the perfect or public monitoring case, public information directly tells the opponents' future action plans. In the private monitoring case, however, each player has to draw statistical inferences about the history of the opponents' private signals to estimate what they are going to do. The inferences quickly become complicated over time, even if players adopt relatively simple strategies.
3. In the perfect or public monitoring case, the set of equilibria has a recursive structure, in the sense that a Nash equilibrium of the repeated game is always played after any history. Now consider a Nash equilibrium of, for example, the repeated Prisoner's Dilemma with imperfect private monitoring. After the equilibrium actions in the first period, say (C, C), players condition their action plans on their private signals ω_1 and ω_2. Hence the continuation play is a *correlated equilibrium*, where it is common knowledge that the probability distribution of the correlation device (ω_1, ω_2) is given by $p(\omega_1, \omega_2|C, C)$. When player 1 deviates to D in the first period, however, the distribution of correlation device is *not* common knowledge: player 1 knows that it is $p(\omega_1, \omega_2|D, C)$, while player 2 keeps the equilibrium expectation $p(\omega_1, \omega_2|C, C)$. Hence, after a deviation, the continuation play is no longer a correlated equilibrium in the usual sense. In addition, the space of the correlation device (the history of private signals) becomes increasingly rich over time. Therefore, the equilibria in the private monitoring case do not have a compact recursive structure; a continuation play is chosen from a different set, depending on the history.

One way to get around these problems is to allow communication (Compte, 1998; Kandori and Matsushima, 1998). In their equilibrium, players truthfully communicate their private signal outcomes in each period. The equilibrium is constructed in such a way that each player's report of her signal is utilized to discipline *other* players and does *not affect one's own continuation payoff*. This implies that each player is indifferent about what to report, and therefore truth telling is *a* best reply. Such an equilibrium, which depends on the history of publicly observable messages, works in much the same way as the equilibria in the public monitoring case. Hence, with communication, the folk theorem is obtained in the private monitoring case.

The remaining issue is to characterize the equilibria in the private monitoring case without communication. From the viewpoint of potential applications, this is important, because collusion or cartel enforcement is a major applied area of repeated games, where communication is explicitly prohibited by the antitrust law.

One may expect that, when players' private information admits sufficient positive correlation, an equilibrium can be constructed in a similar way to the public monitoring case. Sekiguchi (1997) is the first to construct a non-trivial (and nearly efficient) equilibrium in the private monitoring game without communication, and his construction is basically built on such an idea. Strong correlation of private information is, however, not assumed in his model but is derived endogenously. He assumes that private signals provide nearly perfect observability and considered *mixed* strategies. In such a situation, the privately observed random variables, the action-signal pairs, are strongly correlated (because a player's random action is strongly correlated with another player's signal under nearly perfect observability). Mailath and Morris (2002) show that, in general, there is 'continuity' between the public and private but sufficiently correlated monitoring cases, in the sense that any strategy with a *finite memory* works in either case.

Those papers are examples of the *belief-based approach*, which directly addresses the statistical inference problem (see point 2. above). Some other papers follow this approach, and they provide judiciously constructed strategies in rather specific examples, where the inference problem becomes tractable. Aside from the case with near perfect correlation, however, we are yet to have generally applicable results or techniques from this approach.

More successful has been the *belief-free approach*, where an equilibrium is constructed in such a way that the inference problem becomes *irrelevant*. As a leading example, here I explain Ely and Valimaki's work (2002) on the repeated Prisoner's Dilemma with imperfect private monitoring. Each player's strategy is a Markov chain with two states, R (reward) and P (punishment). A specific action is played in each state (C in R, and D in P), and the transition probabilities between the states depend on the realization of the player's private signal. Choose those transition probabilities in such a way that the *opponent* is always indifferent between C and D *no matter which state the player is in*. This requirement can be expressed as a simple system of dynamic programming equations, which has a solution when the discount factor is close to 1 and the private signal is not too uninformative. By construction, any action choice is optimal against this strategy after any history, and in particular this strategy is a best reply to itself (so that it constitutes an equilibrium). Note that one's incentives do not depend on the opponent's state, and therefore one does not have to draw the statistical inferences about the history of the opponent's private signals.

There are certain difficulties, however, in obtaining the folk theorem with such a class of equilibria. First, players may be punished simultaneously in this construction, and our discussion about the public monitoring case shows that some welfare loss is inevitable (unless monitoring is nearly perfect). Second, even if we restrict our attention to the nearly perfect monitoring case, there is a certain set of restrictions imposed on the action profiles that can be sustained by such a belief-free equilibrium.

Those difficulties can be resolved when we consider *block strategies*. Block strategies treat the stage games in T consecutive periods as if they were a single stage game, or a block stage game, and applies the belief-free approach with respect to those block stage games. It is now known that, by using the block strategies, the folk theorem under private monitoring holds in the nearly perfect monitoring case (Hörner and Olszewski, 2006) and for some two-player games where monitoring is far from perfect (Matsushima, 2004). In the former, the block structure of the stage game helps to satisfy the restrictions imposed on the sustainable actions in belief-free equilibria. In the latter, an equilibrium is constructed where players choose constant actions in each block. This means that players have T samples of private signals for the constant actions, so that the observability practically becomes nearly perfect when T is large. With this increased observability and some restrictions on payoff functions, the folk theorem is obtained. For this construction to be feasible, the signals have to satisfy certain strong conditions, such as independence (across players).

The general folk theorem, or a general characterization of equilibria, for the private monitoring case is yet to be obtained, and it remains an important open question in economic theory. A comprehensive technical exposition of the perfect monitoring, imperfect public monitoring, and private monitoring cases can be found in Mailath and Samuelson (2006).

KANDORI MICHIHIRO

See also **cartels; cooperation; reputation; social norms.**

Bibliography

Abreu, D., Dutta, P. and Smith, L. 1994. The folk theorem for repeated games: a NEU condition. *Econometrica* 62, 939–48.

Abreu, D., Pearce, D. and Stacchetti, E. 1990. Towards a theory of discounted repeated games with imperfect monitoring. *Econometrica* 58, 1041–64.

Aumann, R. 1959. Acceptable points in general cooperative N-person games. In *Contributions to the Theory of Games*, vol. 4, ed. R.D. Luce and A.W. Tucker. Princeton: Princeton University Press.

Axelrod, R. 1984. *Evolution of Cooperation*. New York: Basic Books.

Benoit, J.P. and Krishna, V. 1985. Finitely repeated games. *Econometrica* 53, 905–22.

Coase, R. 1937. The nature of the firm. *Economica* n.s. 4, 386–405.

Compte, O. 1998. Communication in repeated games with imperfect private monitoring. *Econometrica* 66, 597–626.

Ely, J. and Valimaki, J. 2002. A robust folk theorem for the Prisoner's Dilemma. *Journal of Economic Theory* 102, 84–105.
Friedman, J. 1971. A non-cooperative equilibrium for supergames. *Review of Economic Studies* 38, 1–12.
Fudenberg, D., Levine, D. and Maskin, E. 1994. The folk theorem with imperfect public information. *Econometrica* 62, 997–1040.
Fudenberg, D. and Maskin, E. 1986. The folk theorem in repeated games with discounting or with incomplete information. *Econometrica* 54, 533–54.
Hörner, J. and Olszewski, W. 2006. The folk theorem for games with private almost-perfect monitoring. *Econometrica* 74, 1499–544.
Kandori, M. and Matsushima, H. 1998. Private observation, communication and collusion. *Econometrica* 66, 627–52.
Macaulay, S. 1963. Non-contractual relations in business: a preliminary study. *American Sociological Review* 28, 55–67.
Mailath, G. and Morris, S. 2002. Repeated games with imperfect private monitoring: notes on a coordination perspective. *Journal of Economic Theory* 102, 189–228.
Mailath, G. and Samuelson, L. 2006. *Repeated Games and Reputations: Long-Run Relationships.* Oxford: Oxford University Press.
Matsushima, H. 2004. Repeated games with private monitoring: two players, *Econometrica* 72, 823–52.
Pearce, D. 1990. Repeated games: cooperation and rationality. In *Advances in Economic Theory*, ed. J. Laffont. Cambridge: Cambridge University Press.
Rubinstein, A. 1979. Equilibrium in supergames with overtaking criterion. *Journal of Economic Theory* 21, 1–9.
Sannikov, Y. 2005. Games with imperfectly observable actions in continuous time. Mimeo, University of California, Berkeley.
Sekiguchi, T. 1997. Efficiency in repeated Prisoner's Dilemma with private monitoring. *Journal of Economic Theory* 76, 345–61.

reputation

In a dynamic setting signals sent now may affect the current and future behaviour of other players; thus, signals can have effects unrelated to their current costs and benefits. It is the interplay between signals and their long-run consequences that is studied in the literature on reputation.

The literature on reputation has two main themes. The first is that introducing a small amount of incomplete information in a dynamic game can dramatically change the set of equilibrium payoffs: introducing something to signal can have big implications in a dynamic model. These kinds of result can also be interpreted as providing a robustness check. Dynamic and repeated games typically have many equilibria, and reputation results allow us to determine which equilibria continue to be played when a game is 'close' to complete information. The second theme of the literature on reputations is that introducing incomplete information in a dynamic game may introduce new and important signalling dynamics in the players' strategies. Thus reputation effects tell us something about behaviour. This theme is particularly important in applications to macroeconomics and to industrial organization, for example. For either of these themes to be relevant it is necessary to have a dynamic game with incomplete information, so work on reputation has been influenced by, and influences, the larger literature on repeated and dynamic games of incomplete information. An excellent detailed treatment of reputation can be found in Mailath and Samuelson (2006).

An example

Most of the results below will be described in the context of a simple infinitely repeated trading game. The row player is a seller who can produce high or low quality. The column player is a buyer. Producing high quality is always expensive for the seller, so she would rather produce low quality; the buyer, however, wants to buy only a high-quality product. The only non-standard element is that the buyer regrets not buying a high-quality product. The trading game (Figure 1) has a unique equilibrium (L, N).

Let us record some facts about this game. The set

$$V \equiv \{(x, y) : x > 0, y > -1/3, y \leq x \text{ and } y \leq 3 - 2x\} \subset \mathbb{R}^2,$$

	Buy (B)	Not Buy (N)
High Quality (H)	(1, 1)	(−1, −1)
Low Quality (L)	(2, −1)	(0, 0)

Figure 1 A trading game

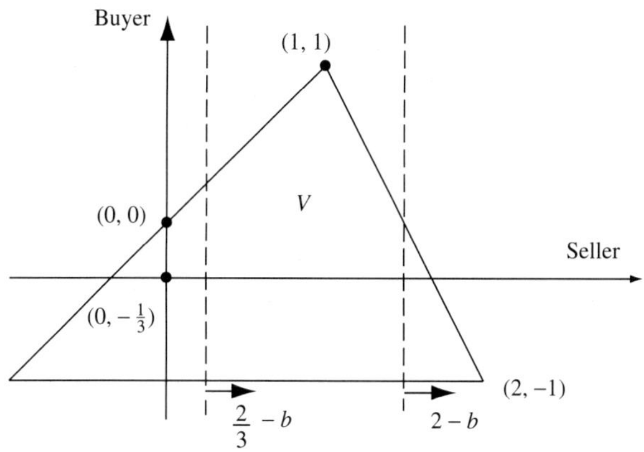

Figure 2 Sets of equilibrium payoffs and reputation bounds

illustrated in Figure 2, is the set of feasible and strictly individually rational payoffs for the trading game. The axes are drawn through the minmax payoffs to make V clear. If the seller could commit to a pure strategy, she would prefer to choose H as the buyer's best response to this is B. However, she could do even better by committing to a mixed strategy; playing (3/4, 1/4) for example would also ensure the buyer played B and give the seller a bigger payoff. Reputation arguments can provide ways for these commitment payoffs to be achieved by sellers who are not actually committed to anything.

The trading game is played in each of the periods $t=1,2,\ldots$ with perfect monitoring; at the end of the period the players get to observe all payoffs and the pure action taken by their opponent. If both players' discount factors, $\delta<1$, were sufficiently large, any point in V could be sustained as an equilibrium payoff. If the seller is long lived but faces an infinite sequence of buyers who each live one period, then any point on the line segment joining (0, 0) to (1, 1) is an equilibrium payoff. (No seller payoff above 1 is achievable if mixed actions are not observable; see Fudenberg, Kreps and Maskin, 1990.)

The stage is now set. To understand how reputation works we will need to introduce something for the seller to signal. Its commitment to high quality? Its low cost of high quality? Its commitment to always ripping off customers...? At this stage it is unnecessary to be specific, and we will concentrate on the general issues of learning. There are two types of sellers, 'strong' and 'normal', that the buyer may face in a game. The seller is told their type by nature at time $t=0$. The buyer, however, is unaware of nature's selection and spends the rest of the game looking at the seller's behaviour and trying to figure out what type she is. The normal seller plays action $a \in \{H,L\}$ with probability $\tilde{\sigma}^t(a)$ at time t, and the strong seller plays a with probability $\hat{\sigma}^t(a)$ at time t. Everything we say in the section below applies to the case where normal and strong sellers follow history-dependent strategies. (These behaviour strategies do depend on the – public – history of play before time t, but let us keep this out of our notation.) An initially uninformed buyer attaches probability p^t to the strong type and $1-p^t$ to the normal type at time t; again this depends on the observed history. Our buyer expects the seller to play a with probability $\sigma^t(a) = p^t\tilde{\sigma}^t(a) + (1-p^t)\hat{\sigma}^t(a)$, and as time passes the buyers observe the outcomes of this strategy and revise their prior accordingly.

Tricks with Bayes's rule and martingales

Now we generate three properties of learning that are extensively used in the reputations literature. We will call them the 'merging' property, the 'right ballpark' property and the 'finite surprises' property. These properties are based on some simple facts about how Bayesian agents revise their beliefs, that is, how uncertainty about the seller's type is processed by the buyers or any other observer of its behaviour. A more advanced treatment of these results can be found in Sorin (1999). We defer any derivation of reputation results to the next section, so a reader could skip this section.

How does the buyer revise his or her beliefs in the light of an observed action a^t? A plain application of Bayes' rule tells us

$$p^{t+1} = \frac{Pr(a^t \cap \text{Strong})}{Pr(a^t)} = \frac{p^t\hat{\sigma}^t(a^t)}{\sigma^t(a^t)}.$$

Or, in terms of the change in the beliefs

$$p^{t+1} - p^t = \frac{p^t[\hat{\sigma}^t(a^t) - \sigma^t(a^t)]}{\sigma^t(a^t)} = \frac{p^t(1-p^t)[\hat{\sigma}^t(a^t) - \tilde{\sigma}^t(a^t)]}{\sigma^t(a^t)}.$$

These equalities are powerful tools when combined with the properties of the priors.

Merging property. This tells us exactly how the long-run behaviour of the sellers is related to the buyer's long run beliefs. Either $p^t(1-p^t) \to 0$ and the buyer eventually learns the type of the seller and can perfectly predict their actions, or all types of the seller end up behaving in the same way $\hat{\sigma}^t(a^t) - \tilde{\sigma}^t(a^t) \to 0$ and again the buyer can perfectly predict their actions. Nothing else can happen!

The stochastic process $\{p^t\}$ is a martingale on $[0, 1]$ with respect to public histories. To see this there is a simple calculation we can do.

$$E(p^{t+1}|h_t) = \sum_{a^t} Pr(a^t)p^{t+1} = \sum_{a^t} \sigma^t(a^t)\frac{p^t\hat{\sigma}^t(a^t)}{\sigma^t(a^t)} = p^t.$$

(The expectation $E(\cdot)$ is taken with respect to the buyer's beliefs about future play.) Bounded martingales converge almost surely (see Williams, 1991, for example), which implies $|p^{t+1} - p^t| \to 0$ almost surely. Applying this to the second equality above (noting that $|\sigma^t(a^t)| \leq 1$), we get

$$p^t(1-p^t)|\hat{\sigma}^t(a^t) - \tilde{\sigma}^t(a^t)| \to 0, \text{ (Merging)}$$

almost surely. This kind of result is extensively used in Hart (1985) and the literature that stems from his work.

Right ballpark property. The strong seller knows that the future will evolve according to the strategy $\hat{\sigma}$ (we use $\hat{Pr}(\cdot)$ and $\hat{E}(\cdot)$ to denote her probability measure and its expectation). This seller might ask, as she plays out an equilibrium, how little probability the buyers can attach to the strong seller, or how low p^t could get when she plays $\hat{\sigma}$. Of course, when the seller is in fact the strong type it is very unlikely that p^t becomes low – beliefs must stay in the right ballpark. (For example, if $\hat{\sigma}$ was actually a pure strategy the strong seller cannot ever believe p^t will decrease. As she plays $\hat{\sigma}$ there will be periods in which the normal type of seller could have done something different, so observing the actions of $\hat{\sigma}$ will cause buyers to revise p^t upwards.)

From the perspective of the strong seller, the likelihood ratio is a martingale:

$$\hat{E}\left(\frac{1-p^{t+1}}{p^{t+1}}|h_t\right) = \frac{1-p^t}{p^t}.$$

(The calculation is just like the earlier one for p^t, where we use $\hat{Pr}(a^t) = \hat{\sigma}(a^t)$.) Let τ be the first time, s say, that $p_s \leq \nu$ and let C^t be the event that $\tau \leq t$. That is, sometime in the first t periods $p_s < \nu$. Then the martingale property combined with the optional stopping theorem (for example, Williams, 1991) implies

$$\frac{1-p^0}{p^0} = \hat{E}\left(\frac{1-p^{t+1}}{p^{t+1}}\right) \geq \hat{Pr}(C^t)E\left(\frac{1-p^{\tau}}{p^{\tau}}|C^t\right) \geq \hat{Pr}(C^t)\frac{1-\nu}{\nu}.$$

The above gives an upper bound on $\hat{Pr}(C^t)$ that is independent of t. Thus it also bounds the probability that p_t is ever below ν:

$$\hat{Pr}(\exists t \text{ s.t. } p_t < \nu) \leq \frac{\nu}{p^0}. \quad \text{(Right Ballpark)}$$

Hence, the strong seller knows that it is very unlikely that the buyer's posterior will ever be close to certain she is actually the normal seller.

Finite surprises property. The strong seller might also ask how many times (as she plays $\hat{\sigma}$) the uninformed buyers will make a big mistake in predicting her strategy, that is, how many periods does $||\hat{\sigma}^t - \sigma^t|| > \nu$ occur when the seller actually plays $\hat{\sigma}$. Here we are helped by the fact that our seller has only two actions, so the variation distance between the mixed actions is just twice the difference in probability of the realized action $||\hat{\sigma}^t - \sigma^t|| = 2|\hat{\sigma}^t(a^t) - \sigma^t(a^t)|$. Let M_N be the event that there are more than N *mistakes*, $||\hat{\sigma}^t - \sigma^t|| > \nu$, before time T. The finite surprises property is that independently of the equilibrium $\hat{Pr}(M_N) \to 0$ as $T, N \to \infty$. Thus, it is very unlikely that there are many periods in which the buyers do not think the seller will play as the strong type if the seller is indeed this type.

Jensen's inequality applied to the likelihood ratio above implies that the prior is a submartingale, that is, $\hat{E}(p^{t+1}|h_t) \geq p^t$. There is a second property of martingales we can now use: they cannot move around very much: $\sum_{t=1}^{T}\hat{E}((p^{t+1} - p^t)^2) \leq 1$.

(A proof of this fact follows from $\hat{E}(p^{t+1} - p^t)^2 \leq \hat{E}((p^{t+1})^2 - (p^t)^2)$.) A substitution from the first Bayes' rule equality above then tells us

$$1 \geq \sum_{t=1}^{T} \hat{E}((p^t[\hat{\sigma}^t(a^t) - \sigma^t(a^t)])^2).$$

It is obvious that only a few of the (non-negative) terms in the sum above can be much above zero, otherwise the upper bound will be violated. The right ballpark property tells us it is very unlikely that $p^t < \nu$. On the event $\{p^t \geq \nu \forall t\} \cap M_N$, the p^t in the above expectation is greater than ν and there are at least N differences that are bigger than $\nu/2$, so the sum is at least $N\nu(\nu/2)^2$, hence

$$1 \geq \sum_{t=1}^{T} \hat{E}(p^t[\hat{\sigma}^t(a^t) - \sigma^t(a^t)])^2 \geq \hat{Pr}(\{p^t \geq \nu \forall t\} \cap M_N)\frac{N\nu^3}{4}.$$

Using the fact that $Pr(A \cap B) \geq Pr(B) - Pr(A^c)$ we now have an upper bound on $\hat{Pr}(M_N)$.

$$\frac{4}{N\nu^3} + \hat{Pr}(\exists t \text{ s.t. } p_t < \nu) \geq \hat{Pr}(M_N).$$

The right ballpark property gives us

$$\hat{Pr}(M_N) \leq \frac{\nu}{p^0} + \frac{4}{N\nu^3}. \quad \text{(Finite Surprises)}$$

As the size of the surprises becomes small $\nu \to 0$ and the number of surprises becomes large $N\nu^3 \to \infty$, the strong seller must attach smaller and smaller probability to M_N. Fudenberg and Levine (1989; 1992), for example, invoke this property.

Basic reputation results: behaviour

The three tools above are sufficient to establish most well-known reputation results. The arguments below are entirely general, and are widely applied, but we use them only in the trading game. To make things simple, suppose that for some reason the strong seller is committed to playing $(b, 1-b)$, that is, in every period t the strong buyer provides high quality with probability b. We reserve the discussion of more complicated types of reputations for a later section.

From the perspective of the buyer, any equilibrium will consist of two phases: an initial phase when there is learning and signalling about the seller's type (this is sometimes called reputation building, although often reputation destruction is what occurs), and a terminal phase when the learning has virtually settled down. It is the merging property that tells us there must be this latter phase. The play in the game moves into this second phase either because the buyer is almost sure he knows the type of the seller (reputation considerations have vanished) or because the sellers are playing in the same way. Thus the equilibria of dynamic signalling games are inherently non-stationary, which is in contrast to much of the work on repeated games. Of course,

Markovian equilibria can be calculated but these too will exhibit the two phases of play. The initial learning, when reputation builds or is destroyed, depends on the particular equilibrium and the game being studied. This phase may last only one period (if a once and for all revealing action is taken by a seller) but frequently it is long and has a random duration (if both types of seller randomize, for example).

Let us first examine reputation destruction in the case where $b \approx 1$, so the strong seller is committed to high quality and only very occasionally slips up. There is an equilibrium of this game where the normal type of seller will offer low quality more often than the strong type, and thereby gradually reveal her type (destroy her reputation for being good). Nevertheless, as this occurs she will enjoy heightened payoffs. The trade-offs our normal seller experiences in this game are what drive the reputation destruction. A seller offering low quality today enjoys the benefit of a higher payoff now, but the observation of low quality typically leads the buyers to revise downwards their probability of the strong seller and buy less in the future, whereas a seller offering high quality will lead the buyer's posterior on the strong seller to be revised upwards and an increased likelihood of buying in the future. Exactly how the normal seller chooses to trade off long-run benefits and short-run costs is unclear. It is possible that pooling dominates and that future buying is so strong that the normal seller prefers to offer high quality today even if it costs something in the short run. However, in this equilibrium the normal seller perceives the long-run benefits to be relatively small and prefers to offer low quality today. The normal seller can be thought of as exploiting, or cashing in, the value of her accumulated reputation. We also know, from the finite surprises property, that there will be finite opportunities for the normal seller to do this. Relatively soon there will come a time where the buyers know the seller is normal and purchase accordingly.

Reputation building (as opposed to destruction) is more likely in a world where there is the possibility that one is thought to be bad, for example, if the strong type is committed to ripping customers off and only occasionally produces a good product ($b \approx 0$). In such a world the normal seller wants to tell buyers she is not this type, because by playing as the strong type she is doomed to never trade. She is building a reputation for *not* being the strong type. To do this the normal type will have to incur the cost of repeatedly offering high quality, even if the buyer is not buying. This is expensive and will drag down the normal seller's equilibrium payoff. But, as above, it will increase the likelihood of future buying by decreasing the likelihood of a strong seller. In contrast to the reputation destruction case, there are short-run costs borne by the normal type to achieve long-run gains. Again, the nature of these costs and benefits rely on the buyers' uncertainty about the seller's type.

Basic reputation results: payoffs

Reputation issues can have an extreme effect on payoffs, and this is what first came to the attention of economists. The general question of how the presence of something to signal in the repeated game affects the equilibrium payoffs could be answered in a number of ways. One way would be to calculate equilibria explicitly. This is usually difficult and would not establish results that hold for *all* equilibria.

Instead, a different approach is taken that is described in the following recipe:

1. If the seller is strong, then in finite time the buyers will believe they face a seller who plays arbitrarily close to $(b, 1-b)$ for ever.
2. Figure out what the buyers will do when the seller is strong.
3. Use step 2 to evaluate the normal seller's payoff if she pretends to be strong for ever.
4. At a Nash equilibrium the answer to step 3 is a lower bound on the normal seller's equilibrium payoff.

Step 1 is independent of the model and is a result of our earlier calculations. The right ballpark property tells us that p^t does not tend to zero when the seller is strong. The merging property then implies either $p^t \to 1$, or eventually all remaining normal types of buyer are also playing arbitrarily close to $(b, 1-b)$. In either of these cases, at a large but finite time the buyers believe that they face a seller who will always play $(b, 1-b)$.

Before proceeding to apply this recipe, we illustrate its power with the remarkable results we expect to get. Let us first consider a world where buyers are short run. We will show that introducing an arbitrarily small probability that there is a strong seller places a lower bound on the normal seller's equilibrium payoffs of $2-b$ (when $b>1/3$). Thus for b close to 1/3 the equilibrium payoffs in the complete information game (the segment joining (0, 0) and (1, 1)) and the incomplete information game are disjoint! Moreover, the normal seller can get almost his maximum feasible payoff at every equilibrium. In the second case, where buyers are also long run, we will get less strong conclusions; nevertheless, we will show that the normal type of seller must get at least $2/3 - b$ when $b>1/3$. These payoffs are illustrated in Figure 2.

The really difficult part of our recipe is step 2, because we have to understand how the buyers will behave in equilibrium. We therefore need to consider as separate cases what happens if buyers are short run or long run. Also, the amount of discounting that the sellers do affects the answer to step 3, so we need to consider different arguments for different amounts of discounting. The following catalog moves from simple to more elaborate arguments and from stronger to weaker reputation effects.

Reputation without discounting: short-term buyers

When a buyer lives only one period he plays a best response to the seller's current action. By step 1 in the very long run this will be B if $b>1/3$ and N if $b>1/3$. Step 3 is simple; by playing $\hat{\sigma}$ for ever the normal seller knows that in a large but finite time she can ensure the buyer will behave as above and so she will receive a stage game payoff approximately $R^*(b)$, where

$$R^*(b) := \begin{cases} 2-b & b>1/3, \\ -b & b<1/3. \end{cases}$$

If there is no discounting, and limits are correctly taken, $R^*(b)$ will equal the normal type's payoff from playing $\hat{\sigma}$ for ever. Thus, at step 4, at any Nash equilibrium the normal type must get at least $R^*(b)$.

In a general game R^* is equal to the seller's payoff from playing the strong type's stage game strategy when the buyer plays his or her unique best response. (If the best response is not unique this is not correct.)

Reputation with discounting: short-term buyers

Step 2 is as above – we still have short-term buyers. When the normal seller discounts payoffs, however, playing $\hat{\sigma}$ and eventually getting $R^*(b)$ every period does not tell us what her payoff discounted to time zero will be. There is an order of limits issue; as the discounting of the seller becomes weaker ($\delta \to 1$), it could be that the equilibria change and there are more and more periods where the seller is not getting $R^*(b)$. It is now that the finite surprises property plays an important role. First notice that, when v is chosen appropriately and $||\sigma - \hat{\sigma}|| < v$, then playing a best response to σ is the same as playing a best response to $\hat{\sigma}$. Hence, it is only when a surprise occurs that the normal seller is not getting $R^*(b)$ from playing $\hat{\sigma}$. But the probability of more than N surprises can be made very small *independently of the discounting.* So, as the discounting becomes weak and N periods have a small effect on total discounted payoff, there is a small probability of the normal seller of getting anything less than $R^*(b)$ when she plays $\hat{\sigma}$. Any Nash equilibrium, therefore, gives the normal seller at least $R^*(b)$. This is the kind of argument first made in specific cases by Kreps and Wilson (1982) and Milgrom and Roberts (1982), and generalized in Fudenberg and Levine (1989; 1992).

Reputation without discounting: one long-run buyer

If the buyer lives for many periods, he will not necessarily play a short-run best response to $(b, 1-b)$ even if he expects it to be played for ever. We can, however, use some weaker information. At an equilibrium the buyer must on average get at least $-1/3$ (his minmax payoff) against $(b, 1-b)$. This implies that the buyer has to buy with at least probability 1/3 when $b > 1/3$ and buy with at most probability 1/3 when $b < 1/3$. There are, consequently, some bounds on the normal seller's payoff when she has played $\hat{\sigma}$ for a sufficiently long time. While playing $(b, 1-b)$ she gets $2-b$ when the buyer buys and $-b$ if not; thus, if the buyer buys with probability greater than 1/3, she expects to receive a payoff of at least $2/3-b$. If the buyer buys with at most probability 1/3, she expects to get at least $-b$. The seller is not discounting, so what she gets in the long run from playing $\hat{\sigma}$ is also what she expects to get at time zero. Our answer to step 3, therefore, is

$$R^{\dagger}(b) := \begin{cases} (2/3) - b & b > 1/3, \\ -b & b < 1/3; \end{cases}$$

and we have a weaker lower bound on the normal type's payoff.

In an arbitrary game $R^{\dagger}$ is equal to the seller's worst payoff from playing as the strong type when the buyer plays a response that gives him more than his minmax payoff. In certain cases this can be a very strong restriction – for example, if the seller has a pure strategy that minmaxed the buyer and there is a unique response for the buyer that ensured he received his minmax payoff. Certain games, known as games of conflicting interests, have the property that the best action for the seller to commit to is pure and minmaxes the buyer. $R^{\dagger}$ is a very tight bound for such games.

Reputation with discounting: one long-run buyer

This final case combines most of the above issues. If the seller discounts the future much less than the buyer, then in the long run the seller must get $R^{\dagger}(b)$ from playing $\hat{\sigma}$. If a normal seller pretends to be strong, the buyers think there are at most N periods when the strong strategy is not played. Imagine now we have a buyer who cares only about what happens in the next t' periods. Such a buyer can think there are at most $t'N$ periods in which $\hat{\sigma}$ is not played for the next t' periods. (This kind of argument is due to Schmidt, 1993.) As the seller becomes very patient Nt' periods become of vanishing importance and the normal seller's payoff is bounded below by $R^{\dagger}(b)$. If the seller and buyer discount equally, however, reputation effects cannot be found except in some very special cases.

Imperfect monitoring: temporary and bad reputation

The analysis of reputation given above presupposes perfect monitoring by the buyers and sellers of each others' actions. In many dynamic and repeated games this is not likely. To what extent do the above results continue to hold when the players are not able to see exactly what their opponent did in any one period? Perhaps reputations are harder to establish if the observed behaviour is noisy? On the other hand, perhaps deviations from the strong type's action are harder to detect and so reputations last longer and are more valuable.

The merging, right ballpark and finite surprises properties all hold true under imperfect monitoring, with a suitable redefinition, provided there is enough statistical information for the buyer to eventually identify the seller's behaviour. (This is a full-rank condition on the players' signals.) As a result, the bounds on payoffs given in the previous section continue to hold.

Under imperfect monitoring with adequate statistical information there is one new behavioural feature of these games – reputation is almost always temporary, that is, the buyer will eventually get to know the seller's type. To see why this is so, let us amend the game in Figure 1 by restricting the buyer to imperfectly observe the seller's action. With probability $1-\varepsilon$ the buyer observes the seller's true action in the current period, but with probability $-\varepsilon$ he observes the reverse action. (We must also assume the buyer does not see his own payoffs, otherwise he can deduce the seller's action from his payoff.) Consider a game where the seller always provides high quality ($b=0$) and suppose that reputation is permanent in such a game. Then p would, at least some of the time, converge to a number that is not zero or one. (Remember beliefs have to converge.) The merging property tells us that, in this case where the limit of beliefs is between zero and one, the buyer will be certain the normal seller is always providing high quality. Such buyers will ignore the occasional low-quality product as just unlucky outcomes, and there will be no loss of seller reputation if the buyer ever receives low quality. The normal type of seller can, therefore, deviate from always providing high quality, gain one unit of profit, and not face any costs in terms of loss of reputation. This cannot be an equilibrium. The initial claim that reputation is permanent has to be false as a result of this contradiction. The details of this argument can be found in Cripps, Mailath and Samuelson (2004).

When the monitoring is not statistically informative, 'bad reputation', due to Ely and Valimaki (2003), is a possibility. Uninformative monitoring is a particular problem in repeated extensive form games, because players do not get to see the actions their opponent would have taken on other branches of the game tree. Bad reputation may arise in our example if the buyer could take an action (such as not to buy) that stopped the seller being able to signal her type. Then, the normal seller might find herself permanently stuck in a situation where she cannot sell. This is not particularly surprising if the buyers were strongly convinced they faced a strong seller that almost always provided low quality. However, in certain circumstances this problem is much more severe: even if the buyers were almost certain the seller were normal, every equilibrium has trade ending in a bounded and finite time. Thus, it is possible that introducing something for the seller to signal has huge negative costs for her equilibrium payoffs. To illustrate this, suppose the seller were a restaurant with imperfect control over quality, although it does have a strategy (for example, doubling the butter and salt content!) that makes it more likely the buyer will think the meal he received is good – but is actually damaging to the buyer. When play has reached the position where just one more bad meal will lead the buyer to permanently avoid the restaurant, then the restaurant will choose to use this unhealthy strategy. Knowing this, the buyer will choose to go elsewhere for his last but one meal too, and there is an unravelling of the putative equilibrium. Buyers eat at the restaurant only if they get very few bad meals, because they know they are in for clogged arteries and high blood pressure after that. Bad reputation arises because the seller cannot resist the temptation of taking actions that are actually unfavourable to the buyer in an effort to regain his good opinion. They actually have the reverse effect of ultimately driving the buyers away.

Reputation for what?

In our discussion we consider a strong type of seller who is committed to playing a particular fixed (random) action in each period. Is this form of uncertainty the only relevant one, or are there other potential types of strong seller that may do even better for our normal seller? There are two alternatives to consider: the strong seller is committed to playing a history-dependent strategy, or the strong player is equipped with a payoff function and her strategy is determined by an equilibrium.

If the seller faces a sequence of short-term buyers, then committing to a fixed stage game action is the best she could ever do, because each buyer's optimization focuses on what the seller does in the current period – the future is irrelevant. Even when the buyers are long lived, there are circumstances where committing to play a fixed action imparts a strategic advantage in repeated play, for example in most coordination or common interest games. However, there are other repeated games, such as the Prisoner's Dilemma, and dynamic games where committing to a fixed stage action is worthless. What the seller would like to do is to commit to a strategy, such as tit-for-tat, which would persuade a sufficiently patient buyer to cooperate with the strong type. Provided some rather strong conditions are satisfied, this is possible.

Our recipe for reputation results will break down when we consider strong sellers with payoffs rather than actions; nevertheless, reputation results are possible. For example, if the strong seller had payoffs of 2 for high quality and zero for low quality he would be strategically identical to a seller who always provided high quality.

Many players: social reputation and other considerations

Thus far we have resolutely stuck to a model of two players, but it is clear that reputation is a pervasive social and competitive phenomenon. Here we sketch some of the issues in many-player reputation. The literature on this area is in its infancy; very little can be said with much certainty now.

The easiest case to deal with is what happens as the number of uninformed players (the buyers in our example) increases. Here the benefit to the seller of building a reputation for high quality increases, as providing a good product today means the seller is more likely to trade with many buyers tomorrow. In a way, increasing the number of buyers is like making the seller more patient, and so we would expect the seller to be more inclined to build a reputation in this case.

A second case would be where there are very large numbers of informed buyers trying to acquire reputations for individual or group characteristics. Models of career concerns are similar to reputation models and have many workers trying to acquire reputations for individual characteristics. Also, there are models of group reputation, such as Tirole (1996), where a particular class of individuals behaves in a particular way to perpetuate the 'group's' reputation. In both these types of model the large numbers assumption allows one individual's reputation decision to be treated as virtually independent of others. Thus they can be analysed using quite simple tools.

A final case is where a few informed agents are in competition or collusion with each other. Collusion in team reputation obviously introduces a public goods issue. If one player contributes to the good name of the group, he or she does not get to enjoy the full benefits of the contribution. Typically, therefore, reputations for such teams are harder to establish. One might conjecture that competition appears to drive a player towards excessive investment in reputation, but there are many effects at work that we do not completely understand. For example, competitors may also act to undermine their rival's reputation and to interfere with its development. This is a fertile region for applied and theoretical investigations.

MARTIN W. CRIPPS

See also **repeated games.**

Bibliography

Cripps, M.W., Mailath, G.J. and Samuelson, L. 2004. Imperfect monitoring and impermanent reputations. *Econometrica* 72, 407–32.

Ely, J. and Valimaki, J. 2003. Bad reputation. *Quarterly Journal of Economics* 118, 785–814.

Fudenberg, D., Kreps, D. and Maskin, E. 1990. Repeated games with long-run and short-run players. *Review of Economic Studies* 57, 555–74.

Fudenberg, D. and Levine, D.K. 1989. Reputation and equilibrium selection in games with a patient player. *Econometrica* 57, 759–78.

Fudenberg, D. and Levine, D.K. 1992. Maintaining a reputation when strategies are imperfectly observed. *Review of Economic Studies* 59, 561–79.

Hart, S. 1985. Nonzero-sum two-person repeated games with incomplete information. *Mathematics of Operations Research* 10, 117–53.

Kreps, D. and Wilson, R. 1982. Reputation and imperfect information. *Journal of Economic Theory* 27, 253–79.

Mailath, G.J. and Samuelson, L. 2006. *Repeated Games and Reputations: Long-Run Relationships.* Oxford: Oxford University Press.

Milgrom, P. and Roberts, J. 1982. Predation, reputation and entry deterrence. *Journal of Economic Theory* 27, 280–312.

Schmidt, K. 1993. Reputation and equilibrium characterization in repeated games of conflicting interests. *Econometrica* 61, 325–51.

Sorin, S. 1999. Merging, reputation, and repeated games with incomplete information. *Games and Economic Behavior* 29, 274–308.

Tirole, J. 1996. A theory of collective reputations (with applications to the persistence of corruption and to firm quality). *Review of Economic Studies* 63, 1–22.

Williams, D. 1991. *Probability with Martingales.* Cambridge: Cambridge University Press.

revelation principle

Communication is central to the economic problem (Hayek, 1945). Opportunities for mutually beneficial transactions cannot be found unless individuals share information about their preferences and endowments. Markets and other economic institutions should be understood as mechanisms for facilitating communication. However, people cannot be expected to reveal information when it is against their interests; for example, a seller may conceal his willingness to sell at a lower price. Rational behaviour in any specific communication mechanism can be analysed using game-theoretic equilibrium concepts, but efficient institutions can be identified only by comparison with all possible communication mechanisms. The revelation principle is a technical insight that allows us, in any given economic situation, to make general statements about all possible communication mechanisms.

The problem of making statements about all possible communication systems might seem intractably complex. Reports and messages may be expressed in rich languages with unbounded vocabulary. Communication systems can include both public announcements and private communication among smaller groups. Communication channels can have noise that randomly distorts messages. A communication mechanism may also specify how contractually enforceable transactions will depend on agents' reports and messages. So a general communication mechanism for any given set of agents may specify (*a*) a set of possible reports that each agent can send, (*b*) a set of possible messages that each agent can receive from the communication system, and (*c*) a probabilistic rule for determining the messages received and the enforceable transactions as a function of the reports sent by the agents. However, the revelation principle tells us that, for many economic purposes, it is sufficient for us to consider only a special class of mechanisms, called 'incentive-compatible direct-revelation mechanisms'.

In these mechanisms, every economic agent is assumed to communicate only with a central mediator. This mediator may be thought of as a trustworthy person or as a computer at the centre of a telephone network. In a direct-revelation mechanism, each individual is asked to report all of his private information confidentially to the mediator. After receiving these reports, the mediator then specifies all contractually enforceable transactions, as a function of these reports. If any individual controls private actions that are not contractually enforceable (such as efforts that others cannot observe), then the mediator also confidentially recommends an action to the individual. A direct-revelation mechanism is any rule for specifying how the mediator determines these contractual transactions and privately recommended actions, as a function of the private-information reports that the mediator receives.

A direct-revelation mechanism is said to be 'incentive compatible' if, when each individual expects that the others will be honest and obedient to the mediator, then no individual could ever expect to do better (given the information available to him) by reporting dishonestly to the mediator or by disobeying the mediator's recommendations. That is, the mechanism is incentive compatible if honesty and obedience is an equilibrium of the resulting communication game. The set of incentive-compatible direct-revelation mechanisms has good mathematical properties that often make it easy to analyse because it can be defined by a collection of linear inequalities, called 'incentive constraints'. Each of these incentive constraints expresses a requirement that an individual's expected utility from using a dishonest or disobedient strategy should not be greater than the individual's expected utility from being honest and obedient, when it is anticipated that everyone else will be honest and obedient.

The analysis of such incentive-compatible direct-revelation mechanisms might seem to be of limited interest, because real institutions rarely use such fully centralized mediation and often generate incentives for dishonesty or disobedience. For any equilibrium of any general communication mechanism, however, there exists an incentive-compatible direct-revelation mechanism that is essentially equivalent. This proposition is the revelation principle. Thus, the revelation principle tells us that, by analysing the set of incentive-compatible direct-revelation mechanisms, we can derive general properties of all equilibria of all coordination mechanisms.

The terms 'honesty' and 'obedience' here indicate two fundamental aspects of the general economic problem of communication. In a general communication system, an individual may send out messages or reports to share information that he knows privately, and he may also receive messages or recommendations to guide actions that he controls privately. The problem of motivating individuals to report their private information honestly is called 'adverse selection', and the problem of motivating individuals to implement their recommended actions obediently is called 'moral hazard'. To describe the intuition behind the revelation principle, let us consider first the special cases where only one or the other of these problems exists.

Pure adverse selection

First, let us formulate the revelation principle for the case of pure adverse selection, as developed in Bayesian social choice theory. In this case we are given a set of individuals, each of whom has some initial private information that may be called the individual's 'type', and there is a planning question of how a social allocation of resources should depend on the individuals' types. Each individual's payoff can depend on the resource allocation and on the types of all individuals according to some given utility function, and each type of each individual has some given probabilistic beliefs about the types of all other individuals. A general communication system would allow each individual i to send a message m_i in some rich language, and then the chosen resource allocation would depend on all these messages according to some rule $\gamma(m_1, \ldots, m_n)$. In any equilibrium of the game defined by this communication system, each individual i must have some strategy σ_i for choosing his message as a function of his type t_i, so that $m_i = \sigma_i(t_i)$.

For the given equilibrium $(\sigma_1, \ldots, \sigma_n)$ of the given social-choice rule γ, the revelation principle is satisfied by a mediation plan in which each individual is asked to confidentially report his type t_i to a central mediator, who then implements the social choice

$$\mu(t_1, \ldots, t_n) = \gamma(\sigma_1(t_1), \ldots, \sigma_n(t_n)).$$

So the mediator computes what message would be sent by the reported type of each individual i under his or her strategy σ_i, and then the mediator implements the resource allocation that would result from these messages under the rule γ. It is easy to see that honesty is an equilibrium under this mediation plan μ. If any individual could gain by lying to this mediator, when all others are expected to be honest, then this individual could have also gained by lying to himself when implementing his equilibrium strategy σ_i under the given mechanism γ, which would contradict the optimality condition that defines an equilibrium. So μ is an incentive-compatible direct-revelation mechanism that is equivalent to the given general mechanism γ with the given equilibrium $(\sigma_1, \ldots, \sigma_n)$.

In this case of pure adverse selection, the revelation principle was introduced by Gibbard (1973), but for a narrower solution concept (dominant strategies, instead of Bayesian equilibrium). The revelation principle for the broader solution concept of Bayesian equilibrium was recognized by Dasgupta, Hammond and Maskin (1979), Harris and Townsend (1981), Holmstrom (1977), Myerson (1979), and Rosenthal (1978).

Pure moral hazard

Next let us formulate the revelation principle for the case of pure moral hazard, as developed in Aumann's (1974) theory of correlated equilibrium. In this case we are given a set of individuals, each of whom controls some actions, and each individual's payoff can depend on the actions $(c_1, \ldots, c_n)$ that are chosen by all individuals, according to some given utility function $u_i(c_1, \ldots, c_n)$. That is, we are given a game in strategic form. In this case of pure moral hazard, nobody has any private information initially, but a communication process could give individuals different information before they choose their actions. In a general communication system, each individual i could get some message m_i in some rich language, with these messages $(m_1, \ldots, m_n)$ being randomly drawn from some joint probability distribution ρ. In any equilibrium of the game generated by adding this communication system, each individual i has some strategy σ_i for choosing his action c_i as a function of his message m_i, so that $c_i = \sigma_i(m_i)$.

For the given equilibrium $(\sigma_1, \ldots, \sigma_n)$ of the game with the given communication system ρ, the revelation principle is satisfied by a mediation plan in which the mediator randomly generates recommended actions in such a way that the probability of recommending actions $(c_1, \ldots, c_n)$ is the same as the probability of the given communication system ρ yielding messages $(m_1, \ldots, m_n)$ that would induce the players to choose $(c_1, \ldots, c_n)$ in the σ equilibrium. That is, the probability

$\mu(c_1, \ldots, c_n)$ of the mediator recommending $(c_1, \ldots, c_n)$ is

$$\mu(c_1, \ldots, c_n) = \rho(\{(m_1, \ldots, m_n)|\sigma_1(m_1) = c_1, \ldots, \sigma_n(m_n) = c_n\}).$$

Then the mediator confidentially tells each individual i only which action c_i is recommended for him. Obedience is an equilibrium under this mediation plan μ because, if any individual could gain by disobeying this mediator when all others are expected to be obedient, then this individual could have also gained by disobeying himself in implementing his equilibrium strategy σ_i in the given game with communication system ρ. So μ is an incentive-compatible direct-revelation mechanism that is equivalent to the given mechanism ρ with the given equilibrium $(\sigma_1, \ldots, \sigma_n)$.

General formulations

Problems of adverse selection and moral hazard can be combined in the framework of Harsanyi's (1967) Bayesian games, where players have both types and actions. The revelation principle for general Bayesian games was formulated by Myerson (1982; 1985). A further generalization of the revelation principle to multistage games was formulated by Myerson (1986). In each case, the basic idea is that any equilibrium of any general communication system can be simulated by a maximally centralized communication system in which, at every stage, each individual confidentially reports all his private information to a central mediator, and then the mediator confidentially recommends an action to each individual, and the mediator's rule for generating recommendations from reports is designed so that honesty and obedience form an equilibrium of the mediated communication game.

The basic assumption here is that, although the motivations of all economic agents are problematic, we can find a mediator who is completely trustworthy and has no costs of processing information. Asking agents to reveal all relevant information to the trustworthy mediator maximizes the mediator's ability to implement any coordination plan. But telling any other agent more than is necessary to guide his choice of action would only increase the agent's ability to find ways of profitably deviating from the coordination plan.

For honesty and obedience to be an equilibrium, the mediation plan must satisfy incentive constraints which say that no individual could ever expect to gain by deviating to a strategy that involves lying to the mediator or disobeying a recommendation from the mediator. In a dynamic context, we must consider that an individual's most profitable deviation from honesty and obedience could be followed by further deviations in the future. So, to verify that an individual could never gain by lying, we must consider all possible deviation strategies in which the individual may thereafter choose actions that can depend disobediently on the mediator's recommendations (which may convey information about others' types and actions).

When we use sequential equilibrium as the solution concept for dynamic games with communication, the set of actions that can be recommended in a sequentially incentive-compatible mechanism must be restricted somewhat. In a Bayesian game, if some action d_i could never be optimal for individual i to use when his type is t_i, no matter what information he obtained about others' types and actions, then obedience could not be sequentially rational in any mechanism where the mediator might ever recommend this action d_i to i after he reports type t_i. Myerson (1986) identified a larger set of *co-dominated actions* that can never be recommended in any sequentially incentive-compatible mechanism. Suppose that, if any individual observed a zero-probability event, then he could attribute this surprise to a mistake by the trembling hand of the mediator. Under this assumption, Myerson (1986) showed that the effect of requiring sequential rationality in games with communication is completely characterized by the requirement that no individuals should ever be expected to choose any co-dominated actions. (See Gerardi and Myerson, 2007.)

Limitations

The revelation principle says that each equilibrium of any communication mechanism is equivalent to the honest-obedient equilibrium of an incentive-compatible direct-revelation mechanism. But this direct-revelation mechanism may have other dishonest equilibria, which might not correspond to equilibria of the original mechanism. So the revelation principle cannot help us when we are concerned about the whole set of equilibria of a communication mechanism. Similarly, a given communication mechanism may have equilibria that change in some desirable way as we change the players' given beliefs about each others' types, but these different equilibria would correspond to different incentive-compatible mechanisms, and so this desirable property of the given mechanism could not be recognized with the revelation principle.

The assumption that perfectly trustworthy mediators are available is essential to the mathematical simplicity of the incentive-compatible set. Otherwise, if individuals can communicate only by making public statements that are immediately heard by everybody, then the set of equilibria may be smaller and harder to compute.

In principal-agent analysis we often apply the revelation principle to find the incentive-compatible mechanism that is optimal for the principal. If the principal would be tempted to use revealed information opportunistically, then there could be loss of generality in assuming that the agents reveal all their private information to the principal. But we should not confuse the principal with the mediator. The revelation principle can still be applied if the principal can get a trustworthy mediator to take the agents' reports and use them according to any specified mechanism.

There are often questions about whether the allocation selected by a mechanism could be modified by subsequent exchanges among the individuals. An individual's right to offer his possessions for sale at some future date could be accommodated in mechanism design by additional moral-hazard constraints.

For example, suppose the principal can sell an object each day, on days 1 and 2. The only buyer's value for such objects is either low \$1 or high \$3, low having probability

0.25. To maximize the principal's expected revenue with the buyer participating honestly, an optimal mechanism would sell both objects for \$3 if the buyer's type is high, but would sell neither if the buyer is low. But if no sale is recommended then the principal could infer that the buyer is low and would prefer to sell for \$1. Suppose now that the principal cannot be prevented from offering to sell for \$1 on either day. With these additional moral-hazard constraints, an optimal mechanism uses randomization by the mediator to conceal information from the principal. If the buyer reports low then the mediator recommends no sale on day 1 and selling for \$1 on day 2. If the buyer reports high, then with probability 1/3 the mediator recommends no sale on day 1 and selling for \$3 on day 2, but with probability 2/3 recommends selling for \$1.50 on both days. A no-sale recommendation on day 1 implies probability 0.5 of low, so that obedience yields the same expected revenue $0.5 \times (0 + 1) + 0.5 \times (0 + 3)$ as deviating to sell for \$1 on both days.

A proliferation of such moral-hazard constraints may greatly complicate the analysis, however. So in practice we often apply the revelation principle with an understanding that we may be overestimating the size of the feasible set, by assuming away some problems of mediator imperfection or moral hazard. When we use the revelation principle to show that a seemingly wasteful mechanism is actually efficient when incentive constraints are recognized, such overestimation of the incentive-feasible set would not weaken the impact of our results (as long as this mechanism remains feasible).

Centralized mediation is emphasized by the revelation principle as a convenient way of characterizing what people can achieve with communication, but this analytical convenience does not imply that centralization is necessarily the best way to coordinate an economy. For fundamental questions about socialist centralization versus free-market decentralization, we should be sceptical about an assumption that centralized control over national resources could not corrupt any mediator. The power of the revelation principle for such questions is instead its ability to provide a common analytical framework that applies equally to socialism and capitalism. For example, a standard result of revelation-principle analysis is that, if only one producer knows the production cost of a good, then efficient incentive-compatible mechanisms must allow this monopolistic producer to take positive informational rents or profits (Baron and Myerson, 1982). Thus the revelation principle can actually be used to support arguments for decentralized multi-source production, by showing that problems of profit-taking by an informational monopolist can be just as serious under socialism as under capitalism.

Nash (1951) advocated a different methodology for analysing communication in games. In Nash's approach, all opportunities for communication should be represented by moves in our extensive model of the dynamic game. Adding such communication moves may greatly increase the number of possible strategies for a player, because each strategy is a complete plan for choosing the player's moves throughout the game. But if all communication will occur in the implementation of these strategies, then the players' initial choices of their strategies must be independent. Thus, Nash argued, any dynamic game can be normalized to a static strategic-form game, where players choose strategies simultaneously and independently, and Nash equilibrium is the general solution for such games.

With the revelation principle, however, communication opportunities are omitted from the game model and are instead taken into account by using incentive-compatible mechanisms as our solution concept. Characterizing the set of all incentive-compatible mechanisms is often easier than computing the Nash equilibria of a game with communication. Thus, by applying the revelation principle, we can get both a simpler model and a simpler solution concept for games with communication. But, when we use the revelation principle, strategic-form games are no longer sufficient for representing general dynamic games, because normalizing a game model to strategic form would suppress implicit opportunities for communicating during the game (see Myerson, 1986). So the revelation principle should be understood as a methodological alternative to Nash's strategic-form analysis.

ROGER B. MYERSON

See also **mechanism design.**

Bibliography

Aumann, R. 1974. Subjectivity and correlation in randomized strategies. *Journal of Mathematical Economics* 1, 67–96.

Baron, D. and Myerson, R. 1982. Regulating a monopolist with unknown costs. *Econometrica* 50, 911–30.

Dasgupta, P., Hammond, P. and Maskin, E. 1979. The implementation of social choice rules: some results on incentive compatibility. *Review of Economic Studies* 46, 185–216.

Gerardi, D. and Myerson, R. 2007. Sequential equilibrium in Bayesian games with communication. *Games and Economic Behavior* 60, 104–34.

Gibbard, A. 1973. Manipulation of voting schemes: a general result. *Econometrica* 41, 587–601.

Harris, M. and Townsend, R. 1981. Resource allocation under asymmetric information. *Econometrica* 49, 1477–99.

Harsanyi, J. 1967. Games with incomplete information played by Bayesian players. *Management Science* 14, 159–82, 320–34, 486–502.

Hayek, F. 1945. The use of knowledge in society. *American Economic Review* 35, 519–30.

Holmstrom, B. 1977. On incentives and control in organizations. Ph.D. thesis, Stanford University.

Myerson, R. 1979. Incentive-compatibility and the bargaining problem. *Econometrica* 47, 61–73.

Myerson, R. 1982. Optimal coordination mechanisms in generalized principal-agent problems. *Journal of Mathematical Economics* 10, 67–81.

Myerson, R. 1985. Bayesian equilibrium and incentive compatibility: an introduction. In *Social Goals and Social Organization*, ed. L. Hurwicz, D. Schmeidler and H. Sonnenschein. Cambridge: Cambridge University Press.

Myerson, R. 1986. Multistage games with communication. *Econometrica* 54, 323–58.

Nash, J. 1951. Non-cooperative games. *Annals of Mathematics* 54, 286–95.

Rosenthal, R. 1978. Arbitration of two-party disputes under uncertainty. *Review of Economic Studies* 45, 595–604.

Shapley value

The *value* of an uncertain outcome (a 'lottery') is an a priori measure, in the participant's utility scale, of what he expects to obtain (this is the subject of 'utility theory'). The question is, how would one evaluate the prospects of a player in a multi-person interaction, that is, in a game?

This question was originally addressed by Lloyd S. Shapley (1953a). The framework was that of n-person games in coalitional form with side-payments, which are given by a set N of 'players', say 1, 2, ..., n, together with a 'coalitional function' v that associates to every subset S of N ('coalition') a real number $v(S)$, the maximal total payoff the members of S can obtain (the 'worth' of S). An underlying assumption of this model is that there exists a medium of exchange ('money') that is freely transferable in unlimited amounts between the players, and moreover every player's utility is additive with respect to it (that is, a transfer of x units from one player to another decreases the first one's utility by x units and increases the second one's utility by x units; the total payoff of a coalition can thus be meaningfully defined as the sum of the payoffs of its members). This requirement is known as existence of 'side payments' or 'transferable utility'. In addition, the game is assumed to be adequately described by its coalitional function (that is, the worth $v(S)$ of each coalition S is well defined, and the abstraction from the extensive structure of the game to its coalitional function leads to no essential loss; such a game is called a '*c*-game'). These assumptions may be interpreted in a broader and more abstract sense. For example, in a voting situation, a 'winning coalition' is assigned worth 1, and a 'losing' coalition, worth 0. The essential feature is that the prospects of each coalition may be summarized by one number.

The *Shapley value* associates to each player in each such game a unique payoff – his 'value'. The value is required to satisfy the following four axioms. (EFF) *Efficiency* or *Pareto optimality:* The sum of the values of all players equals $v(N)$, the worth of the grand coalition of all players (in a superadditive game $v(N)$ is the maximal amount that the players can jointly get); this axiom combines feasibility and efficiency. (SYM) *Symmetry* or *equal treatment:* If two players in a game are substitutes (that is, the worth of no coalition changes when replacing one of the two players by the other one), then their values are equal. (NUL) *Null* or *dummy player:* If a player in a game is such that the worth of every coalition remains the same when he joins it, then his value is zero. (ADD) *Additivity:* The value of the sum of two games is the sum of the values of the two games (equivalently, the value of a probabilistic combination of two games is the same as the probabilistic combination of the values of the two games; this is analogous to 'expected utility'). The surprising result of Shapley is that these four axioms *uniquely determine* the values in *all* games.

Remarkably, the Shapley value of a player in a game turns out to be exactly his *expected marginal contribution to a random coalition*. The marginal contribution of a player i to a coalition S (that does not contain i) is the change in the worth when i joins S, that is, $v(S \cup \{i\}) - v(S)$. To obtain a random coalition S not containing i, arrange the n players in a line (for example, 1, 2, ..., n) and put in S all those that precede i in that order; all $n!$ orders are assumed to be equally likely. The formula for the Shapley value is striking, first, since it is a consequence of very simple and basic axioms and, second, since the idea of marginal contribution is so fundamental in much of economic analysis.

It should be emphasized that the value of a game is an a priori measure, that is, an evaluation before the game is actually played. Unlike other solution concepts (for example, core, von Neumann–Morgenstern solution, bargaining set), it need not yield a 'stable' outcome (the probable final result when the game is actually played). These final stable outcomes are in general not well determined; the value – which is uniquely specified – may be thought of as their expectation or average. Another interpretation of the value axioms regards them as rules for 'fair' division, guiding an impartial 'referee' or 'arbitrator'. Also, as suggested above, the Shapley value may be understood as the utility of playing the game (Shapley, 1953a; Roth, 1977).

In view of both its strong intuitive appeal and its mathematical tractability, the Shapley value has been the focus of much research and many applications. We can only briefly mention some of these here (together with just a few representative references). The reader is referred to the survey of Aumann (1978) and, for more extensive coverage, to the *Handbook of Game Theory* (Aumann and Hart, vol 1: 1992 [HGT1], vol 2: 1994 [HGT2], vol 3: 2002 [HGT3]), especially Chapters 53–58, as well as parts of Chapters 32–34 and 37.

Variations

Following Shapley's pioneering approach, the concept of *value* has been extended, modified and generalized.

Weighted values

Assume that the players are of unequal 'size' (for example, a player may represent a 'group', a 'department', and so on), and this is expressed by given (relative) weights. This setup leads to 'weighted Shapley values' (Shapley, 1953b); in unanimity games, for example, the values of the players are no longer equal but, rather, proportional to their weights [HGT3, Ch. 54].

Semi-values

Abandoning the efficiency axiom (EFF) yields the class of 'semi-values' (Dubey, Neyman and Weber, 1981). An interesting semi-value is the *Banzhaf index* (Penrose, 1946; Banzhaf, 1965; Dubey and Shapley, 1979), originally proposed as a measure of power in voting games. Like the Shapley value, it is also an expected marginal contribution, but here all coalitions not containing player i are equally likely [HGT3, Ch. 54].

Other axiomatizations

There are alternative axiomatic systems that characterize the Shapley value. For instance, one may replace the additivity axiom (ADD) with a *marginality axiom* that requires the value of a player to depend only on his marginal contributions (Young, 1985). Another approach is based on the existence of a *potential* function together with efficiency (EFF) (Hart and Mas-Colell, 1989) [HGT3, Ch. 53].

Consistency

Given a solution concept which associates payoffs to games, assume that a group of players in a game have already agreed to it, are paid off accordingly, and leave the game; consider the 'reduced game' among the remaining players. If the solution of the reduced game is the same as that of the original game, then the solution is said to be *consistent*. It turns out that consistency, together with some elementary requirements for two-player games, characterizes the Shapley value (Hart and Mas-Colell, 1989) [HGT3, Ch. 53], [HGT1, Ch. 18].

Large games

Assume that the number of players increases and individuals become negligible. Such models are important in applications (such as competitive economies and voting), and there is a vast body of work on values of large games that has led to beautiful and important insights (for example, Aumann and Shapley, 1974) [HGT3, Ch. 56].

NTU games

These are games 'without side payments', or 'with non-transferable utility' (that is, the existence of a medium of utility exchange is no longer assumed). The simplest such games, two-person pure bargaining problems, were originally studied by Nash (1950). Values for general NTU games, which coincide with the Shapley value in the side payments case, and with the Nash bargaining solution in the two-person case, have been introduced by Harsanyi (1963), Shapley (1969), Maschler and Owen (1992) [HGT3, Ch. 55].

Non-cooperative foundations

Bargaining procedures whose non-cooperative equilibrium outcome is the Shapley value have been proposed by Gul (1989) (see Hart and Levy, 1999; Gul, 1999) and Winter (1994) for strictly convex games, and by Hart and Mas-Colell (1996) for general games [HGT3, Ch. 53].

Other extensions

This includes games with communication graphs (Myerson, 1977), coalition structures (Aumann and Drèze, 1974; Owen, 1977; Hart and Kurz, 1983), and others [HGT2, Ch. 37], [HGT3, Ch. 53].

Economic applications

Perfect competition

In the classical economic model of perfect competition, the commodity prices are determined by the requirement that total demand equals total supply; this yields a *competitive* (or *Walrasian*) *equilibrium*. A different approach in such setups looks at the cooperative 'market game' where the members of each coalition can freely exchange among themselves the commodities they own. A striking phenomenon occurs: various game-theoretic solutions of the market games yield precisely the competitive equilibria. In particular, in perfectly competitive economies every Shapley value allocation is competitive and, if the utilities are smooth, then every competitive allocation is also a value allocation. This result, called the *value equivalence principle*, is remarkable since it joins together two very different approaches: competitive prices arising from supply and demand on the one hand, and marginal contributions to trading coalitions on the other. The value equivalence principle has been studied in a wide range of models (for example, Shapley, 1964; Aumann, 1975). While it is undisputed in the TU case, its extension to the general NTU case seems less clear (it holds for the Shapley NTU value, but not necessarily for other NTU values) [HGT3, Ch. 57].

Cost allocation

Consider the problem of allocating joint costs in a 'fair' manner. Think of the various 'tasks' (or 'projects', 'departments', and so on) as players, and let $v(S)$ be the total cost of carrying out the set S of tasks (Shubik, 1962). It turns out that the axioms determining the Shapley value are easily translated into postulates appropriate for solving cost allocation problems (for example, the efficiency axiom becomes 'total-cost-sharing'). Two notable applications are airport landing fees (a task here is an aircraft landing; Littlechild and Owen, 1973) and telephone billing (each time unit of a phone call is a player; the resulting cost allocation scheme was put into actual use at Cornell University; Billera, Heath and Raanan, 1978) [HGT2, Ch. 34].

Other applications

The value has been applied to various economic models; for example, models of taxation where a political power structure is given in addition to the economic data (Aumann and Kurz, 1977). Further references to economic applications can be found in Aumann (1985) [HGT3, Ch. 58], [HGT2, Ch. 33].

Political applications

What is the 'power' of an individual or a group in a voting situation? A trivial observation – though not always remembered in practice – is that the political power need not be proportional to the number of votes (see Shapley, 1981, for some interesting examples). It is therefore important to find an objective method of measuring power in such situations. The Shapley value (known in this setup as the *Shapley–Shubik index*; Shapley and Shubik, 1954) is, by its very nature, a most appropriate candidate. Indeed, consider a simple political game, described by specifying whether each coalition is 'winning' or 'losing'. The Shapley value of a player i turns out to be the probability that i is the 'pivot' or 'key' player, namely, that in a random order of all players those preceding i are losing, whereas together with i they are winning. For example, in a 100-seat parliament with simple majority (that is, 51 votes are needed to win), assume there is one large party having 33 seats and the rest are divided among many small parties; the value of the large party is then close to 50%, considerably more than its voting weight (that is, its 33% share of the seats). In contrast, when there are two large parties each having 33 seats and a large number of small parties, the value of each large party is close to 25% – much less than its voting weight of 33%. To understand this, think of the competition between the two large parties to attract the small parties to form a winning coalition; in contrast, when there is only one large party, the competition is between the small parties (to join the large party).

The Shapley value has also been used in more complex models, where 'ideologies' and 'issues' are taken into account (thus, not all arrangements of the voters are equally likely; an 'extremist' party, for example, is less likely to be the pivot than a 'middle-of-the-road' one; Owen, 1971; Shapley, 1977).

References to political applications of the Shapley value may be found in Shapley (1981); these include various parliaments (USA, France, Israel), the United Nations Security Council, and others [HGT2, Ch. 32].

SERGIU HART

Bibliography

Aumann, R.J. 1975. Values of markets with a continuum of traders. *Econometrica* 43, 611–46.

Aumann, R.J. 1978. Recent developments in the theory of the Shapley Value. *Proceedings of the International Congress of Mathematicians*, Helsinki.

Aumann, R.J. 1985. On the non-transferable utility value: a comment on the Roth–Shafer examples. *Econometrica* 53, 667–78.

Aumann, R.J. and Drèze, J.H. 1974. Cooperative games with coalition structures. *International Journal of Game Theory* 3, 217–37.

Aumann, R.J. and Hart, S., eds. 1992 [HGT1], 1994 [HGT2], 2002 [HGT3]. *Handbook of Game Theory, with Economic Applications*, vols 1–3. Amsterdam: North-Holland.

Aumann, R.J. and Kurz, M. 1977. Power and taxes. *Econometrica* 45, 1137–61.

Aumann, R.J. and Shapley, L.S. 1974. *Values of Non-atomic Games*. Princeton: Princeton University Press.

Banzhaf, J.F. 1965. Weighted voting doesn't work: a mathematical analysis. *Rutgers Law Review* 19, 317–43.

Billera, L.J., Heath, D.C. and Raanan, J. 1978. Internal telephone billing rates: a novel application of non-atomic game theory. *Operations Research* 26, 956–65.

Dubey, P., Neyman, A. and Weber, R.J. 1981. Value theory without efficiency. *Mathematics of Operations Research* 6, 122–8.

Dubey, P. and Shapley, L.S. 1979. Mathematical properties of the Banzhaf Power Index. *Mathematics of Operations Research* 4, 99–131.

Gul, F. 1989. Bargaining foundations of Shapley value. *Econometrica* 57, 81–95.

Gul, F. 1999. Efficiency and immediate agreement: a reply to Hart and Levy. *Econometrica* 67, 913–18.
Harsanyi, J.C. 1963. A simplified bargaining model for the *n*-person cooperative game. *International Economic Review* 4, 194–220.
Hart, S. and Kurz, M. 1983. Endogenous formation of coalitions. *Econometrica* 51, 1047–64.
Hart, S. and Levy, Z. 1999. Efficiency does not imply immediate agreement. *Econometrica* 67, 909–12.
Hart, S. and Mas-Colell, A. 1989. Potential, value and consistency. *Econometrica* 57, 589–614.
Hart, S. and Mas-Colell, A. 1996. Bargaining and value. *Econometrica* 64, 357–80.
Littlechild, S.C. and Owen, G. 1973. A simple expression for the Shapley value in a special case. *Management Science* 20, 370–72.
Maschler, M. and Owen, G. 1992. The consistent Shapley value for games without side payments. In *Rational Interaction: Essays in Honor of John Harsanyi*, ed. R. Selten. New York: Springer.
Myerson, R.B. 1977. Graphs and cooperation in games. *Mathematics of Operations Research* 2, 225–29.
Nash, J.F. 1950. The bargaining problem. *Econometrica* 18, 155–62.
Owen, G. 1971. Political games. *Naval Research Logistics Quarterly* 18, 345–55.
Owen, G. 1977. Values of games with a priori unions. In *Essays in Mathematical Economics and Game Theory*, ed. R. Henn and O. Moeschlin. New York: Springer.
Penrose, L.S. 1946. The elementary statistics of majority voting. *Journal of the Royal Statistical Society* 109, 53–7.
Roth, A.E. 1977. The Shapley value as a von Neumann–Morgenstern utility. *Econometrica* 45, 657–64.
Shapley, L.S. 1953a. A value for *n*-person games. In *Contributions to the Theory of Games, II*, ed. H.W. Kuhn and A.W. Tucker. Princeton: Princeton University Press.
Shapley, L.S. 1953b. Additive and non-additive set functions. Ph.D. thesis, Princeton University.
Shapley, L.S. 1964. Values of large games VII: a general exchange economy with money. Research Memorandum 4248-PR. Santa Monica, CA: RAND Corp.
Shapley, L.S. 1969. Utility comparison and the theory of games. In *La Décision: agrégation et dynamique des ordres de préférence*. Paris: Editions du CNRS.
Shapley, L.S. 1977. A comparison of power indices and a nonsymmetric generalization. Paper No. P–5872. Santa Monica, CA: RAND Corp.
Shapley, L.S. 1981. Measurement of power in political systems. *Game Theory and its Applications*, Proceedings of Symposia in Applied Mathematics, vol. 24. Providence, RI: American Mathematical Society.
Shapley, L.S. and Shubik, M. 1954. A method for evaluating the distribution of power in a committee system. *American Political Science Review* 48, 787–92.
Shubik, M. 1962. Incentives, decentralized control, the assignment of joint costs and internal pricing. *Management Science* 8, 325–43.
Winter, E. 1994. The demand commitment bargaining and snowballing cooperation. *Economic Theory* 4, 255–73.
Young, H.P. 1985. Monotonic solutions of cooperative games. *International Journal of Game Theory* 14, 65–72.

stochastic adaptive dynamics

Stochastic adaptive dynamics require analytical methods and solution concepts that differ in important ways from those used to study deterministic processes. Consider, for example, the notion of asymptotic stability: in a deterministic dynamical system, a state is locally asymptotically stable if any sufficiently small deviation from the original state is self-correcting. We can think of this as a first step toward analysing the effect of stochastic shocks; that is, a state is locally asymptotically stable if, after the impact of a small, one-time shock, the process evolves back to its original state.

This idea is not entirely satisfactory, however, because it treats shocks as if they were isolated events. Economic systems are usually composed of large numbers of interacting agents whose behaviour is *constantly* being buffeted by perturbations from various sources. These *persistent* shocks have substantially different effects from *one-time* shocks; in particular, persistent shocks can accumulate and tip the process out of the basin of attraction of an asymptotically stable state. Thus, in a stochastic setting, conventional notions of dynamic stability – including *evolutionarily stable strategies* – are inadequate to characterize the long-run behaviour of the process. Here we shall outline an alternative approach that is based on the theory of large deviations in Markov processes (Freidlin and Wentzell, 1984; Foster and Young, 1990; Young, 1993a).

Types of stochastic perturbations

Before introducing formal definitions, let us consider the various kinds of stochastic shocks to which a system of interacting agents may be exposed. First, there is the interaction process itself whereby agents randomly encounter other agents in the population. Second, the agents' behaviour will be *intentionally* stochastic if they are employing mixed strategies. Third, their behaviour may be *unintentionally* stochastic if their payoffs are subject to unobserved utility shocks. Fourth, mutation processes may cause one type of agent to change spontaneously into another type. Fifth, in and out-migration can introduce new behaviours into the population or extinguish existing ones. Sixth, the system may be hit by aggregate shocks that change the *distribution* of behaviours. This list is by no means exhaustive, but it does convey some sense of the range of stochastic influences that arise quite naturally in economic (and biological) contexts.

Stochastic stability

The early literature on evolutionary game dynamics tended to sidestep stochastic issues by appealing to the law of large numbers. The reasoning is that, when a population is large, random influences at the individual level will tend to average out, and the aggregate state variables will evolve according to the *expected* (hence deterministic) direction of motion. While this approximation may be reasonable in the short and medium run, however, it can be quite misleading when extrapolated over longer periods of time. The difficulty is that, even when the stochastic shocks have very small probability, their accumulation can have dramatic long-run effects that push the process far away from its deterministic trajectory.

The key to analysing such processes is to observe that, when the aggregate stochastic effects are 'small' and the resulting process is ergodic, the long-run distribution will often be concentrated on a very small subset of states – possibly, in fact, on a *single* state. This leads to the idea of *stochastic stability*, a solution concept first proposed for general stochastic dynamical systems by Foster and Young (1990, p. 221): 'the stochastically stable set (SSS) is the set of states S such that, in the long run, it is nearly certain that the system lies within every open set containing S as the noise tends slowly to zero.' The analytical technique for computing these states relies on the theory of large deviations first developed for continuous-time processes by Freidlin and Wentzell (1984), and subsequently extended to general finite-state Markov chains by Young (1993a). It is in the latter form that the theory is usually applied in economic contexts.

An illustrative example

The following simple model illustrates the basic ideas. Consider a population of n agents who are playing the 'Stag Hunt' game:

	A	B
A	10, 10	0, 7
B	7, 0	7, 7

The *state* of the process at time t is the current number of agents playing A, which we shall denote by $z_t \in Z = \{0, 1, 2, \ldots, n\}$. Time is discrete. At the start of period $t+1$, one agent is chosen at random. Strategy A is a best response if $z_t \leq .7n$ and B is a best response if $z_t \geq .7n$. (We assume that the player includes herself in assessing the current distribution, which simplifies the computations.) With high probability, say $1-\varepsilon$, the agent chooses a best response to the current distribution of strategies; while with probability ε she chooses A or B at random (each with probability $\varepsilon/2$).

We can interpret such a departure from best response behaviour in various ways: it might be a form of experimentation, it might be a behavioural 'mutation', or it might simply be a form of ignorance – the agent may not know the current state. Whatever the explanation, the result is a *perturbed best response process* in which individuals choose (myopic) best responses to the current state with high probability and depart from best response behaviour with low probability.

This process is particularly easy to visualize because it is one-dimensional: the states can be viewed as points on a line, and in each period the process moves to the left by one step, to the right by one step, or stays put. Figure 1 illustrates the situation when the population consists of ten players.

The transitions indicated by solid arrows have high probability and represent the direction of best response, that is, the main flow of the process. The dashed arrows go against the flow and have low probability, which is the same order of magnitude as ε. (The process can also loop by staying in a given state with positive probability; these loops are omitted from the figure to avoid clutter.)

In this example the transition probabilities are easy to compute. Consider any state z to the left of the critical value $z^* = 7$. The process moves right if and only if one more agent plays A. This occurs if and only if an agent currently playing B is drawn (an event with probability $1 - z/10$) and this agent mistakenly chooses A (an event with probability $\varepsilon/2$). In other words, if $z<7$ the probability of moving right is $R_z = (1 - z/10)\ (\varepsilon/2)$. Similarly, the probability of moving left is $L_z = (z/10)\ (1 - \varepsilon/2)$. The key point is that the right transitions have much smaller probability than the left transitions when ε is small. Exactly the reverse is true for those states $z>7$. In this case the probability of moving right is $R_z = (1 - z/10)(1 - \varepsilon/2)$, whereas the probability of moving left is $L_z = (z/10)(\varepsilon/2)$. (At $z=7$ the process moves left with probability .15, moves right with probability .35, and stays put with probability .50.)

Computing the long-run distribution

Since this finite-state Markov chain is irreducible (each state is reachable from every other via a finite number of transitions), the process has a unique long-run distribution. That is, with probability 1, the relative frequency of being in any given state z equals some number μ_z *independently of the initial state*. Since the process is one-dimensional, the equations defining μ are particularly transparent, namely, it can be shown that for every $z<n$, $\mu_z R_z = \mu_{z+1} L_{z+1}$. This is known as the *detailed balance condition*. It has a simple interpretation: in the long run, the process transits from $z+1$ to z as often as it transits from z to $z+1$.

The solution in this case is very simple. Given any state z, consider the directed tree T_z consisting of all right transitions from states to the left of z and all left transitions from states to the right of z. This is called a *z-tree* (see Figure 2).

An elementary result in Markov chain theory says that, for one-dimensional chains, the long-run probability of being in state z is proportional to the product of the probabilities on the edges of T_z:

$$\mu_z \propto \prod_{y<z} R_y \prod_{y>z} L_y. \tag{1}$$

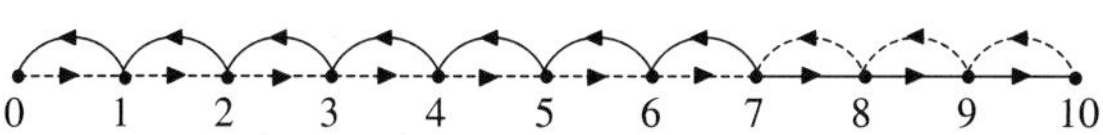

Figure 1 Transitions for the perturbed best response process in the Stag Hunt game and a population of ten agents. *Note*: Each vertex represents the number of agents playing action A at a given time. Solid arrows are transitions with high-probability, dotted arrows are transitions with low (order-ε) probability.

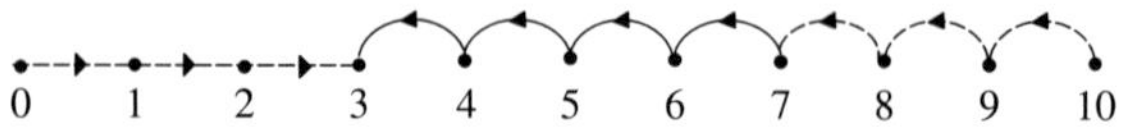

Figure 2 The unique 3-tree

This is a special case of the *Markov chain tree theorem*, which expresses the stationary distribution of any finite chain in terms of the probabilities of its z-trees. (Versions of this result go back at least to Kirchhoff's work in the 1840s; see Haken, 1978, s. 4.8. Freidlin and Wentzell, 1984, use it to study large deviations in continuous-time Wiener processes.)

Formula (1) allows us to compute the order-of-magnitude probability of each state without worrying about its exact magnitude. Figure 2 shows, for example, that μ_3, the long-run probability of state $z = 3$, must be proportional to ε^6, because the 3-tree has six dotted arrows, each of which has probability of order ε. Using this method we can easily compute the relative probabilities of each state.

Stochastic stability and equilibrium selection

This example illustrates a general property of adaptive processes with small persistent shocks. That is, the persistent shocks act as a *selection mechanism*, and the *selection strength increases the less likely the shocks are*. The reason is that the long-run distribution depends on the probability of escaping from various states, and the critical escape probabilities are *exponential* in ε. Figure 1 shows, for example, that the probability of all-B (the left endpoint) is larger by a factor of $1/\varepsilon$ than the probability of any other state, and it is larger by a factor of $1/\varepsilon^4$ than the probability of all-A (the right endpoint). It follows that, as ε approaches zero, the long-run distribution of the process is concentrated entirely on the all-B state. It is the unique stochastically stable state.

While stochastic stability is defined in terms of the limit as the perturbation probabilities go to zero, sharp selection can in fact occur when the probabilities are quite large. To illustrate, suppose that we take $\varepsilon = .20$ in the above example. This defines a very noisy adjustment process, but in fact the long-run distribution is still strongly biased in favour of the all-B state. It can be shown, in fact, that the all-B state is nearly 50 times as probable as the all-A state. (See Young, 1998b, ch. 4, for a general analysis of stochastic selection bias in one-dimensional evolutionary models.)

A noteworthy feature of this example is that the stochastically stable state (all-B) does not correspond to the Pareto optimal equilibrium of the game, but rather to the risk dominant equilibrium (Harsanyi and Selten, 1988). The connection between stochastic stability and risk dominance was first pointed out by Kandori, Mailath and Rob (1993). Essentially their result says that, in any symmetric 2×2 game with a uniform mutation process, the risk dominant equilibrium is stochastically stable provided the population is sufficiently large. The logic of this connection can be seen in the above example. In the pure best response process ($\varepsilon = 0$) there are two absorbing states: all-B and all-A. The basin of attraction of all-B is the set of states to the left of the critical point, while the basin of attraction of the all-A is the set of states to the right of the critical point. The left basin is bigger than the right basin. To go from the left endpoint into the opposite basin therefore requires more 'uphill' motion than to go the other way around. In any symmetric 2×2 coordination game the risk dominant equilibrium is the one with the widest basin, hence it is stochastically stable under uniform stochastic shocks of the above type.

How general is this result? It depends in part on the nature of the shocks. On the one hand, if we change the probabilities of left and right transitions in an arbitrary way, then we can force any given state – including non-equilibrium states – to have the highest long-run probability; indeed this follows readily from formula (1). (See Bergin and Lipman, 1996.) On the other hand, there are many natural perturbations that do lead to the risk dominant equilibrium in 2×2 games. Consider the following class of perturbed best response dynamics. In state z, let $\Delta(z)$ be the expected payoff from playing A against the population minus the payoff from playing B against the population. Assume that in state z the probability of choosing A divided by the probability of choosing B is well-approximated by a function of form $e^{h(\Delta(z))/\beta}$ where $h(\Delta)$ is non-decreasing in Δ, strictly increasing at $\Delta = 0$, and skew-symmetric ($h(\Delta) = - h(-\Delta)$). The positive scalar β is a measure of the noise level. In this set-up, a state is *stochastically stable* if its long-run probability is bounded away from zero as $\beta \rightarrow 0$. Subject to some minor additional regularity assumptions, it can be shown that, in any symmetric 2×2 coordination game, if the population is large enough, the unique stochastically stable state is the one in which everyone plays the risk-dominant equilibrium (Blume, 2003).

Unfortunately, the connection between risk dominance and stochastic stability breaks down – even for uniform mutation rates – in games with more than two strategies per player (Young, 1993a). The difficulty stems from the fact that comparing 'basin sizes' works only in special situations. To determine the stochastically stable states in more general settings requires finding the path of least resistance – the path of greatest probability – from every absorbing set to every other absorbing set, and then constructing a rooted tree from these critical paths (Young, 1993a). (An absorbing set is a minimal set of states from which the *unperturbed* process cannot escape.) What makes the one-dimensional situation so special is that there are only two absorbing sets – the left endpoint and the right endpoint – and there is a unique directed path going from left to right and another unique path going from right to left. (For other situations in which the analysis can be simplified, see Ellison, 2000; Kandori and Rob, 1995.)

There are many games of economic importance in which this theory has powerful implications for equilibrium selection. In the non-cooperative Nash bargaining model, for example, the Nash bargaining solution is essentially the unique stochastically stable outcome (Young, 1993b). Different assumptions about the one-shot bargaining process lead instead to the selection of the Kalai–Smorodinsky solution (Young, 1998a; for further variations see Binmore, Samuelson and Young, 2003). In a standard oligopoly framework, marginal cost pricing turns out to be the stochastically stable solution (Vega-Redondo, 1997).

Speed of adjustment

One criticism that has been levelled at this approach is that it may take an exceedingly long time for the evolutionary process to reach the stochastically stable states when it starts from somewhere else. The difficulty is that, when the shocks have very small probability, it takes a long time (in expectation) before enough of them accumulate to tip the process into the stochastically stable state(s). While this is correct in principle, the waiting time can be very sensitive to various modelling details. First, it depends on the size and probability of the shocks themselves. As we have already noted, the shocks need not be small for sharp selection to occur, in which case the waiting time need not be long either. (In the above example we found that an error rate of 20 per cent still selects the all-B state with high probability.) Second, the expected waiting time depends crucially on the *topology* of interaction. In the above example we assumed that each agent reacts to the distribution of actions in the whole population. If instead we suppose that people respond only to actions of those in their immediate geographic (or social) neighbourhood, the time to reach the stochastically stable state is greatly reduced (Ellison, 1993; Young, 1998b, ch. 6). Third, the waiting time is reduced if the stochastic perturbations are not independent, either because the agents act in a coordinated fashion, or because the utility shocks among agents are statistically correlated (Young, 1998b, ch. 9; Bowles, 2004).

Path dependence

The results discussed above rely on the assumption that the adaptive process is ergodic, that is, its long-run behaviour is almost surely independent of the initial state. Ergodicity holds if, for example, the number of states is finite, the transition probabilities are time-homogeneous, and there is a positive probability of transiting from any state to any other state within a finite number of periods. One way in which these conditions may fail is that the weight of history grows indefinitely. Consider, for example, a two-person game G together with a population of potential row players and another population of potential column players. Assume that an initial history of plays is given. In each period, one row player and one column player are drawn at random, and each of them chooses an ε-trembled best reply to the opposite population's previous actions (alternatively, to a random sample of fixed size drawn from the opponent's previous actions). This is a stochastic form of fictitious play (Fudenberg and Kreps, 1993; Kaniovski and Young, 1995). The *proportion* of agents playing each action evolves according to a stochastic difference equation in which the magnitude of the stochastic term decreases over time; in particular it decreases at the rate $1/t$.

This type of process is not ergodic. It can be shown, in fact, that the long-run proportions converge almost surely either to a neighbourhood of all-A or to a neighbourhood of all-B, where the relative probabilities of these two events depend on the initial state (Kaniovski and Young, 1995). Processes of this type require substantially different techniques of analysis from the ergodic processes discussed earlier; see in particular Arthur, Ermoliev and Kaniovski (1987), Benaim and Hirsch (1999) and Hofbauer and Sandholm (2002).

Summary

The introduction of persistent random shocks into models with large numbers of interacting agents can be handled using methods from stochastic dynamical systems theory; moreover, there is virtually no limit on the dimensionality of the systems that can be analysed using these techniques. Such processes can exhibit path dependence if the weight of history is allowed to grow indefinitely. If instead past actions fade away or are forgotten, the presence of persistent random shocks makes the process ergodic and its long-run behaviour is often easier to analyse. An important feature of such ergodic models is that some equilibrium states are much more likely to occur in the long run than others, and this holds independently of the initial state. The length of time that it takes to reach such states from out-of-equilibrium conditions depends on key structural properties of the model, including the size and frequency of the stochastic shocks, the extent to which they are correlated among agents, and the network topology governing agents' interactions with one another.

H. PEYTON YOUNG

Bibliography

Arthur, W.B., Ermoliev, Y. and Kaniovski, Y. 1987. Adaptive growth processes modelled by urn schemes. *Kybernetika* 6, 49–57; English trans in *Cybernetics* 23, 779–89.

Benaim, M. and Hirsch, M.W. 1999. Mixed equilibria and dynamical systems arising from fictitious play in perturbed games. *Games and Economic Behavior* 29, 36–72.

Bergin, J. and Lipman, B.L. 1996. Evolution with state-dependent mutations. *Econometrica* 64, 943–56.

Binmore, K., Samuelson, L. and Young, H.P. 2003. Equilibrium selection in bargaining models. *Games and Economic Behavior* 45, 296–328.

Blume, L.E. 1995. The statistical mechanics of best-response strategy revision. *Games and Economic Behavior* 11, 111–45.

Blume, L.E. 2003. How noise matters. *Games and Economic Behavior* 44, 251–71.

Bowles, S. 2004. *Microeconomics: Behavior, Institutions, and Evolution.* Princeton, NJ: Princeton University Press.

Ellison, G. 1993. Learning, local interaction, and coordination. *Econometrica* 61, 1047–71.

Ellison, G. 2000. Basins of attraction, long run equilibria, and the speed of step-by-step evolution. *Review of Economic Studies* 67, 17–45.

Foster, D. and Young, H.P. 1990. Stochastic evolutionary game dynamics. *Theoretical Population Biology* 38, 219–32.

Freidlin, M. and Wentzell, A. 1984. *Random Perturbations of Dynamical Systems.* Berlin: Springer-Verlag.

Fudenberg, D. and Kreps, D. 1993. Learning mixed equilibria. *Games and Economic Behavior* 5, 320–67.

Haken, H. 1978. *Synergetics.* Berlin: Springer-Verlag.

Harsanyi, J. and Selten, R. 1988. *A General Theory of Equilibrium Selection in Games.* Cambridge MA: MIT Press.

Hofbauer, J. and Sandholm, W. 2002. On the global convergence of stochastic fictitious play. *Econometrica* 70, 2265–94.
Kandori, M., Mailath, G. and Rob, R. 1993. Learning, mutation, and long-run equilibria in games. *Econometrica* 61, 29–56.
Kandori, M. and Rob, R. 1995. Evolution of equilibria in the long run: a general theory and applications. *Journal of Economic Theory* 65, 383–414.
Kaniovski, Y. and Young, H.P. 1995. Learning dynamics in games with stochastic perturbations. *Games and Economic Behavior* 11, 330–63.
Karlin, S. and Taylor, H.M. 1975. *A First Course in Stochastic Processes*. New York: Academic Press.
Vega-Redondo, F. 1997. The evolution of Walrasian behaviour. *Econometrica* 65, 375–84.
Young, H.P. 1993a. The evolution of conventions. *Econometrica* 61, 57–84.
Young, H.P. 1993b. An evolutionary model of bargaining. *Journal of Economic Theory* 59, 145–68.
Young, H.P. 1998a. Conventional contracts. *Review of Economic Studies* 65, 776–92.
Young, H.P. 1998b. *Individual Strategy and Social Structure*. Princeton. NJ: Princeton University Press.

strategic and extensive form games

1. Introduction

Game theory is a collection of models designed to understand situations in which decision-makers interact. This article discusses models that focus on the behaviour of individual decision-makers. These models are sometimes called 'non-cooperative'.

2. Strategic games

2.1 Definition

The basic model of decision-making by a single agent consists of a set of possible actions and a preference relation over this set. The simplest theory of the agent's behaviour is that she chooses a member of the set that is best according to the preference relation.

The model of a *strategic game* extends this model to many agents, who are referred to as *players*. Each player has a set of possible actions and a preference relation over action *profiles* (lists of actions, one for each player).

Definition 1 *A* ***strategic game with deterministic preferences*** *consists of*

- *a set* N *(the set of* ***players****)*

and for each player $i \in N$

- *a set* A_i *(the set of player i's possible* ***actions****)*
- *a preference relation* $\succsim_i$ *over the set* $\times_{j \in N} A_j$ of action profiles.

A strategic game $\langle N, (A_i), (\succsim_i) \rangle$ *is* ***finite*** *if the set N of players and the set* A_i *of actions of each player i are finite.*

The fact that each player's preferences are defined over the set of action profiles allows for the possibility that each player cares not only about her own action but also about the other players' actions, distinguishing the model from a collection of independent single-agent decision problems.

Notice that the model does not have a temporal dimension. An assumption implicit in the solution notions applied to a game is that each player independently commits to an action before knowing the action chosen by any other player. Notice also that no structure is imposed on the players' sets of actions. In the simplest cases, a player's set of actions may consist of two elements; in more complex cases, it may consist, for example, of an interval of real numbers, a set of points in a higher dimensional space, a set of functions from one set to another, or a combination of such sets. In particular, an action may be a contingent plan, specifying a player's behaviour in a variety of possible circumstances, so that the model is not limited to 'static' problems (see Section 3.1.1). Thus, although the model has no temporal dimension, it may be used to study 'dynamic' situations under the assumption that each player chooses her plan of action once and for all.

A few examples give an idea of the range of situations that the model encompasses. The most well-known strategic game is the *Prisoner's Dilemma*. In this game, there are two players ($N = \{1, 2\}$, say), each player has two actions, *Quiet* and *Fink*, and each player's preference relation ranks the action pair in which she chooses *Fink* and the other player chooses *Quiet* highest, then (*Quiet*, *Quiet*), then (*Fink*, *Fink*), and finally the action profile in which she chooses *Quiet* and the other player chooses *Fink*. In this example, as in most examples, working with payoff representations of the players' preference relations is simpler than working with the preference relations themselves. Taking a payoff function for each player that assigns the payoffs 3, 2, 1, and 0 to the four outcomes, we may conveniently represent the game in the table in Figure 1. (Any two-player strategic game in which each player has finitely many actions may be represented in a similar table.)

This game takes its name from the following scenario. The two players are suspected of joint involvement in a major crime. Sufficient evidence exists to convict each one of a minor offence, but conviction of the major crime requires at least one of them to confess, thereby implicating the other (that is, one player 'finks'). Each suspect may stay quiet or may fink. If a single player finks, she is rewarded by being set free, whereas the other player is convicted of the major offence. If both players fink, then each is convicted but serves only a moderate sentence.

The game derives its interest not from this specific interpretation but because the structure of the players' preferences fits many other social and economic situations. The combination of the desirability of the players' coordinating on an outcome and the incentive on the part of each player individually to deviate from this outcome is present in situations as diverse as duopolists setting prices and countries involved in an arms race.

Another example of a strategic game models oligopoly as suggested by Cournot (1838). The players are the n firms, each player's set of actions is the set of possible outputs (the set of non-negative real numbers), and the preference relation of player i is represented by its profit, given by the payoff function u_i defined by

$$u_i(q_1, \ldots, q_n) = q_i P\left(\sum_{j=1}^{n} q_j\right) - C_i(q_i),$$

		Player 2 *Quiet*	Player 2 *Fink*
Player 1	*Quiet*	2, 2	0, 3
Player 1	*Fink*	3, 0	1, 1

Figure 1 The Prisoner's Dilemma

where q_i is player i's output, C_i is its cost function, and P is the inverse demand function, giving the market price for any total output. Another strategic game that models oligopoly, associated with the name of Bertrand, differs from Cournot's model in taking the set of actions of each player to be the set of possible prices (which requires profit to be defined as a function of prices).

A strategic game that models competition between candidates for political office was suggested by Hotelling (1929). The set of players is a finite set of candidates; each player's set of actions is the same subset X of the line, representing the set of possible policies. Each member of a continuum of citizens (who are not players in the game) has single-peaked preferences over X. Each citizen votes for the candidate whose position is closest to her favourite position. A density function on X represents the distribution of the citizens' favourite policies. The total number of votes obtained by any player is the integral with respect to this density over the subset of X consisting of points closer to the player's action (chosen policy) than to the action of any other player. A player's preferences are represented by the payoff function that assigns 1 to any action profile in which she obtains more votes than every other player, $1/k$ to any action profile in which she obtains at least as many votes as any other player and $k \geq 2$ players tie for the highest number of votes, and 0 to any action profile in which she obtains fewer votes than some other player.

2.2 Nash equilibrium

Which action profile will result when a strategic game is played? Game theory provides two main approaches to answering this question. One isolates action profiles that correspond to stable 'steady states'. This approach leads to the notion of Nash equilibrium, discussed in this section. The other approach, discussed in Section 2.5, isolates action profiles that are consistent with each player's reasoning regarding the likely actions of the other players, taking into account the other players' reasoning about each other and the player in question.

Fix an n-player strategic game and suppose that for each player in the game there exists a population of K individuals, where K is large. Imagine that, in each of a long sequence of periods, K sets of n individuals are randomly selected, each set consisting of one individual from each population. In each period, each set of n individuals plays the game, the individual from population i playing the role of player i, for each value of i. The selected sets change from period to period; because K is large, the chance that an individual will play the game with the same opponent twice is low enough not to enter her strategic calculations. If play settles down to a steady state in which each individual in each population i chooses the same action, say a_i^*, whenever she plays the game, what property must the profile a^* satisfy?

In such a (deterministic) steady state, each individual in population i knows from her experience that every individual in every other population j chooses a_j^*. Thus we can think of each such individual as being involved in a single person decision problem in which the set of actions is A_i and the preferences are induced by player i's preference relation in the game when the action of every other player j is fixed at a_j^*.

That is, a_i^* maximizes i's payoff in the game given the actions of all other players. Or, looked at differently, a^* has the property that no player i can increase her payoff by changing her action a_i^* *given the other players' actions*. An action profile with this property is a *Nash equilibrium*. (The notion is due to Nash, 1950; the underlying idea goes back at least to Cournot, 1838.) For any action profile b, denote by (a_i, b_{-i}) the action profile in which player i's action is a_i and the action of every other player j is b_j.

Definition 2 *A* ***Nash equilibrium*** *of the strategic game* $\langle N, (A_i), (\succsim_i)\rangle$ *is an action profile* a^* *for which*

$$a^* \succsim_i (a_i, a_{-i}^*) \text{ for all } a_i \in A_i$$

for every player $i \in N$.

By inspection of the four action pairs in the *Prisoner's Dilemma* (Figure 1) we see that the action pair (*Fink*, *Fink*) is the only Nash equilibrium. For each of the three other action pairs, a player choosing *Quiet* can increase her payoff by switching to *Fink*, given the other player's action.

The games in Figure 2 immediately answer three questions. Does every strategic game necessarily have a Nash equilibrium? Can a strategic game have more than one Nash equilibrium? Is it possible that every player is better off in one Nash equilibrium than she is in another Nash equilibrium? The left-hand game, which models the game 'Matching pennies', has no Nash equilibrium. The right-hand game has two Nash equilibria, (*B*,*B*) and (*C*, *C*), and both players are better off in (*C*, *C*) than they are in (*B*, *B*).

In some games, especially ones in which each player has a continuum of actions, Nash equilibria may most easily be found by first computing each player's best action for every configuration of the other players' actions. For each player i, let u_i be a payoff function that represents player i's preferences. Fix a player i and define, for each list a_{-i} of the other players' actions, the set of actions that maximize i's payoff:

$$B_i(a_{-i}) = \{a_i \in A_i : a_i \text{ maximizes } u_i(a_i, a_{-i}) \text{ over } a_i \in A_i\}.$$

Each member of $B_i(a_{-i})$ is a *best response* of player i to a_{-i}; the function B_i is called player i's *best response function*. (Note that it is set-valued.) An action profile a^* is a Nash equilibrium if and only if

$$a_i^* \in B_i(a_{-i}^*) \text{ for every player } i.$$

In some games, the set $B_i(a_{-i})$ is a singleton for every player i and every list a_{-i}. For such a game, denote the single element by $b_i(a_{-i})$. Then the condition for the action

	B	*C*
B	1, −1	−1, 1
C	−1, 1	1, −1

	B	*C*
B	1, 1	0, 0
C	0, 0	2, 2

Figure 2 Two strategic games

profile a^* to be a Nash equilibrium may be written as

$$a_i^* = b_i(a_{-i}^*) \text{ for every player } i,$$

a collection of n equations in n unknowns.

Consider, for example, a two-player game in which each player's set of actions is the set of non-negative real numbers and the preference relation of each player i is represented by the payoff function u_i defined by

$$a_i(c + a_j - a_i)$$

where $c > 0$ is a constant. In this game each player i has a unique best response to every action a_j of the other player (j), given by $b_i(a_j) = \frac{1}{2}(c + a_j)$. The two equations $a_1 = \frac{1}{2}(c + a_2)$ and $a_2 = \frac{1}{2}(c + a_1)$ immediately yield the unique solution $(a_1, a_2) = (c, c)$, which is thus the only Nash equilibrium of the game.

2.3 Mixed strategy Nash equilibrium

In a steady state modelled by the notion of Nash equilibrium, all individuals who play the role of a given player choose the same action whenever they play the game. We may generalize this notion. In a *stochastic steady state*, the rule used to select an action by individuals in the role of a given player is probabilistic rather than deterministic. In a *polymorphic steady state*, each individual chooses the same action whenever she plays the game, but different individuals in the role of a given player choose different deterministic actions.

In both of these generalized steady states an individual faces uncertainty: in a stochastic steady state because the individuals with whom she plays the game choose their actions probabilistically, and in a polymorphic steady state because her potential opponents, who are chosen probabilistically from their respective populations, choose different actions. Thus, to analyse the players' behaviour in such steady states, we need to specify their preferences regarding lotteries over the set of action profiles. The following extension of Definition 1 assumes that these preferences are represented by the expected value of a payoff function. (The term 'vNM preferences' refers to von Neumann and Morgenstern, 1944, pp. 15–31; 1947, pp. 204–221, who give conditions on preferences under which such a representation exists.)

Definition 3 *A **strategic game** (with vNM preferences) consists of*

- *a set N (the set of **players**)*

and for each player $i \in N$

- *a set A_i (the set of player i's possible **actions**)*
- *a function $u_i : \times_{j \in N} A_j \to \mathbb{R}$ (player i's **payoff function**, the expected value of which represents i's preferences over the set of lotteries over action profiles).*

A probability distribution over A_i, the set of actions of player i, is called a *mixed strategy* of player i. The notion of a *mixed strategy Nash equilibrium* corresponds to a stochastic steady state in which each player chooses her mixed strategy to maximize her expected payoff, given the other players' mixed strategies.

Definition 4 *A **mixed strategy Nash equilibrium** of the strategic game $\langle N, (A_i), (u_i)\rangle$ is a profile α^* in which each component α_i^* is a probability distribution over A_i that satisfies*

$$U_i(\alpha^*) \geq U_i(\alpha_i, \alpha_{-i}^*) \text{ for every probability distribution } \alpha_i \text{ on } A_i$$

for every player $i \in N$, where $U_i(\alpha)$ is the expected value of $u_i(a)$ under α.

Suppose that each player's set of actions is finite and fix the mixed strategy of every player $j \neq i$ to be α_j. Then player i's expected payoff when she uses the mixed strategy α_i is a weighted average of her expected payoffs to each of the actions to which α_i assigns positive probability. Thus, if α_i maximizes player i's expected payoff given α_{-i}, then so too do all the actions to which α_i assigns positive probability. This observation has two significant consequences. First, a mixed strategy Nash equilibrium corresponds not only to a stochastic steady state but also to a polymorphic steady state. (The equilibrium probability $\alpha_i^*(a_i)$ is the fraction of individuals in population i that choose a_i.) Second, the fact that in a mixed strategy Nash equilibrium each player is indifferent between all the actions to which her mixed strategy assigns positive probability is sometimes useful when computing mixed strategy Nash equilibria.

To illustrate the notion of a mixed strategy Nash equilibrium, consider the games in Figure 2. In the game on the left, a player's expected payoff is the same (equal to 0) for her two actions when the other player chooses each action with probability $\frac{1}{2}$, so that the game has a mixed strategy Nash equilibrium in which each player chooses each action with probability $\frac{1}{2}$. The game has no other mixed strategy Nash equilibrium because each player's best response to any mixed strategy other than the one that assigns probability $\frac{1}{2}$ to each action is either the action B or the action C, and we know that the game has no equilibrium in which neither player randomizes.

The game on the right of Figure 2 has three mixed strategy Nash equilibria. Two correspond to the Nash equilibria of the game in which randomization is not allowed: each player assigns probability 1 to B, and each player assigns probability 1 to C. In the third equilibrium, each player assigns probability $\frac{2}{3}$ to B and probability $\frac{1}{3}$ to C. This strategy pair is a mixed strategy Nash equilibrium because each player's expected payoff to each of her actions is the same (equal to $\frac{2}{3}$ for both players).

The notion of mixed strategy Nash equilibrium generalizes the notion of Nash equilibrium in the following sense.

- If a^* is a Nash equilibrium of the strategic game $\langle N, (A_i), (\succsim_i)\rangle$, then the mixed strategy profile in which each player i assigns probability 1 to a_i^* is a mixed strategy Nash equilibrium of any strategic game with vNM preferences $\langle N, (A_i), (u_i)\rangle$ in which, for each player i, u_i represents $\succsim_i$.
- If a^* is a mixed strategy Nash equilibrium of the strategic game with vNM preferences $\langle N, (A_i), (u_i)\rangle$ in which for each player i there is an action a_i^* such that $\alpha_i^*(a_i^*) = 1$, then a^* is a Nash equilibrium of the strategic game $\langle N, (A_i), (\succsim_i)\rangle$ in which, for each player i, $\succsim_i$ is the preference relation represented by u_i.

The following result gives a sufficient condition for a strategic game to have a mixed strategy Nash equilibrium.

Definition 5 A strategic game with vNM preferences $\langle N, (A_i), (u_i) \rangle$ in which the set N of players is finite has a mixed strategy Nash equilibrium if either (a) the set A_i of actions of each player i is finite or (b) the set A_i of actions of each player i is a compact convex subset of a Euclidean space and the payoff function u_i of each player i is continuous.

Part *(a)* of this result is due to Nash (1950, 1951) and part *(b)* is due to Glicksberg (1952).

In many games of economic interest the players' payoff functions are not continuous. Several results giving conditions for the existence of a mixed strategy Nash equilibrium in such games are available; see, for example, Section 5 of Reny (1999).

As I have noted, in any mixed strategy Nash equilibrium in which some player chooses an action with positive probability less than 1, that player is indifferent between all the actions to which her strategy assigns positive probability. Thus, she has no positive reason to choose her equilibrium strategy: any other strategy that assigns positive probability to the same actions is equally good. This fact shows that the notion of a mixed strategy equilibrium lacks robustness. A result of Harsanyi (1973) addresses this issue. For any strategic game G, Harsanyi considers a game in which the players' payoffs are randomly perturbed by small amounts from their values in G. In any play of the perturbed game, each player knows her own payoffs, but not (exactly) those of the other players. (Formally the perturbed game is a Bayesian game, a model described in Section 2.6.) Typically, a player has a unique optimal action in the perturbed game, and this game has an equilibrium in which no player randomizes. (Each player's equilibrium action depends on the value of her own payoffs.) Harsanyi shows that the limit of these equilibria as the perturbations go to zero defines a mixed strategy Nash equilibrium of G, and almost any mixed strategy Nash equilibrium of G is associated with the limit of such a sequence. Thus we can think of the players' strategies in a mixed strategy Nash equilibrium as approximations to collections of *strictly* optimal actions.

2.4 Correlated equilibrium

One interpretation of a mixed strategy Nash equilibrium is that each player conditions her action on the realization of a random variable, where the random variable observed by each player is independent of the random variable observed by every other player. This interpretation leads naturally to the question of how the theory changes if the players may observe random variables that are not independent.

To take a simple example, consider the game at the right of Figure 2. Suppose that the players observe random variables that are perfectly correlated, each variable taking one value, say x, with some probability p, and another value, say y, with probability

$1 - p$. Consider the strategy that chooses the action B if the realization of the player's random variable is x and the action C if the realization is y. If one player uses this strategy, the other player optimally does so too: if the realization is x, for example, she knows the other player will choose B, so that her best action is B. Thus the strategy pair is an equilibrium.

More generally, the players may observe random variables that are partially correlated. Equilibria in which they do so exist for the game at the right of Figure 2, but the game in Figure 3 is more interesting.

Consider the random variable that takes the values x, y and z, each with probability $\frac{1}{3}$. Player 1 observes only whether the realization is in $\{x, y\}$ or is z (but not, in the first case, whether it is x or y), and player 2 observes only whether it is in $\{x, z\}$ or is y. Suppose that player 1 chooses B if she observes $\{x, y\}$ and C if she observes z, and player 2 chooses B if she observes $\{x, z\}$ and C if she observes y. Then neither player has an incentive to change her action, whatever she observes. If, for example, player 1 observes $\{x, y\}$, then she infers that x and y have each occurred with probability $\frac{1}{2}$, so that player 2 will choose each of her actions with probability $\frac{1}{2}$. Thus her expected payoff is 4 if she chooses B and $\frac{7}{2}$ if she chooses C, so that B is optimal. Similarly, if player 1 observes z, she infers that player 2 will choose B, so that C is optimal for her. The outcome is (B, B) with probability $\frac{1}{3}$, (B, C) with probability $\frac{1}{3}$, and (C, B) with probability $\frac{1}{3}$, so that each player's expected payoff is 5.

An interesting feature of this equilibrium is that both players' payoffs exceed their payoffs in the unique mixed strategy Nash equilibrium (in which each player chooses B with probability $\frac{2}{3}$ and obtains the expected payoff $\frac{14}{3}$).

In general, a *correlated equilibrium* of a strategic game with vNM preferences consists of a probability space and, for each player, a partition of the set of states and a function associating an action with each set in the partition (the player's *strategy*) such that, for each player and each set in the player's partition, the action assigned by her strategy to that set maximizes her expected payoff given the probability distribution over the other players' actions implied by her information. (The notion of correlated equilibrium is due to Aumann, 1974.)

The appeal of a correlated equilibrium differs little from the appeal of a mixed strategy equilibrium. In one respect, in fact, most correlated equilibria are more appealing: the action specified by each player's strategy for each member of her partition of the set of states is strictly optimal (she is not indifferent between that action and any others). Nevertheless, the notion of correlated equilibria has found few applications.

	B	C
B	6, 6	2, 7
C	7, 2	0, 0

Figure 3 A strategic game

2.5 Rationalizability

The outcome (*Fink*, *Fink*) of the *Prisoner's Dilemma* is attractive not only because it is a Nash equilibrium (and hence consistent with a steady state). In addition, for each player, *Fink* is optimal and *Quiet* is suboptimal *regardless* of the other player's action. That is, we may argue solely on the basis of a player's rationality that she will select *Fink*; no reference to her belief about the other player's action is necessary.

We say that the mixed strategy α_i of player i is *rational* if there exists a probability distribution over the other players' actions to which it is a best response. (The probability distribution may entail correlation between the other players' actions; we do not require it to be derived from independent mixed strategies.) With this terminology, the only rational action for each player in the *Prisoner's Dilemma* is *Fink*.

This definition of rationality puts no restriction on the probability distribution over the other players' actions that justifies a player's mixed strategy. In particular, an action is rational even if it is a best response only to a belief that assigns positive probability to the other players' not being rational. For example, in the game on the left of Figure 4, Q is rational for player 1, but all the mixed strategies of player 2 to which Q is a best response for player 1 assign probability of at least $\frac{1}{2}$ to Q, which is not rational for player 2. Such beliefs are ruled out if we assume that each player is not only rational, but also believes that the other players are rational. In the game on the left of Figure 4 this assumption means that player 1's beliefs must assign positive probability only to player 2's action F, so that player 1's only optimal action is F. That is, in this game the assumptions that each player is rational and that each player believes the other player is rational isolate the action pair (F, F).

We may take this argument further. Consider the game on the right of Figure 4. Player 1's action Q is consistent with player 1's rationality and also with a belief that player 2 is rational (because both actions of player 2 are rational). It is not, however, consistent with player 1's believing that player 2 believes that player 1 is rational. If player 2 believes that player is rational, her belief must assign probability 0 to player 1's action X (which is not a best response to any strategy of player 2), so that her only optimal action is F. But if player 2 assigns positive probability only to F, then player 1's action Q is not optimal.

In all of these games — the *Prisoner's Dilemma* and the two in Figure 4 — player 1's action F survives any number of iterations of the argument: it is consistent with player 1's rationality, player 1's belief that player 2 is rational, player 1's belief that player 2 believes that player 1 is rational, and so on. An action with this property is called

	Q	F
Q	3, 2	0, 3
F	2, 0	1, 1

	Q	F
Q	4, 2	0, 3
X	1, 1	1, 0
F	3, 0	2, 2

Figure 4 Two variants of the *Prisoner's Dilemma*

rationalizable, a notion developed independently by Bernheim (1984) and Pearce (1984). (Both Bernheim and Pearce discuss a slightly different notion, in which players are restricted to beliefs that are derived from *independent* probability distributions over each of the other player's actions. Their notion does not have the same properties as the notion described here.)

The set of action profiles in which every player's action is rationalizable may be given a simple characterization. First define a strictly dominated action.

Definition 6 *Player i's action a_i in the strategic game with vNM preferences $\langle N,(A_i),(u_i)\rangle$ is* ***strictly dominated*** *if for some mixed strategy α_i of player i we have*

$$U_i(\alpha_i, a_{-i}) > u_i(a_i, a_{-i}) \text{ for every } a_{-i} \in \times_{j\in N\setminus\{i\}} A_j,$$

where $U_i(\alpha_i, a_{-i})$ is player i's expected payoff when she uses the mixed strategy α_i and the other players' actions are given by a_{-i}.

Note that the fact that α_i in this definition is a mixed strategy is essential: some strictly dominated actions are not strictly dominated by any action. For example, in the variant of the game at the left of Figure 4 in which player 1 has an additional action, say Z, with $u_1(Z,Q) = 0$ and $u_1(Z,F) = 5$, the action F is not strictly dominated by any action, but is strictly dominated by the mixed strategy that assigns probability $\frac{3}{4}$ to Q and probability $\frac{1}{4}$ to Z.

We may show that an action in a finite strategic game is not rational if and only if it is strictly dominated. Given this result, it is not surprising that actions are rationalizable if they survive the iterated elimination of strictly dominated actions, defined precisely as follows.

Definition 7 *Let $G = \langle N,(A_i),(u_i)\rangle$ be a strategic game. For each $j \in N$, let $X_j^1 = A_j$, and for each $j \in N$ and each $t \geq 1$, let X_j^{t+1} be a subset of X_j^t with the property that every member of $X_j^t \setminus X_j^{t+1}$ is strictly dominated in the game $\langle N,(X_i^t),(u_i^t)\rangle$, where u_i^t denotes the restriction of the function u_i to $\times_{j\in N} X_j^t$. If no member of X_j^T for any $j \in N$ is strictly dominated, then the set $\times_{j\in N} X_j^T$* ***survives iterated elimination of strictly dominated actions****.*

The procedure specified in this definition does not pin down exactly which actions are eliminated at each step. Only strictly dominated actions are eliminated, but not *all* such actions are necessarily eliminated. Thus the definition leaves open the question of the uniqueness of the set of surviving action profiles. In fact, however, this set is unique; it coincides with the set of profiles of rationalizable actions.

Proposition 8 *In a finite strategic game the set of action profiles that survives iterated elimination of strictly dominated actions is unique and is equal to the set of profiles of rationalizable actions.*

Every action of any player used with positive probability in a correlated equilibrium is rationalizable. Thus the set of profiles of rationalizable actions is the largest 'solution' for a strategic game that we have considered. In many games, in fact, it is very large. (If no player has a strictly dominated action, all actions of every player are rationalizable, for example.) However, in several of the games mentioned in the previous sections, each player has a single rationalizable action, equal to her unique Nash equilibrium action. This property holds, with some additional assumptions, for Cournot's and Bertrand's oligopoly games with two firms and Hotelling's model of electoral competition with two candidates. The fact that in other games the set of rationalizable actions is large has limited applications of the notion, but it remains an important theoretical construct, delineating exactly the conclusion we may reach by assuming that the players take into account each others' rationality.

2.6 Bayesian games

In the models discussed in the previous sections, every player is fully informed about all the players' characteristics: their actions, payoffs, and information. In the model of a Bayesian game, players are allowed to be uncertain about these characteristics. We call each configuration of characteristics a *state*. The fact that each player's information about the state may be imperfect is modelled by assuming that each player does not observe the state but rather receives a *signal* that may depend on the state. At one extreme, a player may receive a different signal in every state; such a player has perfect information. At another extreme, a player may receive the same signal in every state; such a player has no information about the state. In between these extremes are situations in which a player is partially informed; she may receive the same signal in states ω_1 and ω_2, for example, and a different signal in state ω_3.

To make a decision, given her information, a player needs to form a belief about the probabilities of the states between which she cannot distinguish. We assume that she starts with a *prior belief* over the set of states, and acts upon the posterior belief derived from this prior, given her signal, using Bayes's Law. If, for example, there are three states, ω_1, ω_2, and ω_3, to which her prior belief assigns probabilities $\frac{1}{2}$, $\frac{1}{4}$, and $\frac{1}{4}$, and she receives the same signal, say X, in states ω_1 and ω_2, and a different signal, say Y, in state ω_3, then her posterior belief assigns probability $\frac{2}{3}$ to ω_1 and probability $\frac{1}{3}$ to ω_2 when she receives the signal X and probability 1 to ω_3 when she receives the signal Y.

In summary, a Bayesian game is defined as follows. (The notion is due to Harsanyi, 1967/68.)

Definition 9 *A **Bayesian game** consists of*

- *a set N (the set of **players**)*
- *a set Ω (the set of **states**)*

and for each player $i \in N$

- *a set A_i (the set of player i's possible **actions**)*
- *a set T_i (the set of **signals** that player i may receive) and a function $\tau_i : \Omega \to T_i$, associating a signal with each state (player i's **signal function**)*
- *a probability distribution p_i over Ω (player i's **prior belief**), with $p_i(\tau_i^{-1}(t_i)) > 0$ for all $t_i \in T_i$*

- *a function* $u_i : (\times_{j \in N} A_j) \times \Omega \to \mathbb{R}$ *(player i's* ***payoff function****, the expected value of which represents i's preferences over the set of lotteries on the set* $(\times_{j \in N} A_j) \times \Omega$*).*

This definition allows the players to hold different prior beliefs. In many applications every player is assumed to hold the same prior belief.

A widely studied class of Bayesian games models auctions. An example is a single-object auction in which each player knows her own valuation of the object but not that of any other player and believes that every player's valuation is independently drawn from the same distribution. In a Bayesian game that models such a situation, the set of states is the set of profiles of valuations and the signal received by each player depends only on her own valuation, not on the valuation of any other player. Each player holds the same prior belief, which is derived from the assumption that each player's valuation is drawn independently from the same distribution.

The desirability for a player of each of her actions depends in general on the signal she receives. Thus a candidate for an equilibrium in a Bayesian game is a profile of functions, one for each player; the function for player i associates an action (member of A_i) with each signal she may receive (member of T_i). We refer to player i after receiving the signal t_i as *type* t_i of player i. A Nash equilibrium of a Bayesian game embodies the same principle as does a Nash equilibrium of a strategic game: each player's action is optimal given the other players' actions. Thus, in an equilibrium, the action of each type of each player maximizes the payoff of that type given the action of every other type of every other player. That is, a Nash equilibrium of a Bayesian game is a Nash equilibrium of the strategic game in which the set of players is the set of pairs (i,t_i), where i is a player in the Bayesian game and t_i is a signal that she may receive.

Definition 10 *A* ***Nash equilibrium of a Bayesian game*** $\langle N, \Omega, (a_i), (T_i), (\tau_i), (p_i), (u_i)\rangle$ *is a Nash equilibrium of the following strategic game.*

- *The set of players is the set of all pairs* (i, t_i) *such that* $i \in N$ *and* $t_i \in T_i$.
- *The set of actions of player* (i, t_i) *is* A_i.
- *The payoff of player* (i, t_i) *when each player* (j, t_j) *chooses the action* $a(j, t_j)$ *is*

$$\sum_{\omega \in \Omega} Pr(\omega | t_i) u_i((a_i, \hat{a}_{-i}(\omega)), \omega),$$

where $\hat{a}_j(\omega) = a(j, \tau_j(\omega))$ *for each* $j \in N$.

To illustrate this notion, consider the two-player Bayesian game in which there are two states, each player has two actions (B and C), player 1 receives the same signal in both states, player 2 receives a different signal in each state, each player's prior belief assigns probability $\frac{1}{3}$ to state 1 and probability $\frac{2}{3}$ to state 2, and the payoffs are those shown in Figure 5. A Nash equilibrium of this Bayesian game is a Nash equilibrium of the three-player game in which the players are player 1 and the two types of player 2 (one for each state). I claim that the strategy profile in which player 1 chooses B, type 1 of player 2 (that is, player 2 after receiving the signal that the state is 1) chooses C, and

	State 1 (probability $\frac{1}{3}$)			State 2 (probability $\frac{2}{3}$)	
	B	C		B	C
B	1, 0	0, 1	B	1, 1	0, 0
C	1, 1	1, 0	C	1, 0	0, 1

Figure 5 A Bayesian game

type 2 of player 2 chooses B is a Nash equilibrium. The actions of the two types of player 2 are best responses to the action B of player 1. Given these actions, player 1's expected payoff to B is $\frac{2}{3}$ (because with probability $\frac{1}{3}$ the state is 1 and player 2 chooses C and with probability $\frac{2}{3}$ the state is 2 and player 2 chooses B) and her expected payoff to C is $\frac{1}{3}$. Thus player 1's action B is a best response to the actions of the two types of player 2.

3. Extensive games

Although situations in which players choose their actions sequentially may be modelled as strategic games, they are more naturally modelled as extensive games. In Section 3.1 I discuss a model in which each player, when choosing an action, knows the actions taken previously. In Section 3.2 I discuss a more complex model that allows players to be imperfectly informed. (The notion of an extensive game is due to von Neumann and Morgenstern, 1944, and Kuhn, 1950; 1953.) The formulation in terms of histories is due to Ariel Rubinstein.)

3.1 Extensive games with perfect information

An *extensive game with perfect information* describes the sequential structure of the players' actions. It does so by specifying the set of sequences of actions that may occur and the player who chooses an action after each subsequence. A sequence that starts with an action of the player who makes the first move and ends when no move remains is called a *terminal history*.

Definition 11 *An* ***extensive game with perfect information*** *consists of*

- *a set N (the set of* ***players****)*
- *a set H of sequences (the set of* ***terminal histories****) with the property that no sequence is a proper sub-history of any other sequence*
- *a function P (the* ***player function****) that assigns a player to every proper subsequence of every terminal history*

and for each player $i \in N$

- *a preference relation $\succsim_i$ over the set H of terminal histories.*

The restriction on the set H is necessary for its members to be interpreted as terminal histories: if (x, y, z) is a terminal history then (x, y) is not a terminal history, because z may be chosen after (x, y). We refer to subsequences of terminal histories as *histories*.

The sets of actions available to the players when making their moves, while not explicit in the definition, may be deduced from the set of terminal histories. For any history h, the set of actions available to $P(h)$, the player who moves after h, is the set of actions a for which (h, a) is a history. We denote this set $A(h)$.

Two simple examples of extensive games with perfect information are shown in Figure 6. In the game on the left, the set of terminal histories is $\{(X, w), (X, x), (Y, y), (Y, z)\}$ and the player function assigns player 1 to the empty history (a subsequence of every terminal history) and player 2 to the histories X and Y. The game begins with player 1's choosing either X or Y. If she chooses X, then player 2 chooses either w or x; if she chooses Y, then player 2 chooses either y or z. In the game on the right, the set of terminal histories is $\{(W, x, Y), (W, x, Z), (W, y), X\}$ and the player function assigns player 1 to the empty history and the history (W, x), and player 2 to the history W.

Another example of an extensive game with perfect information is a sequential variant of Cournot's model of oligopoly in which firm 1 chooses an output, then firm 2 chooses an output, and so on. In this game, the set of terminal histories is the set of all sequences $(q_1, \ldots, q_n)$ of outputs for the firms; the player function assigns player 1 to the empty history and, for $k = 1, \ldots, n-1$, player $k+1$ to every sequence $(q_1, \ldots, q_k)$. (Because a continuum of actions is available after each non-terminal history, this game cannot easily be represented by a diagram like those in Figure 6.)

A further example is the bargaining game of alternating offers. This game has terminal histories of infinite length (those in which every offer is rejected).

3.1.1 Strategies

A key concept in the analysis of an extensive game is that of a *strategy*. The definition is very simple: a strategy of any player j is a function that associates with *every* history h after which player j moves a member of $A(h)$, the set of actions available after h.

Definition 12 *A* ***strategy*** *of player j in an extensive game with perfect information $\langle N, H, P, (\succsim_i)\rangle$ is a function that assigns to every history h (subsequence of H) for which $P(h) = j$ an action in $A(h)$.*

In the game at the left of Figure 6, player 1 has two strategies, X and Y. Player 2 has *four* strategies, which we may represent by *wy*, *wz*, *wy*, and *xz*, where the first

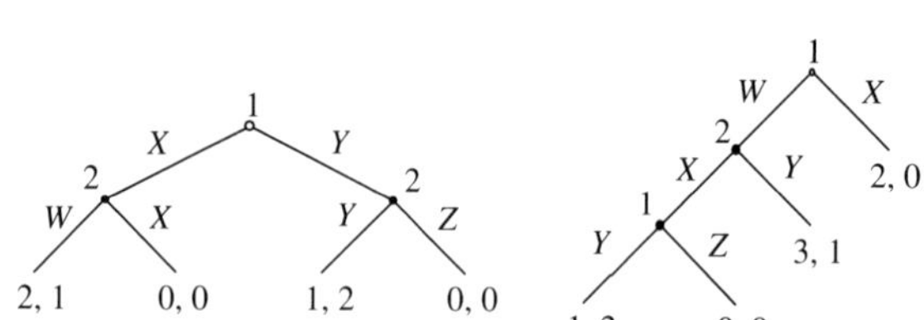

Figure 6 Two extensive games with perfect information. *Note*: Player 1's payoff is the first number in each pair.

component in each pair is the action taken after the history X and the second component is the action taken after the history Y. This example illustrates that a strategy is a complete plan of action, specifying the player's action in every eventuality. Before the game begins, player 2 does not know whether player 1 will choose X or Y; her strategy prepares her for both eventualities.

The game at the right of Figure 6 illustrates another aspect of the definition. Player 1 in this game has *four* strategies, WY, WZ, XY, and XZ. In particular, XY and XZ are distinct strategies. (Remember that a player's strategy assigns an action to *every* history after which she moves.) I discuss the interpretation of strategies like these in Section 3.1.3.

3.1.2 Nash equilibrium

A Nash equilibrium of an extensive game with perfect information is defined in the same way as a Nash equilibrium of a strategic game: it is a strategy profile with the property that no player can increase her payoff by changing her strategy, given the other players' strategies. Precisely, first define the *outcome* $O(S)$ of a strategy profile s to be the terminal history that results when the players use s. (The outcome $O(X, wy)$ of the strategy pair (X, wy) in the game on the left of Figure 6, for example, is the terminal history (X, w).)

Definition 13 *A* ***Nash equilibrium*** *of the extensive game with perfect information* $\langle N, H, P, (\succsim_i)\rangle$ *is a strategy profile* s^* *for which*

$$O(s^*) \succsim_i O(s_i, s^*_{-i}) \text{ for all } s_i \in S_i$$

for every player $i \in N$*, where* S_i *is player* i*'s set of strategies.*

As an example, the game on the left of Figure 6 has three Nash equilibria, (X, wy), (X, wz) and (Y, xy). (One way to find these equilibria is to construct a table like the one in Figure 1 in which each row is a strategy of player 1 and each column is a strategy of player 2.)

For each of the last two equilibria, there exists a history h such that the action specified by player 2's strategy after h is not optimal for her in the rest of the game. For example, in the last equilibrium, player 2's strategy specifies that she will choose x after the history X, whereas only w is optimal for her after this history. Why is such a strategy optimal? Because player 1's strategy calls for her to choose Y, so that the action player 2 plans to take after the history X has no effect on the outcome: the terminal history is (Y, y) regardless of player 2's action after the history X.

I argue that this feature of the strategy pair (Y, xy) detracts from its status as an equilibrium. Its equilibrium status depends on player 1's believing that if she deviates to X then player 2 will choose x. Given that only w is optimal for player 2 after the history X, such a belief seems unreasonable.

Suppose that player 1 forms her belief on the basis of her experience. If she *always* chooses Y, then no amount of experience will enlighten her regarding player 2's choice after the history X. However, in a slightly perturbed steady state in which she very occasionally erroneously chooses X at the start of the game and player 2 chooses her optimal action whenever called upon to move, player 1 knows that player 2 chooses w, not x, after the history X.

If player 1 bases her belief on her reasoning about player 2's rational behaviour (in the spirit of rationalizability), she reaches the same conclusion. (Note, however, that this reasoning process is straightforward in this game only because the game has a finite horizon and one player is indifferent between two terminal histories if and only if the other player is also indifferent.)

In either case, we conclude that player 1 should believe that player 2 will choose w, not x, after the history X. Similarly, the Nash equilibrium (X, wz) entails player 1's unreasonable belief that player 2 will choose z, rather than y, after the history Y. We now extend this idea to all extensive games with perfect information.

3.1.3 Subgame perfect equilibrium

A *subgame perfect equilibrium* is a strategy profile in which each player's strategy is optimal not only at the start of the game, but also after every history. (The notion is due to Selten, 1965.)

Definition 14 *A* ***subgame perfect equilibrium*** *of the extensive game with perfect information* $\langle N, H, P, (\succsim_i)\rangle$ *is a strategy profile* s^* *for which*

$$O_h(s^*) \succsim_i O_h(s_i, s^*_{-i}) \text{ for all } s_i \in S_i$$

for every player $i \in N$ *and every history* h *after which it is player* i*'s turn to move (that is,* $P(h) = i$*), where* S_i *is player* i*'s set of strategies and* $O_h(s)$ *is the terminal history consisting of* h *followed by the sequence of actions generated by* s *after* h.

For any non-terminal history h, define the *subgame* following h to be the part of the game that remains after h has occurred. With this terminology, we have a simple result: a strategy profile is a subgame perfect equilibrium if and only if it induces a Nash equilibrium in every subgame. Note, in particular, that a subgame perfect equilibrium is a Nash equilibrium of the whole game. (The function O in Definition 2 is the same as the function $O_\emptyset$ in Definition 3, where $\emptyset$ denotes the empty history.) The converse is not true, as we have seen: in the game at the left of Figure 6, player 2's only optimal action after the history X is w and her only optimal action after the history Y is y, so that the game has a single subgame perfect equilibrium, (X, wy), whereas it has three Nash equilibria.

Now consider the game at the right of Figure 6. Player 1's only optimal action after the history (W, x) is Y; given that player 1 chooses Y after (W, x), player 2's only optimal action after the history W is x; and given that player 2 chooses x after the history W, player 1's only optimal action at the start of the game is X. Thus the game has a unique subgame perfect equilibrium, (XY, x).

Note, in particular, that player 1's strategy XZ, which generates the same outcome as does her strategy XY regardless of player 2's strategy, is *not* part of a subgame perfect equilibrium. That is, the notion of subgame perfect equilibrium differentiates between these two strategies even though they correspond to the same 'plan of action'. This observation brings us back to a question raised in Section 3.1.1: how should the strategies *XZ* and *XY* be interpreted?

If we view a subgame perfect equilibrium as a model of a perturbed steady state in which every player occasionally makes mistakes, the interpretation of player 1's strategy *XY* is that she chooses *X* at the start of the game, but, if she erroneously chooses *W* and player 2 subsequently chooses *x*, she chooses *Y*. More generally, a component of a player's strategy that specifies an action after a history *h* precluded by the other components of the strategy is interpreted to be the action the player takes if, after a series of mistakes, the history *h* occurs. Note that this interpretation is strained in a game in which some histories occur only after a long series of mistakes, and thus are extremely unlikely.

In some finite horizon games, we may alternatively interpret a subgame perfect equilibrium to be the outcome of the players' calculations about each other's optimal actions. If no player is indifferent between any two terminal histories, then every player can deduce the actions chosen in every subgame of length 1 (at the end of the game); she can use this information to deduce the actions chosen in every subgame of length 2; and she can similarly work back to the start of every subgame at which she has to choose an action. Under this interpretation, the component *Y* of the strategy *XY* in the game at the right of Figure 6 is player 2's belief about player 1's action after the history (*W*, *x*) and also player 1's belief about the action player 2 believes player 1 will choose after the history (*W*, *x*). (This interpretation makes sense also under the weaker condition that, whenever one player is indifferent between the outcomes of two actions, every other player is also indifferent – a sufficient condition for each player to be able to deduce her payoff when the other players act optimally, even if she cannot deduce the other players' strategies.)

This interpretation, like the previous one, is strained in some games. Consider the game that differs from the one at the right of Figure 6 only in that player 1's payoff of 3 after the history (*W*, *y*) is replaced by 1. The unique subgame perfect equilibrium of this game is (*XY*, *x*) (as for the original game). The equilibrium entails player 2's belief that player 1 will choose *Y* if player 2 chooses *x* after player 1 chooses *W*. But choosing *W* is inconsistent with player 1's acting rationally: she guarantees herself a payoff of 2 if she chooses *X*, but can get at most 1 if she chooses *W*. Thus it seems that player 2 should either take player 1's action *W* as an indication that player 1 believes the game to differ from the game that player 2 perceives, or view the action as a mistake. In the first case the way in which player 2 should form a belief about player 1's action after the history (*W*, *x*) is unclear. The second case faces difficulties in games with histories that occur only after a long series of mistakes, as for the interpretation of a subgame perfect equilibrium as a perturbed steady state.

The subgame perfect equilibria of the games in Figure 6 may be found by working back from the end of the game, isolating the optimal action after any history given the optimal actions in the following subgame. This procedure, known as *backward induction*, may be used in any finite horizon game in which no player is indifferent between any two terminal histories. A modified version that deals appropriately with indifference may be used in any finite horizon game.

3.2 Extensive games with imperfect information

In an extensive game with perfect information, each player, when taking an action, knows all actions chosen previously. To capture situations in which some or all players are not perfectly informed of past actions we need to extend the model. A general extensive game allows arbitrary gaps in players' knowledge of past actions by specifying, for each player, a partition of the set of histories after which the player moves. The interpretation of this partition is that the player, when choosing an action, knows only the member of the partition in which the history lies, not the history itself. Members of the partition are called *information sets*. When choosing an action, a player has to know the choices available to her; if the choices available after different histories in a given information set were different, the player would know the history that had occurred. Thus for an information partition to be consistent with a player's not knowing which history in a given information set has occurred, for every history h in any given information set, the set $A(h)$ of available actions must be the same. We denote the set of actions available after the information set I_i by $A(I_i)$.

Definition 15 *An* ***extensive game*** *consists of*

- *a set N (the set of* ***players****)*
- *a set H of sequences (the set of* ***terminal histories****) with the property that no sequence is a proper subhistory of any other sequence*
- *a function P (the* ***player function****) that assigns a player to every proper subsequence of every terminal history*

and for each player $i \in N$

- *a partition $\mathcal{I}_i$ of the set of histories assigned to i by the player function (player i's* ***information partition****) such that for every history h in any given member of the partition (****information set****), the set $A(h)$ of actions available is the same*
- *a preference relation $\succsim_i$ over the set H of terminal histories.*

(A further generalization of the notion of an extensive game allows for events to occur randomly during the course of play. This generalization involves no significant conceptual issue, and I do not discuss it.)

An example is shown in Figure 7. The dotted line indicates that the histories X and Y are in the same information set: player 2, when choosing between x and y, does not know whether the history is X or Y. (Formally, player 2's information partition is $\{\{X, Y\}, \{Z\}\}$. Notice that $A(X) = A(Y)(= \{x, y\})$, as required by the definition.)

A strategy for any player j in an extensive game associates with each of her information sets I_j a member of $A(I_j)$.

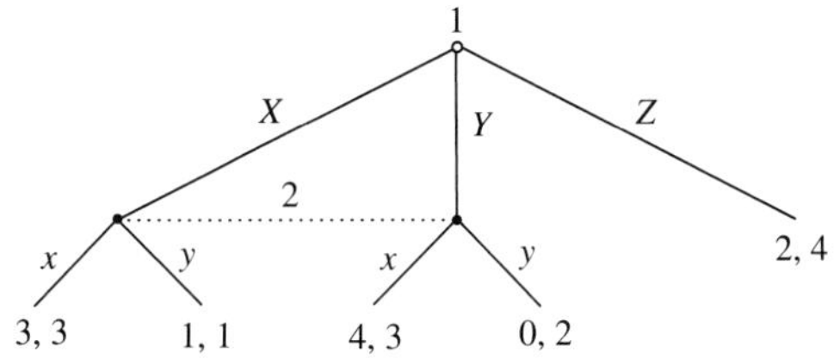

Figure 7 An extensive game with imperfect information. *Note*: The dotted line indicates that the histories *X* and *Y* are in the same information set.

Definition 16 *A* ***strategy*** *of player j in an extensive game* $\langle N, H, P, (\mathcal{I}_i), (\succsim_i)\rangle$ *is a function that assigns to every information set* $I_j \in \mathcal{I}_j$ *of player j an action in* $A(I_j)$.

Given this definition, a Nash equilibrium is defined exactly as for an extensive game with perfect information (Definition 2) — and, as before, is not a satisfactory solution. Before discussing alternatives, we need to consider the possibility of players' randomizing.

In an extensive game with perfect information, allowing players to randomize does not significantly change the set of equilibrium outcomes. In an extensive game with imperfect information, the same is not true. A straightforward way of incorporating the possibility of randomization is to follow the theory of strategic games and allow each player to choose her strategy randomly. That is, we may define a mixed strategy to be a probability distribution over (pure) strategies. An approach more directly suited to the analysis of an extensive game is to allow each player to randomize independently at each information set. This second approach involves the notion of a behavioural strategy, defined as follows.

Definition 17 *A* ***behavioural strategy*** *of player j in an extensive game* $\langle N, H, P, (\mathcal{I}_i), (\succsim_i)\rangle$ *is a function that assigns to each information set* $I_j \in \mathcal{I}_j$ *a probability distribution over the actions in* $A(I_j)$, *with the property that each probability distribution is independent of every other distribution.*

For a large class of games, mixed strategies and behavioural strategies are equivalent: for every mixed strategy there exists a behavioural strategy that yields the same outcome regardless of the other players' strategies, and vice versa. (This result is due to Kuhn, 1950; 1953.) In the discussion that follows, I work with behavioural strategies.

The notion of subgame perfect equilibrium for an extensive game with perfect information embodies two conditions: whenever a player takes an action, (*a*) this action is optimal given her belief about the other players' strategies and (*b*) her belief about the other players' strategies is correct. In such a game, each player needs to form a belief only about the other players' future actions. In an extensive game with imperfect information, players need also to form beliefs about the other player's *past* actions. Thus, in order to impose condition *b* on a strategy profile in an extensive game with imperfect information, we need to consider how a player choosing an action at an information set containing more than one history forms a belief about which history has occurred and what it means for such a belief to be correct.

Consider the game in Figure 7. If player 1's strategy is X or Y, then the requirement that player 2's belief about the history be correct is easy to implement: if player 1's strategy specifies X then she believes X has occurred, whereas if player 1's strategy specifies Y then she believes Y has occurred. If player 1's strategy is Z, however, this strategy gives player 2 no basis on which to form a belief — we cannot derive from player 1's strategy a belief of player 2 about player 1's action. The main approach to defining equilibrium avoids this difficulty by specifying player 1's belief as a component of an equilibrium. Precisely, we define a belief system and an assessment as follows.

Definition 18 *A **belief system** is a function that assigns to every information set a probability distribution over the set of histories in the set. An **assessment** is a pair consisting of a profile of behavioral strategies and a belief system.*

We may now define an equilibrium to be an assessment satisfying conditions a and b. To do so, we need to decide exactly how to implement b. One option is to require consistency of beliefs with strategies only at information sets reached if the players follow their strategies, and to impose no conditions on beliefs at information sets not reached if the players follow their strategies. The resulting notion of equilibrium is called a weak sequential equilibrium. (The name 'perfect Bayesian equilibrium' is sometimes used, although the notion with this name defined by Fudenberg and Tirole, 1991, covers a smaller class of games and imposes an additional condition on assessments.)

Definition 19 *An assessment* (β, μ), *where* β *is a behavioural strategy profile and* μ *is a belief system, is a **weak sequential equilibrium** if it satisfies the following two conditions.*

Sequential rationality. *Each player's strategy is optimal in the part of the game that follows each of her information sets, given the other players' strategies and her belief about the history in the information set that has occurred. Precisely, for each player i and each information set* I_i *of player i, player i's expected payoff to the probability distribution over terminal histories generated by her belief* μ_i *at* I_i *and the behaviour prescribed subsequently by the strategy profile* β *is at least as large as her expected payoff to the probability distribution over terminal histories generated by her belief* μ_i *at* I_i *and the behaviour prescribed subsequently by the strategy profile* (γ_i, β_{-i}), *for each of her behavioural strategies* γ_i.

Weak consistency of beliefs with strategies. *For every information set* I_i *reached with positive probability given the strategy profile* β, *the probability assigned by the belief system to each history h in* I_i *is the probability of h occurring conditional on* I_i *being reached, as given by Bayes' law.*

Consider the game in Figure 7. Notice that player 2's action x yields her a higher payoff than does y regardless of her belief. Thus in any weak sequential equilibrium she chooses x with probability 1. Given this strategy, player 1's only optimal strategy

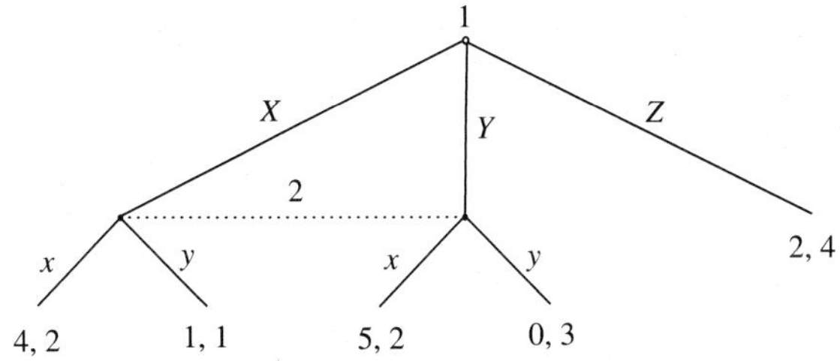

Figure 8 An extensive game with imperfect information

assigns probability 1 to Y. Thus the game has a unique weak sequential equilibrium, in which player 1's strategy is Y, player 2's strategy is x, and player 2's belief assigns probability 1 to the history Y.

Now consider the game in Figure 8. I claim that the assessment in which player 1's strategy is $(\frac{1}{2}, \frac{1}{2}, 0)$, player 2's strategy is $(\frac{1}{2}, \frac{1}{2})$, and player 2's belief assigns probability $\frac{1}{2}$ to X and probability $\frac{1}{2}$ to Y is a weak sequential equilibrium. Given her beliefs, player 2's expected payoffs to x and y are both 2, and given player 2's strategy, player 1's expected payoffs to X and Y are both $\frac{5}{2}$ and her payoff to Z is 2. Thus each player's strategy is sequentially rational. Further, player 2's belief is consistent with player 1's strategy. This game has an additional weak sequential equilibrium in which player 1's strategy is Z, player 2's strategy is y, and player 2's belief assigns probability 1 to the history Y. Note that the consistency condition does not restrict player 2's belief in this equilibrium, because player 1 chooses neither X nor Y with positive probability.

In some games the notion of weak sequential equilibrium yields sharp predictions, but in others it is insufficiently restrictive. Some games, for example, have weak sequential equilibria that do not satisfy a natural generalization of the notion of subgame perfect equilibrium. In response to these problems, several 'refinements' of the notion of a weak sequential equilibrium have been studied, including sequential equilibrium (due to Kreps and Wilson, 1982) and perfect Bayesian equilibrium (due to Fudenberg and Tirole, 1991).

MARTIN J. OSBORNE

See also **bargaining; epistemic game theory: an overview; epistemic game theory: complete information; epistemic game theory: incomplete information; Nash equilibrium, refinements of; non-cooperative games (equilibrium existence).**

I am grateful to Jean Guillaume Forand for comments and to the Social Sciences and Humanities Research Council of Canada for financial support.

Bibliography

Aumann, R. 1974. Subjectivity and correlation in randomized strategies. *Journal of Mathematical Economics* 1, 67–96.

Bernheim, B. 1984. Rationalizable strategic behavior. *Econometrica* 52, 1007–28.

Cournot, A. 1838. Recherches sur les principes mathématiques de la théorie des richesses. Paris: Hachette. Trans. N. Bacon as Researches into the Mathematical Principles of the Theory of Wealth. New York: Macmillan, 1897.

Fudenberg, D. and Tirole, J. 1991. Perfect Bayesian equilibrium and sequential equilibrium. *Journal of Economic Theory* 53, 236–60.

Glicksberg, I. 1952. A further generalization of the Kakutani fixed point theorem, with application to Nash equilibrium points. *Proceedings of the American Mathematical Society* 3, 170–4.

Harsanyi, J. 1967/68. Games with incomplete information played by 'Bayesian' players, Parts I, II, and III. *Management Science* 14, 159–82, 320–34, and 486–502.

Harsanyi, J. 1973. Games with randomly disturbed payoffs: a new rationale for mixed-strategy equilibrium points. *International Journal of Game Theory* 2, 1–23.

Hotelling, H. 1929. Stability in competition. *Economic Journal* 39, 41–57.

Kreps, D. and Wilson, R. 1982. Sequential equilibria. *Econometrica* 50, 863–94.

Kuhn, H. 1950. Extensive games. *Proceedings of the National Academy of Sciences of the United States of America* 36, 570–6.

Kuhn, H. 1953. Extensive games and the problem of information. In *Contributions to the Theory of Games, Volume II (Annals of mathematics studies, 28)*, ed. H. Kuhn and A. Tucker. Princeton, NJ: Princeton University Press.

Nash, Jr., J. 1950. Equilibrium points in n-person games. *Proceedings of the National Academy of Sciences of the United States of America* 36, 48–9.

Nash, Jr., J. 1951. Non-cooperative games. *Annals of Mathematics* 54, 286–95.

Pearce, D. 1984. Rationalizable strategic behavior and the problem of perfection. *Econometrica* 52, 1029–50.

Reny, P. 1999. On the existence of pure and mixed strategy Nash equilibria in discontinuous games. *Econometrica* 67, 1029–56.

Selten, R. 1965. Spieltheoretische Behandlung eines Oligopolmodells mit Nachfragetrgheit. *Zeitschrift für die gesamte Staatswissenschaft* 121, 301–24, 667–89.

von Neumann, J. and Morgenstern, O. 1944. *Theory of Games and Economic Behavior.* Princeton, NJ: Princeton University Press.

von Neumann, J. and Morgenstern, O. 1947. *Theory of Games and Economic Behavior*, 2nd edn. Princeton, NJ: Princeton University Press.

strategy-proof allocation mechanisms

An allocation mechanism is a function mapping agents' preferences into final allocations. For example, the competitive allocation mechanism calculates market-clearing prices to select a feasible, Pareto optimal, final allocation that varies with agents' preferences. This simple view, however, of the mechanism as a map from preferences to final allocations is inadequate because it ignores agents' propensity to maximize. The problem is that each agent can misrepresent his preferences in reporting them to the mechanism because they are private to him and not verifiable. Therefore, he will tend to misrepresent whenever he realizes that misreporting his true preferences will result in a more preferable allocation than a truthful report. If, as is the case with the competitive mechanism, agents do have incentives to misrepresent their preferences, then the resulting allocation is optimal only with respect to their reported preferences, not to their true preferences.

Strategy-proof allocation mechanisms neutralize the complications that strategic misrepresentation creates. Informally, an allocation mechanism is strategy-proof if each agent's maximizing choice of what preference ordering to report depends only on his own preferences and not on the preferences that other agents report. If the mechanism is strategy-proof, then each agent's incentives are to disregard his expectations concerning other agents' reports and truthfully report his own preference ordering because, no matter what, it secures the allocation that is maximal among those that his choice of reported preferences could secure. That is, truth telling is every agent's dominant strategy if the mechanism is strategy-proof. Strategy-proofness makes understanding the strategic choices of agents trivial: they always play their dominant strategy of reporting their true preferences. Whenever a mechanism fails strategy-proofness, game theoretic methods with all their complications become essential to understand its true optimality properties.

Gibbard–Satterthwaite theorem

Strategy-proof mechanisms are desirable, but do they exist? That most voting procedures and the competitive mechanism fail strategy-proofness suggests the conjecture that many, if not all, attractive allocation mechanisms also fail strategy-proofness. This conjecture, which Dummett and Farquharson (1961, p. 34) first made, turns out to be true. Gibbard (1973) and Satterthwaite (1975) independently proved that for general environments no strategy-proof allocation mechanisms exist that satisfy minimal requirements for responsiveness to agents' preferences.

Precise statement of this fundamental result requires some notation. Let $J = \{1, 2, \ldots, n\}$, $n \geq 2$, be a fixed set of agents who must select a single alternative from a set $X = \{x, y, z, \ldots\}$ of $|X|$ distinct, final allocations. Each agent $i \in J$ has transitive preferences P_i over the allocations X. Let P_i represent strict preference and I_i

represent indifference. Thus for each $x, y \in X$ and each $i \in J$, only three possibilities exist: xP_iy (agent i strictly prefers x over y), yP_ix (agent i strictly prefers y over x), or xI_iy (agent i is indifferent between x and y). Not every transitive ordering X is necessarily admissible as a preference ordering P_i. For example, for a particular $x, y \in X$, xP_iy might be the only admissible ordering because allocation x dominates allocation y in terms of the usual non-satiation axiom of consumer demand theory. Therefore, let Σ represent all possible transitive preference orderings over X and let $\Omega_i \subset \Sigma$ represent the set of all transitive preference orderings over X that are admissible. Thus P_i is an admissible ordering for agent i only if $P_i \in \Omega_i$. Let $\Omega = \Omega_1 \times \Omega_2 \times \cdots \times \Omega_n$ be the product of all agents' sets of admissible preference orderings. If every transitive ordering is admissible, then preferences are said to be unrestricted and $\Omega = \Sigma^n$. Call the triple $\langle J, X, \Omega \rangle$ the environment.

An n-tuple $P = (P_1, \ldots, P_n) \in \Omega$ is called a preference profile. An allocation mechanism is a function $f : \Omega \to X$ that maps each admissible preference profile into a final allocation. Agent i can manipulate allocation function f at profile $P \in \Omega$ if an admissible ordering $P_i' \in \Omega_i$ exists such that:

$$f(P_1, \ldots, P_{i-1}, P_i', P_{i+1}, \ldots, P_n)\ P_i\, f(P). \tag{1}$$

The interpretation of (1) is this. Preference ordering P_i is agent i's true preferences. The other agents report preference orderings $P_{-i} = (P_1, \ldots, P_{i-1}, P_{i+1}, \ldots, P_n)$. If agent i reports his preferences truthfully, then the outcome is $f(P_i, P_{-i}) \equiv f(P)$. If he misrepresents his preferences to be P_i', then the outcome is $f(P_i', P_{-i}) \equiv f(P_1, \ldots, P_{i-1}, P_i', P_{i+1}, \ldots, P_n)$. Relation (1) states that agent i prefers $f(P_i', P_{-i})$ to $f(P_i, P_{-i})$. Therefore agent i has an incentive to manipulate f at profile P by misrepresenting his preferences to be P_i' rather than P_i.

An allocation mechanism f is strategy-proof if no admissible profile $P \in \Omega$ exists at which f is manipulable. This means that even if, for instance, agent i has perfect foresight about the preferences the other $n-1$ agents will report, agent i can never do better than to report his true preferences P_i. Truth is always every agent's dominant strategy. Presumably this is sufficient to induce every agent always to report his preferences truthfully. Gibbard (1973) and Satterthwaite (1975) show that strategy-proof allocation mechanisms generally do not exist.

Theorem. If admissible preferences are unrestricted ($\Omega = \Sigma^n$) and at least three possible allocations exist ($|X| \geq 3$), then no strategy-proof allocation mechanism f exists that is non-dictatorial and Pareto optimal.

An allocation mechanism is dictatorial if, for and all profiles $P \in \Omega^n$, an agent i exists such that $f(P) \in \max_X P_i$ where

$$\max_x P_i = \{x : x \in X \text{ and, for all } y \in X, \text{ not } yP_ix\}.$$

That is, a dictatorial mechanism always gives the dictator one of the allocations that he most prefers. A non-dictatorial mechanism is a mechanism that is not dictatorial. A mechanism satisfies Pareto optimality if and only if, for any profile $P \in \Omega^n$ and any

$x, y \in X, xP_iy$ for all $i \in I$ implies $f(P) \neq y$. Pareto optimality implies that, if unanimity exists among the agents that an allocation x is the most preferred feasible allocation, then the mechanism picks x.

The theorem can be proved either through appeal to Arrow's impossibility theorem (1951) for social-welfare functions or through a direct argument. Reny (2001) constructs parallel direct proofs for both the Arrow and the Gibbard–Satterthwaite theorems that make transparent their identical foundations.

The theorem is an impossibility theorem because mechanisms that violate either non-dictatorship or Pareto optimality are unattractive. If, however, the environment $\langle J, X, \Sigma \rangle$ contains only two allocations ($|X| = 2$), then impossibility no longer obtains: majority rule is an attractive, non-dictatorial, Pareto optimal, and strategy-proof mechanism. Normally, however, for economic environments $|X| > 2$. Therefore the theorem implies that existence can be obtained (if at all) only for environments $\langle J, X, \Omega \rangle$ where admissible preferences Ω are restricted to a strict subset of Σ^n.

Environments with restricted admissible preferences

In classical economic environments, whether for private goods or for public goods, admissible preferences are naturally constrained so that for all agents admissible preference orderings Ω_i are a strict subset of Σ. Consequently the Gibbard–Satterthwaite theorem does not rule out strategy-proof, non-dictatorial, and Pareto optimal mechanisms. Nevertheless, with one significant exception, classical economic environments do not admit useful strategy-proof mechanisms.

Consider an exchange economy with private goods first. It involves an enormous restriction in Ω relative to Σ^n: each agent cares only about his private allocation from the economy's endowment and, over his possible private allocations, has smooth preferences that generate classical indifference surfaces. Despite this, building on a long series of papers that includes Hurwicz and Walker (1990), Serizawa and Weymark (2003) show that no strategy-proof allocation mechanism exists that always prescribes allocations that are Pareto optimal and, additionally, provide every agent with a consumption bundle that exceeds some positive, minimum level. Thus strategy-proofness and optimality is incompatible with an acceptable distribution of real income.

Turn next to the canonical spatial model of public goods. Let there be $m \geq 2$ public goods so that allocations lie in the non-negative orthant of R^m_+. Each agent's admissible preferences Ω_i includes preference orderings characterized by a most preferred allocation $x \in R^m_+$ surrounded by convex indifference surfaces of less preferred allocations. Zhou (1991) shows that every strategy-proof mechanism in this environment is dictatorial provided that its range within R^m_+ has dimension at least two. This, as Zhou points out, is precisely analogous to the Gibbard–Satterthwaite theorem since Zhou's requirement that the range have dimension two parallels Gibbard and Satterthwaite's twin requirements that f *is* Pareto optimal and X includes at least three allocations.

If, however, in this set-up there is only one public good ($m = 1$) so that all feasible allocations lie on the non-negative half line R^1_+, then the set of admissible preference profiles reduces to profiles of 'single-peaked preferences' and impossibility switches to possibility. A preference ordering $P_i \in \Omega_i$ is single-peaked if it is characterized by a most preferred allocation $x \in R^1_+$ with the desirability of other allocations y strictly decreasing as the distance $|x - y|$ increases. Given that for all agents Ω_i contains only single-peaked preferences, strategy-proof and Pareto optimal mechanisms exist. Ching (1997) describes the full set of these mechanisms. The simplest of these rules is the generalization of majority rule that picks the median of the agents' most preferred allocations. Other, more complicated rules in this set are augmented median voter rules that make use of 'phantom voters'.

Even though useful strategy-proof mechanisms do not exist in general for classical economic environments, they do exist in some cases for the particular environment of a small-scale allocation problem. Two important examples are briefly discussed here. The first concerns a committee J whose n members are considering a set $K = \{a_1, \ldots, a_m\}$ of $m \geq 2$ proposals and must decide which one of its subsets $B \subset K$ should be approved and implemented. Each member has preferences P_i over the 2^m subsets of K; thus if $A, B \subset K$and AP_iB, this means that member i prefers that the set A of proposals be approved and implemented rather than the set B of proposals. For each $P_i \in \Omega_i$, let $G(P_i) = \{a \in K|\{a\}\ P_i\ \emptyset\}$. Denote member i's 'good' proposals. These are just the proposals a that i would vote to approve if the committee were restricted to choosing between approving the single proposal {a} and no proposals at all. A member i's preferences are separable if, for each ordering $P_i \in \Omega_i$ and each subset $A \subset K$,

$$A \cup \{a\}\ P_i\ Aa \in G(P_i).$$

In words, adding proposal a to the set A of approved proposals improves the outcome in i's eyes if and only if a is one of the proposals he deems good.

Barbera, Sonnenschein and Zhou (1991) fully characterize all useful strategy-proof voting rules when preferences are restricted to be separable. The simplest member of this class of strategy-proof rules is 'voting by quota'. Under that rule, which is defined by a positive integer Q, each committee member casts a ballot listing the proposals that he judges to be good. Any proposal that is listed on at least Q of the ballots is declared approved. For instance, for election to a club that currently has 100 members, each proposed member might need to be named on the ballots of at least 60 current members in order to be allowed to join.

The second example concerns the Vickrey–Clarke–Groves combinatorial auction in which a seller has a set $K = \{a_1, \ldots, a_m\}$ of $m \geq 2$ of unique objects that he is selling to a set J of $n \geq 2$ buyers. An allocation is a vector $x = \{x_1, x_2, \ldots, x_m\}$ where the value of $x_j \in \{0, 1, 2, \ldots, n\}$ indicates to which buyer i object j is assigned (with the convention that the seller is labelled as buyer 0). Thus if $x_3 = 2$, then the third object is assigned to the second buyer.

Buyer i has quasi-linear utility if his utility (in dollars) for an allocation x and monetary transfer t_i to the seller takes the form $v_i(x) = u_i(x) - t_i$ where t_i is the payment buyer i makes to the seller and $u_i(x)$ is the value that he places on allocation x. Let Ω_i consist of all possible quasi-linear utility functions. The efficient allocation x^* assigns the objects to those buyers who most highly them:

$$x^* \in \arg\max_{x \in X} \sum_{k \in J} u_k(x)$$

where X is the set of all possible allocations. If buyer i declines to participate and only the other the other $n-1$ buyers bid, then $x^*_{-i} \in \arg\max_{x \in X} \Sigma_{k \in J \backslash i}\, u_i(x)$ is the efficient allocation for the remaining buyers $k \in J \backslash i$. The value of the m objects if they are optimally allocated to all n buyers is $V^* = \Sigma_{k \in J}\, u_k(x^*)$ and their value if they are optimally allocated to the $n-1$ buyers excluding i is $V^*_{-i} = \Sigma_{k \in J \backslash i}\, u_k(x^*_{-i})$.

With this notation in place, the Vickrey–Clarke–Groves auction consists of three steps. First, each buyer $i \in J$ sends the seller a list of bids for each possible allocation: $u_i(x)$ for all $x \in X$. Second, the seller computes and implements the optimal allocation y^*. Third, the seller collects the Vickrey–Clarke–Groves payments from each buyer for that bundle of goods the buyer is allocated under y^*. Buyer i's payment is $V^*_{-i} - (V^* - u_i(x^*))$: the total value the other buyers accrue if i does not participate less their total value if i does participate. This is the opportunity cost of the bundle agent i receives in the optimal allocation. Given that Ω_i contains only quasi-linear utility functions, this auction is strategy-proof: each buyer's dominant strategy is to report truthfully his $u_i(\cdot)$ to the seller. The reason it is strategy-proof is that (*a*) truthful reporting guarantees that he receives a bundle of goods if and only if the value he places on it exceeds its opportunity cost and (*b*) his bundle's opportunity cost is independent of his report. Groves and Loeb (1975) derive the efficiency and incentive properties of this mechanism, Green and Laffont (1979) develop its full theory, and de Vries and Vohra (2003) review its application to combinatorial auctions.

MARK A. SATTERTHWAITE

Bibliography

Arrow, K. 1951. *Social Choice and Individual Values*. New York: Wiley.

Barbera, S., Sonnenschein, H. and Zhou, L. 1991. Voting by committees. *Econometrica* 59, 595–609.

Ching, S. 1997. Strategy-proofness and 'median voters'. *International Journal of Game Theory* 26, 473–90.

de Vries, S. and Vohra, R.V. 2003. Combinatorial auctions: a survey. *INFORMS Journal of Computing* 15, 284–309.

Dummett, M. and Farquharson, R. 1961. Stability in voting. *Econometrica* 29, 33–44.

Gibbard, A. 1973. Manipulation of voting schemes: a general result. *Econometrica* 41, 587–602.

Green, J. and Laffont, J.-J. 1979. *Incentives in Public Decision-making*. Studies in Public Economics, vol. 1. Amsterdam: North-Holland.

Groves, T. and Loeb, M. 1975. Incentives and public inputs. *Journal of Public Economics* 4, 211–26.

Hurwicz, L. and Walker, M. 1990. On the generic nonoptimality of dominant-strategy allocation mechanisms: a general theorem that includes pure exchange economies. *Econometrica* 58, 683–704.

Reny, P. 2001. Arrow's theorem and the Gibbard–Satterthwaite theorem: a unified approach. *Economics Letters* 70, 99–105.

Satterthwaite, M. 1975. Strategy-proofness and Arrow's conditions: existence and correspondence theorems for voting procedures and social welfare functions. *Journal of Economic Theory* 10, 187–217.

Serizawa, S. and Weymark, J. 2003. Efficient strategy-proof exchange and minimum consumption guarantees. *Journal of Economic Theory* 109, 246–63.

Zhou, L. 1991. Impossibility of strategy-proof mechanisms in economies with pure public goods. *Review of Economic Studies* 58, 107–19.

supermodularity and supermodular games

The concept of complementarity is well established in economics at least since Edgeworth (1881). The basic idea of complementarity is that the marginal value of an action increases with the level of other actions available. The mathematical concept of supermodularity formalizes the idea of complementarity. The theory of monotone comparative statics and supermodular games provides the toolbox to deal with complementarities. This theory, developed by Topkis (1978; 1979), Vives (1985; 1990) and Milgrom and Roberts (1990a), in contrast to classical convex analysis, is based on order and monotonicity properties on lattices (see Topkis, 1998; Vives, 1999; and Vives, 2005, for detailed accounts of the theory and applications). Monotone comparative statics analysis provides conditions under which optimal solutions to optimization problems change monotonically with a parameter. The theory of supermodular games exploits order properties to ensure that the best response of a player to the actions of rivals increases with their level. Indeed, this is the characteristic of games of strategic complementarities (the term was coined in Bulow, Geanakoplos and Klemperer, 1983). The power of the approach is that it clarifies the drivers of comparative statics results and the need of regularity conditions; it allows very general strategy spaces, including indivisibilities and functional spaces such as those arising in dynamic or Bayesian games; it establishes the existence of equilibrium in pure strategies (without requiring quasi-concavity of payoffs, smoothness assumptions, or interior solutions); it allows a global analysis of the equilibrium set when there are multiple equilibria, which has an order structure with largest and smallest elements; and, finally, it finds that those extremal equilibria have strong stability properties and there is an algorithm to compute them.

We will provide an introduction to the approach and some definitions, move on to the basic monotone comparative statics results, and provide the basic results for supermodular games.

Preliminaries and definitions

A binary relation $\geq$ on a nonempty set X is a *partial order* if $\geq$ is reflexive, transitive, and antisymmetric (a binary relation is antisymmetric if $x \geq y$ and $y \geq x$ implies that $x = y$). A partially ordered set $(S, \geq)$ is *completely ordered* if for x and y in S either $x \geq y$ or $y \geq x$. An upper bound on a subset $A \subset X$ is $z \in X$ such that $z \geq x$ for all $x \in A$. A greatest element of A is an element of A that is also an upper bound on A. Lower bounds and least elements are defined analogously. The greatest and least elements of A, when they exist, are denoted max A and min A, respectively. A *supremum* (resp., *infimum*) of A is a least upper bound (resp., greatest lower bound); it is denoted sup A (resp., inf A). A *lattice* is a partially ordered set $(X, \geq)$ in which any two elements have a *supremum* and an *infimum*. Any interval of the real line with the usual order is a lattice since any two points have a *supremum* and an *infimum* in the interval. However, the set in $\mathbb{R}^2$ $\{(1, 0), (0, 1)\}$ is not a lattice with the vector ordering (the usual component-wise ordering), since (1, 0) and (0, 1) have no joint upper bound in the set. However, if we add the points (0, 0) and (1, 1) the set becomes a lattice with the vector ordering (see Figure 1 and let $x=(0, 1)$ and $y=(1, 0)$). A lattice $(X, \geq)$ is *complete* if every non-empty subset has a *supremum* and an *infimum*. Any compact interval of the real line with the usual order, or product of compact intervals with the vector order, is a complete lattice. Open intervals are lattices but they are not complete (for example, the *supremum* of the interval (a, b) does not belong to (a, b)). A subset L of the lattice X is a *sublattice* of X if the *supremum* and *infimum* of any two elements of L belong also to L. A lattice is always a sublattice of itself, but a lattice need not be a sublattice of a larger lattice. Let $(X, \geq)$ and $(T, \geq)$ be partially ordered sets. A function $f: X \rightarrow T$ is *increasing* if, for x, y in X, $x \geq y$ implies that $f(x) \geq f(y)$.

Supermodular functions

A function $g: X \rightarrow \mathbb{R}$ on a lattice X is *supermodular* if, for all x, y in X, $g(\inf(x, y)) + g(\sup(x, y)) \geq g(x) + g(y)$. It is *strictly supermodular* if the inequality is strict for all pairs x, y in X that neither $x \geq y$ nor $y \geq x$ holds. A function f is *(strictly) submodular* if $-f$ is (strictly) supermodular; a function f is *(strictly) log-supermodular* if $\log f$ is (strictly) supermodular. Let X be a lattice and T a partially ordered set. The function $g: X \times T \rightarrow R$ has *(strictly) increasing differences* in (x, t) if $g(x', t) - g(x, t)$ is (strictly) increasing in t for $x' > x$ or, equivalently, if $g(x, t') - g(x, t)$ is (strictly) increasing in x for $t' > t$. Decreasing differences are defined analogously.

Supermodularity is a stronger property than increasing differences: if T is also a lattice and if g is (strictly) supermodular on $X \times T$, then g has (strictly) increasing

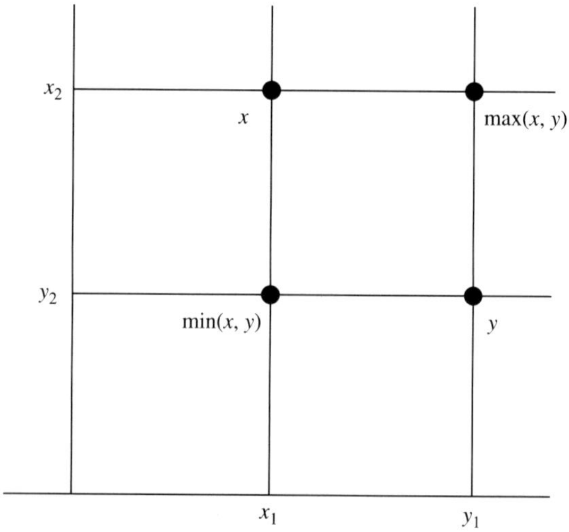

Figure 1 Illustration of supermodular payoff on lattice. *Source*: Vives (1999, p. 25).

differences in (x, t). However, the two concepts coincide on the product of completely ordered sets: in such case a function is supermodular if and only if it has increasing differences in any pair of variables. Both concepts formalize the idea of complementarity: increasing one variable raises the return to increase another variable. For example, the Leontieff utility function $U(x) = \min\{a_1x_1, ..., a_nx_n\}$ with $a_i \geq 0$ for all i is supermodular on $\mathbb{R}^n$. The complementarity idea can be made transparent by thinking of the rectangle in $\mathbb{R}^n$ with vertices $\{\min(x, y), y, \max(x, y), x\}$ and rewriting the definition of supermodularity as $g(\max(x, y)) - g(x) \geq g(y) - g(\min(x, y))$. Consider, for example, points in $\mathbb{R}^2$ $x = (x_1, x_2)$ and $y = (y_1, y_2)$ with the usual order. Then going from $\min(x, y) = (x_1, y_2)$ to y, for given y_2, increases the payoff less than going from x to $\max(x, y) = (y_1, x_2)$, for given $x_2 \geq y_2$ (see Figure 1).

If X is a convex subset of $\mathbb{R}^n$ and if g: $X \to R$ is twice-continuously differentiable, then g has increasing differences in (x_i, x_j) if and only if $\partial^2 g(x)/\partial x_i \partial x_j \geq 0$ for all x and $i \neq j$. For decreasing differences (or submodularity) we would have $\partial^2 g(x)/\partial x_i \partial x_j \leq 0$. This characterization has a direct counterpart with the concept of (weak) cost complementarities if g is a cost function and $x \geq 0$ the production vector. If $\partial^2 g(x)/\partial x_i \partial x_j > 0$ for all x and $i \neq j$, then g is strictly supermodular. The differential characterization of supermodularity can be motivated by the figure as before. As an example consider assortative matching when types x and y in $[0, 1]$ produce $f(x, y)$ when matched and nothing otherwise. If $\partial^2 f/\partial x \partial y > 0$ then in a core allocation matching is positively assortative, that is, matched partners are identical (Becker, 1973; see Shimer and Smith, 2000, for a dynamic model with search where it is required also that $\log \partial f/\partial x$ and $\log \partial^2 f/\partial x \partial y$ are supermodular).

Positive linear combinations and pointwise limits preserve the complementarity properties (supermodularity/increasing differences) of a family of functions g_n: $X \times T \to \mathbb{R}$. Supermodularity is also preserved under integration. This has important consequences for comparative statics under uncertainty and games of incomplete information (see Vives, 1990; and Athey, 2001). Supermodularity is also preserved under the maximization operation. Supermodularity is unrelated to convexity, concavity or returns to scale. Indeed, any real-valued function on a completely ordered set (say the reals) is both supermodular and submodular. This fact also makes clear that supermodularity in Euclidean spaces, in contrast to concavity or convexity, has no connection with continuity or differentiability properties. Note also that if g is a twice-continuously differentiable function, supermodularity only puts restrictions on the cross partials of g while the other concepts impose restrictions also on the diagonal of the matrix of second derivatives.

Monotone comparative statics

Let X be a compact rectangle in Euclidean space and let T be a partially ordered set. Let $g : X \times T \to \mathbb{R}$ be a function that (a) is supermodular and continuous on X for each $t \in T$ and (b) has increasing differences in (x, t). Let $\varphi(t) = \arg\max_{x \in X} g(x, t)$. Then (Topkis, 1978):

1. $\varphi(t)$ is a non-empty compact sublattice for all t;
2. φ is increasing in the sense that, for $t'>t$ and for $x' \in \varphi(t')$ and $x \in \varphi(t)$, we have $\sup(x', x) \in \varphi(t')$ and $\inf(x', x) \in \varphi(t)$; and
3. $t \mapsto \sup\varphi(t)$ and $t \mapsto \inf \varphi(t)$ are increasing selections of φ.

Several remarks are in order: (i) The continuity requirement of g can be relaxed. In more general spaces the requirement is for X to be a complete lattice and for g to fulfil an appropriate order continuity property. (ii) If g has strictly increasing differences in (x, t), then all selections of φ are increasing. (iii) If solutions are interior, and $\partial g/\partial x_i$ is strictly increasing in t for some i, then all selections of φ are strictly increasing (Edlin and Shannon, 1998). (iv) Milgrom and Shannon (1994) relax the complementarity conditions to ordinal complementarity conditions (quasi-supermodularity and a single crossing property), and develop necessary and sufficient conditions for monotone comparative statics.

Let us illustrate the result when $T \subset \mathbb{R}$ and g is twice-continuously differentiable on $X \times T$. Suppose first that $X \subset \mathbb{R}$, g is strictly quasi-concave in x (with $\partial g/\partial x = 0$ implying that $\partial^2 g/(\partial x)^2 < 0$), and that the solution to the maximization problem $\varphi(t)$ is interior. Then, using the implicit function theorem on the interior solution, for which $\partial g(\varphi(t), t)/\partial x = 0$, we obtain that φ is continuously differentiable and $\varphi' = -\frac{\partial^2 g}{\partial x \partial t} / \frac{\partial^2 g}{(\partial x)^2}$. Obviously, sign $\varphi' =$ sign $\frac{\partial^2 g}{\partial x \partial t}$. The solution is increasing (decreasing) in t if there are increasing, $\frac{\partial^2 g}{\partial x \partial t} \geq 0$, (decreasing, $\frac{\partial^2 g}{\partial x \partial t} \leq 0$) differences. The monotone comparative statics result asserts that the solution $\varphi(t)$ will be monotone increasing in t even if g is not strictly quasi-concave in x, in which case $\varphi(t)$ need not be a singleton or convex-valued, provided that $\frac{\partial^2 g}{\partial x \partial t}$ does not change sign. For example, if we consider a single-product monopolist with revenue function $R(x)$ and cost function $C(x, t)$, where x is the output of the firm and t a cost efficiency parameter, we have $g(x, t) = R(x) - C(x, t)$. If $C(\cdot)$ is smooth and $\frac{\partial^2 C}{\partial x \partial t} \leq 0$, an increase in t reduces marginal costs. Then if $R(\cdot)$ is continuous the comparative static result applies, and the largest $\varphi(t)$ and the smallest $\underline{\varphi}(t)$ monopoly outputs are increasing in t. If $\frac{\partial^2 C}{\partial x \partial t} < 0$, then all selections of the set of monopoly outputs are increasing in t. It is worth noting that the comparative statics result is obtained with no concavity assumption on the profit of the firm.

Suppose now that $X \subset \mathbb{R}^k$. If g is strictly concave in x (with the Jacobian of $(\frac{\partial g}{\partial x_1}, \ldots, \frac{\partial g}{\partial x_k})$ with respect to x, H_x, negative definite) and the solution to the optimization problem $\varphi(t) = (\varphi_1(t), ..., \varphi_k(t))$ is interior, then $\varphi(t)$ is continuously differentiable, and $(\frac{\partial \varphi_1}{\partial t}, \ldots, \frac{\partial \varphi_k}{\partial t}) = -H_x^{-1}(\frac{\partial^2 g}{\partial x_1 \partial t}, \ldots, \frac{\partial^2 g}{\partial x_k \partial t})$. If the off-diagonal elements of H_x are nonnegative $\frac{\partial^2 g}{\partial x_i \partial x_j} \geq 0$, $j \neq i$, then all the elements of $-H_x^{-1}$ are nonnegative and the diagonal elements are positive (McKenzie, 1959). A sufficient condition for $\frac{\partial \varphi_i}{\partial t} \geq 0$ for all i is that $\frac{\partial^2 g}{\partial x_i \partial t} \geq 0$ for all i (the statement also holds with strict inequalities). As before, even if H_x is not negative definite, the assumptions that $\frac{\partial^2 g}{\partial x_i \partial x_j} \geq 0$, $j \neq i$, and that $\frac{\partial^2 g}{\partial x_i \partial t} \geq 0$ imply that the solution set $\varphi(t)$ has the monotonicity properties stated in the monotone comparative statics result. Note that when X is multidimensional, the restriction that g be supermodular on X, ensuring that for any components i and j an increase in the variable x_j raises the marginal return of variable x_i, when coupled with increasing differences on $X \times T$ is needed to guarantee the monotonicity of the solution. For example, let us consider a multiproduct monopolist. If the revenue function $R(\cdot)$ is continuous and supermodular on X, the cost function $C(\cdot)$ continuous and submodular on X, and $C(\cdot)$ displays decreasing differences in (x, t), the comparative static result follows. That is, the largest $\varphi(t)$and the smallest $\underline{\varphi}(t)$ monopoly output vectors are increasing in t. In the differentiable case, $\frac{\partial^2 R}{\partial x_i \partial x_j} \geq 0$ and $\frac{\partial^2 C}{\partial x_i \partial x_j} \leq 0$, for all $i \neq j$, and $\frac{\partial^2 C}{\partial x_i \partial t} \leq 0$ for all i. The result hinges on revenue and cost complementarities among outputs, and the impact of the efficiency parameter on marginal costs, and not concavity of profits.

Now let us consider a team problem. Suppose that n persons share a common objective $g(x_1, \ldots, x_n, t)$ where the action of player i x_i is in the rectangle $X_i \subset \mathbb{R}^{k_i}$for each i and t is a payoff relevant parameter. If g is supermodular on $X = \prod_{i=1}^{n} X_i$ and has strictly increasing differences in (x, t), then any optimal solution is increasing in the level of the parameter. For example, the optimal production $g(x, t)$ of the firm (seen as a team problem) is increasing in the level of information technology t (which raises the marginal productivity of any worker of the firm).

Supermodular games

Consider the game $(A_i, \pi_i; i \in N)$ where for each $i = 1, \ldots, n$ in the set of players N, A_i is the strategy set, a subset of Euclidean space, and π_i the payoff of the player (defined on the cross product of the strategy spaces of the players A). Let $a_i \in A_i$ and $a_{-i} \in \Pi_{j \neq i} A_j$ (that is, denote by a_{-i} the strategy profile $(a_1, \ldots, a_n)$ except the ith element). The strategy profiles are endowed with the usual component-wise order. We will say that the game $(A_i, \pi_i; i \in N)$ is *(strictly) supermodular* if for each i, A_i is a compact rectangle of Euclidean space, π_i is continuous and (i) supermodular in a_i for fixed a_{-i} and (ii) displays (strictly) increasing differences in (a_i, a_{-i}). We will say that the game $(A_i, \pi_i;$ $i \in N)$ is *smooth (strictly) supermodular* if furthermore π_i (a_i, a_{-i}) is twice continuously differentiable with (i) $\partial^2 \pi_i / \partial a_{ih} \partial a_{ik} \geq 0$ for all $k \neq h$, and (ii) $\partial^2 \pi_i / \partial a_{ih} \partial a_{jk} \geq (>)\ 0$ for all $j \neq i$ and for all h and k, where a_{ih} denotes the hth component of the strategy a_i of player i. Condition (i) is the strategic complementarity property in own strategies a_i. Condition (ii) is the strategic complementarity property in rivals' strategies a_{-i} .

In a more general formulation strategy spaces need only be complete lattices and this includes functional spaces such as those arising in dynamic or incomplete information games. The complementarity conditions can be weakened to define an 'ordinal supermodular' game (see Milgrom and Shannon, 1994). Furthermore, the application of the theory can be extended by considering increasing transformations of the payoff (which do not change the equilibrium set of the game). For example, we will say that the game is log-supermodular if π_i is nonnegative and log π_i fulfils conditions (i) and (ii). This is the case of a Bertrand oligopoly with differentiated substitutable products, where each firm produces a different variety and marginal costs are constant, whenever the own-price elasticity of demand for firm i is decreasing in the prices of rivals, as with constant elasticity, logit, or constant expenditure demand systems.

In the duopoly case ($n = 2$) the case of strategic substitutability can also be covered. Indeed, suppose that there is strategic complementarity or supermodularity, in own strategies ($\partial^2 \pi_i / \partial a_{ih} \partial a_{ik} \geq 0$ for all $k \neq h$, in the smooth version) and strategic substitutability in rivals' strategies or decreasing differences in (a_i, a_{-i}) ($\partial^2 \pi_i / \partial a_{ih} \partial a_{jk} \leq 0$ for all $j \neq i$ and for all h and k, in the smooth version). Then the game obtained by reversing the order in the strategy space of one of the players, say player 2, is supermodular (Vives, 1990). Cournot competition with substitutable products displays typically strategic substitutability between the (output) strategies of the firms.

In a supermodular game best responses are monotone increasing even when π_i is not quasi-concave in a_i. Indeed, in a supermodular game each player has a largest, $\Psi_i(a_{-i}) = \sup \Psi_i(a_{-i})$, and a smallest, $\underline{\Psi}_i(a_{-i}) = \inf \Psi_i(a_{-i})$, best reply, and they are increasing in the strategies of the other players. Let $\Psi = (\Psi_1, \ldots, \Psi_n)$ and $\underline{\Psi} = (\underline{\Psi}_1, \ldots, \underline{\Psi}_n)$ denote the extremal best reply maps.

Result 1 In a supermodular game there always *exist* a largest $a = \sup\{a \in A : \Psi(a) \geq a\}$ and a smallest $\underline{a} = \inf\{a \in A: \underline{\Psi}(a) \leq a\}$ equilibrium (Topkis, 1979).

The result is shown applying Tarski's fixed point theorem to the extremal selections of the best-reply map, Ψ and $\underline{\Psi}$, which are monotone because of the strategic complementarity assumptions. Tarski's theorem (1955) states that if A is a complete lattice (for example, a compact rectangle in Euclidean space) and $f: A \to A$ an increasing function then f has a largest $\sup\{a \in A : f(a) \geq a\}$ and a smallest $\inf\{a \in A : a \geq f(a)\}$ fixed point. There is no reliance on quasi-concave payoffs and convex strategy sets to deliver convex-valued best replies as required when showing existence using Kakutani's fixed point theorem. The equilibrium set can be shown also to be a complete lattice (Vives, 1990; Zhou, 1994). The result proves useful in a variety of circumstances to get around the existence problem highlighted by Roberts and Sonnenschein (1976). This is the case, for example, of the (log-)supermodular Bertrand oligopoly with differentiated substitutable products.

Result 2 In a *symmetric* supermodular game (that is, a game with payoffs and strategy sets exchangeable against permutations of the players) the extremal equilibria a and $\underline{a}$ are symmetric and, if strategy spaces are completely ordered and the game is strictly supermodular, then *only* symmetric equilibria exist (see Vives, 1999).

The result is useful to show uniqueness since if there is a unique symmetric equilibrium then the equilibrium is unique. For example, in a symmetric version of the Bertrand oligopoly model with constant elasticity of demand and constant marginal costs, it is easy to check that there exists a unique symmetric equilibrium. Since the game is (strictly) log-supermodular, we can conclude that the equilibrium is unique. The existence result of symmetric equilibria is related to the classical results of McManus (1962; 1964) and Roberts and Sonnenschein (1976).

Result 3 In a supermodular game if there are positive spillovers (that is, the payoff to a player is increasing in the strategies of the other players) then the largest (smallest) equilibrium point is the Pareto best (worst) equilibrium (Milgrom and Roberts, 1990a; Vives, 1990).

Indeed, in many games with strategic complementarities equilibria can be Pareto ranked. In the Bertrand oligopoly example the equilibrium with higher prices is Pareto dominant for the firms. This has proved particularly useful in applications in macroeconomics (for example, Cooper and John, 1988) and finance (for example, Diamond and Dybvig, 1983).

Result 4 In a supermodular game:

(*a*) Best-reply dynamics approach the interval $[\underline{a}, a]$ defined by the smallest and the largest equilibrium points of the game. Therefore, if the equilibrium is unique it is globally stable. Starting at any point $A^+(A^-)$ in the intersection of the upper (lower) contour sets of the largest (smallest) best replies of the players, best-reply dynamics lead monotonically downwards (upwards) to an equilibrium (Vives, 1990; 1999). This provides an iterative procedure to find the largest (smallest) equilibrium (Topkis, 1979) starting at sup A (inf A) (see Figure 2).

(*b*) The extremal equilibria correspond to the largest and smallest serially undominated strategies (Milgrom and Roberts, 1990a). Therefore, if the equilibrium is unique, the game is dominance solvable. Rationalizable (Bernheim, 1984; Pearce, 1984) or mixed strategy outcomes must lie in the interval $[\underline{a}, a]$.

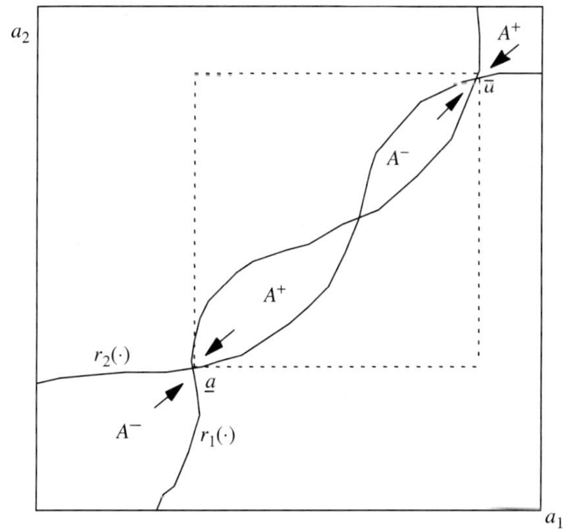

Figure 2 Best-reply dynamics in a supermodular game (with best-reply functions $r_1(\cdot)$, $r_2(\cdot)$). *Source*: Vives (1999, p. 54).

In the Bertrand oligopoly example with linear, constant elasticity, or logit demands, the equilibrium is unique and therefore it is globally stable, and the game is dominance solvable. In the team example it is clear that an optimal solution will be a Nash equilibrium of the game among team members. If the equilibrium is unique, then best-reply dynamics among team members will converge to the optimal solution. This need not be the case if there are multiple equilibria (see Milgrom and Roberts, 1990b, for an application to the theory of the firm).

Result 5 Let us consider a supermodular game with parameterized payoffs $\pi_i\,(a_i, a_{-i}; t)$ with t in a partially ordered set T. If $\pi_i\,(a_i, a_{-i}; t)$ has increasing differences in (a_i, t) (in the smooth version $\partial^2\pi_i/\partial a_{ih}\partial t \geq 0$ for all h and i), then the largest and smallest equilibrium points increase with an increase in t, and starting from any equilibrium, best reply dynamics lead to a (weakly) larger equilibrium following the parameter change. The latter result can be extended to adaptive dynamics, which include fictitious play and gradient dynamics (see Lippman, Mamer and McCardle, 1987; and Sobel, 1988, for early results; and Milgrom and Roberts, 1990a; Milgrom and Shannon, 1994; and Vives, 1999, for extensions). It is worth noting that continuous equilibrium selections that do not increase monotonically with t predict unstable equilibria (Echenique, 2002). The result yields immediately that an increase in an excise tax in a (log-)supermodular Bertrand oligopoly raises prices at an extremal equilibrium.

The basic intuition for the comparative statics result is that an increase in the parameter increases the actions for one player, for given actions or rivals, and this reinforces the desire of all other players to increase their actions because of strategic complementarity. This initiates a mutually reinforcing process that leads to larger equilibrium actions. This is a typical positive feedback in games of strategic complementarities. In this class of games, unambiguous monotone comparative statics obtain if we concentrate on stable equilibria. We can understand this as a multidimensional version of Samuelson's (1947) correspondence principle, which was obtained with standard calculus methods applied to interior and stable one-dimensional models.

A patent race

Let us consider n firms engaged in a memory-less patent race that have access to the same R&D technology. The winner of the patent obtains the prize V and losers obtain nothing. The (instantaneous) probability of innovating is given by $h(x)$ if a firm spends x continuously, where h is a smooth function with $h(0)=0$, $h'>0$, $\lim_{x\to\infty} h'(x) = 0$, and $h'(0) = \infty$. It is assumed also that h is concave but a region of increasing returns for small x may be allowed. If no patent is obtained the (normalized) profit of a firm is zero. The expected discounted profits (at rate r) of firm i investing x_i if rival $j \neq i$ invests x_j is given by

$$\pi_i = \frac{h(x_i)\,V - x_i}{h(x_i) + \Sigma_{j\neq i} h(x_j) + r}.$$

Lee and Wilde (1980) restrict attention to symmetric Nash equilibria of the game and show that, under a uniqueness and stability condition at a symmetric equilibrium x^* expenditure intensity increases with n. The classical approach requires assumptions to ensure a unique and stable symmetric equilibrium and cannot rule out the existence of asymmetric equilibria. Suppose that there are potentially multiple symmetric equilibria and that going from n to $n+1$ new equilibria appear. What comparative static result can we infer then? Using the lattice approach we obtain a more general comparative statics result that allows for the presence of multiple symmetric equilibria (Vives, 1999, Exercise 2.20; and 2005, Section 5.2). Let $h(0)=0$ with h strictly increasing in$[0, x]$, with $h(x)\ V - x<0$ for $x \geq x>0$. Under the assumptions the game is strictly log-supermodular and from Result 2 only symmetric equilibria exist. Let $x_i=x$ and $x_j=y$ for $j \neq i$. Then $\log \pi_i$ has (strictly) increasing differences in (x, n) for all y $(y>0)$ and, according to Result 5, the expenditure intensity x^* at extremal equilibria is increasing in n. Furthermore, starting at any equilibrium, an increase in n will raise research expenditure with out-of-equilibrium adjustment according to best-reply dynamics. This will be so even if new equilibria appear, or some disappear, as a result of increasing n. Finally, if h is smooth with $h'>0$ and $h'(0) = \infty$, then $\partial \log \pi_i / \partial x_i$ is strictly increasing in n and (at extremal equilibria) x^* is *strictly* increasing in n. This follows because, under our assumptions, equilibria are interior and must fulfil the first-order conditions.

The results can be applied to dynamic and incomplete information games, which have complex strategy spaces. For example, in an incomplete information game, if for given types of players the *ex* post game is supermodular, then the Bayesian game is also supermodular and therefore there exist Bayesian equilibria in pure strategies (Vives, 1990). If, furthermore, payoffs to any player have increasing differences between the actions of the player and types, and higher types believe that other players are also of a higher type (according to first-order stochastic dominance), then extremal equilibria of the Bayesian game are monotone increasing in types (Van Zandt and Vives, 2007). This defines a class of monotone supermodular games. An example is provided by global games, introduced by Carlsson and Van Damme (1993) and developed by Morris and Shin (2002) and others with the aim of equilibrium selection. Global games are games of incomplete information with type space determined by each player observing a noisy private signal of the underlying state. The result is obtained applying iterated elimination of strictly dominated strategies. From the perspective of monotone supermodular games we know that extremal equilibria are the outcome of iterated elimination of strictly dominated strategies, that they are monotone in type (and therefore in binary action games there is no loss of generality in restricting attention to threshold strategies), and the conditions put to pin down a unique equilibrium in the global game amount to a lessening of the strength of strategic complementarities (see Vives, 2005, Section 7.2).

XAVIER VIVES

See also **epistemic game theory: incomplete information; global games.**

Bibliography

Athey, S. 2001. Single crossing properties and the existence of pure strategy equilibria in games of incomplete information. *Econometrica* 69(4), 861–89.

Becker, G. 1973. A theory of marriage: part I. *Journal of Political Economy* 81, 813–46.

Bernheim, B.D. 1984. Rationalizable strategic behavior. *Econometrica* 52, 1007–28.

Bulow, J., Geanakoplos, J. and Klemperer, P. 1985. Multimarket oligopoly: strategic substitutes and complements. *Journal of Political Economy* 93, 488–511.

Carlsson, H. and van Damme, E. 1993. Global games and equilibrium selection. *Econometrica* 61, 989–1018.

Cooper, R. and John, A. 1988. Coordinating coordination failures in Keynesian models. *Quarterly Journal of Economics* 103, 441–63.

Diamond, D.W. and Dybvig, P.H. 1983. Bank runs, deposit insurance, and liquidity. *Journal of Political Economy* 91, 401–19.

Echenique, F. 2002. Comparative statics by adaptive dynamics and the correspondence principle. *Econometrica* 70, 833–44.

Edlin, A.S. and Shannon, C. 1998. Strict monotonicity in comparative statics. *Journal of Economic Theory* 81, 201–19.

Lee, T. and Wilde, L. 1980. Market structure and innovation: a reformulation. *Quarterly Journal of Economics* 94, 429–36.

Lippman, S., Mamer, J. and McCardle, K. 1987. Comparative statics in non-cooperative games via transfinitely iterated play. *Journal of Economic Theory* 41, 140–66.

McKenzie, L. 1959. Matrices with dominant diagonals and economic theory. In *Mathematical Methods in the Social Sciences*, ed. K. Arrow, S. Karlin and P. Suppes. Stanford: Stanford University Press.

McManus, M. 1962. Numbers and size in Cournot oligopoly. *Yorkshire Bulletin of Economic and Social Research* 14, 14–22.

McManus, M. 1964. Equilibrium, numbers and size in Cournot oligopoly. *Yorkshire Bulletin of Economic and Social Research* 16, 68–75.

Milgrom, P. and Roberts, J. 1990a. Rationalizability, learning, and equilibrium in games with strategic complementarities. *Econometrica* 58, 1255–77.

Milgrom, P. and Roberts, J. 1990b. The economics of modern manufacturing: technology, strategy, and organization. *American Economic Review* 80, 511–28.

Milgrom, P. and Shannon, C. 1994. Monotone comparative statics. *Econometrica* 62, 157–80.

Morris, S. and Shin, H.S. 2002. Global games: theory and applications: advances in economics and econometrics. In *Proceedings of the Eighth World Congress of the Econometric Society*, ed. M. Dewatripont, L. Hansen, and S. Turnovsky. Cambridge: Cambridge University Press.

Pearce, D. 1984. Rationalizable strategy behavior and the problem of perfection. *Econometrica* 52, 1029–50.

Roberts, J. and Sonnenschein, H. 1976. On the existence of Cournot equilibrium without concave profit functions. *Journal of Economic Theory* 13, 112–17.

Samuelson, P. 1947. *Foundations of Economic Analysis*. Cambridge, MA: Harvard University Press.

Shimer, R. and Smith, L. 2000. Assortative matching and search. *Econometrica* 68, 343–70.

Sobel, M. 1988. Isotone comparative statics in supermodular games. Mimeo, State University of New York.

Tarski, A. 1955. A lattice-theoretical fixpoint theorem and its applications. *Pacific Journal of Mathematics* 5, 285–309.

Topkis, D. 1978. Minimizing a submodular function on a lattice. *Operations Research* 26, 305–21.

Topkis, D. 1979. Equilibrium points in non-zero sum n-person submodular games. *SIAM Journal of Control and Optimization* 17, 773–87.

Topkis, D. 1998. *Supermodularity and Complementarity.* Princeton: Princeton University Press.
Van Zandt, T. and Vives, X. 2007. Monotone equilibria in Bayesian games of strategic complementarities. *Journal of Economic Theory* 134, 339–50.
Vives, X. 1985. Nash equilibrium in oligopoly games with monotone best responses. Working Paper No. 85-10, CARESS, Unibversity of Pennsylvania.
Vives, X. 1990. Nash equilibrium with strategic complementarities. *Journal of Mathematical Economics* 19, 305–21.
Vives, X. 1999. *Oligopoly Pricing: Old Ideas and New Tools.* Boston: MIT Press.
Vives, X. 2005. Complementarities and games: new developments. *Journal of Economic Literature* 43, 437–79.
Zhou, L. 1994. The set of Nash equilibria of a supermodular game is a complete lattice. *Games and Economic Behavior* 7, 295–300.

Index